TJ 223 .M53 M39 1995

McGo   ny J., 1950-

Di

NEW ENGLAND INSTITUTE
OF TECHNOLOGY
LEARNING RESOURCES CENTER

Direct Digital Control:

A Guide to Distributed Building Automation

Direct Digital Control:

A Guide to Distributed Building Automation

John J. McGowan, CEM

Published by
THE FAIRMONT PRESS, INC.
700 Indian Trail
Lilburn, GA 30247

Library of Congress Cataloging-in-Publication Data

McGowan, John J., 1950-
    Direct digital control: a guide to distributed building automation / John J. McGowan.
    p. cm.
    Includes index.
    ISBN 0-88173-166-8
    1. Digital control systems. 2. Buildings--Automation. I. Title.

TJ223.M53M39   1995   696--dc20   94-32120
                                                                                                              CIP

Direct Digital Control: A Guide To Distributed Building Automation
by John J. McGowan.
©1995 by The Fairmont Press, Inc. All rights reserved. No part of this publication may be reproduced or transmitted in any form or by any means, electronic or mechanical, including photocopy, recording, or any information storage and retrieval system, without permission in writing from the publisher.

Published by The Fairmont Press, Inc.
700 Indian Trail
Lilburn, GA 30247

Printed in the United States of America

10 9 8 7 6 5 4 3 2 1

ISBN 0-88173-166-8 FP

ISBN 0-13-206269-0 PH

While every effort is made to provide dependable information, the publisher, authors, and editors cannot be held responsible for any errors or omissions.

Distributed by Prentice Hall PTR
Prentice-Hall, Inc.
A Simon & Schuster Company
Englewood Cliffs, NJ 07632

Prentice-Hall International (UK) Limited, London
Prentice-Hall of Australia Pty. Limited, Sydney
Prentice-Hall Canada Inc., Toronto
Prentice-Hall Hispanoamericana, S.A., Mexico
Prentice-Hall of India Private Limited, New Delhi
Prentice-Hall of Japan, Inc., Tokyo
Simon & Schuster Asia Pte. Ltd., Singapore
Editora Prentice-Hall do Brasil, Ltda., Rio de Janeiro

DEDICATION

*To Dan and Kate
for the foundation of a lifetime*

*To Judy, Dustin and Kendall
with the love of a lifetime*

Table Of Contents

Chapter 1 1
    Distributed Direct Digital Control (DDC)
    and Effective Building Operation

Chapter 2 5
    Why Automate Buildings: Using Networked DDC
    As a Management Tool
    By Dan Agne, Johnson Controls, Inc.

Chapter 3 15
    Integrating Automated Buildings

Chapter 4 47
    Energy 2001 - A View of DDC in the Future
    By George Owens, P.E., The Rouse Company

Chapter 5 57
    Ensuring Successful DDC System Projects

Chapter 6 71
    Developing a Strategic DDC Plan

Chapter 7 79
    Conducting a Building Control Survey

Chapter 8 105
    State of the Art Distributed Direct Digital Control

Chapter 9 127
    Building Wide Sequence of Operation

Chapter 10 189
  Equipment Level DDC Control

Chapter 11 231
  Zone Level DDC Control

Chapter 12 255
  Communication and Networking with
  Distributed DDC Systems

Chapter 13 279
  DDC Operator Interface: Hand Held and Central

Chapter 14 325
  Financing Energy Saving DDC and Other Retrofits

Chapter 15 337
  Specifying Building Automation Systems – Graphic Display
  By Harris D. Bynum, Honeywell, Inc.

Chapter 16 361
  Procurement Practices for High Performance DDC Systems
  By Ken Sinclair, Sinclair Energy Services

Chapter 17 397
  Managing DDC Systems for Results

Chapter 18 415
  DDC System Operator Training
  By Jack Meredith, P.E., British Columbia
  Buildings Corporation

Chapter 19 421
  DDC Industrial Facility Case Study –
  Energy Conservation is a Process
  By Paul Krippner, P.E., Ethicon, Incorporated
  Alan Stewart, P.E., Honeywell, Incorporated

Chapter 20 435
  DDC Commercial Building Case Study: The Museum of New
  Mexico and Energy Management
  By Brian K. Johnson, New Mexico Energy Minerals and Natural
  Resources Department

Chapter 21 455
    DDC New Application Case Study - Monitoring and
    Control of Large Computers and Support Systems
    By John P. Cilia, IBM Corporation

Appendix A 467
    Glossary

Appendix B 481
    List of DDC Manufacturers

Index 485

Chapter 21 455
    *DDC New Application Case Study - Monitoring and
Control of Large Computers and Support Systems
By John P. Cilia, IBM Corporation*

Appendix A 467
    Glossary

Appendix B 481
    List of DDC Manufacturers

Index 485

Chapter 1

Distributed Direct Digital Control (DDC) and Effective Building Operation

Distributed Direct Digital Control Systems have gone under a number of names over the years. They have been called Energy Management and Building Automation Systems to name just two. Yet more important than the name, is the fact that DDC remains one of the most universally applied and successful technologies installed in buildings.

As will be outlined later in the book, the major reason that people began installing computerized controls in their buildings was to save energy cost. This impetus started the trend, but the many benefits that these systems have offered users over the years have spurred growth of this business to nearly one billion dollars per year. An industry of this magnitude creates a significant number of jobs and further involves significant capital investment on the part of a wide variety of users. Application of these systems is cost effective in commercial and industrial facilities. Inspiring success stories have been documented in the largest factory and in simple commercial facilities. Schools for example, have been consistently successful with DDC systems.

Given the opportunity for DDC in any type of building, the title of this chapter states the essential premise of this book. As the reader you have purchased this book for a specific reason. It is the author's hope that your expectations will be met, and to that end this chapter will allow comparison of your goals with a brief statement of the goals for this book. Beyond initial motivation however, ultimately any reader who is goaled by improving the effectiveness of buildings for their employer, client or customer will benefit from this book.

Effective building operation is the central theme of the book. This can be defined as an efficient building that meets the occupants needs at the lowest possible operating cost. An effective building also focuses on meeting the occupants needs in a proactive manner. This means optimizing comfort and building system function.

Two decades ago when the author got involved with these systems it was acceptable to sacrifice occupant comfort or productivity to save energy and cost. Since that time energy cost has varied in importance for the user. Chapter 8 will outline the energy market and take a look at the motivation for buying systems. In that vein, energy cost savings are still important, but not to the exclusion of those using the building.

As a result of the importance of effective buildings, this book is absolutely committed to outlining a technology that enhances this operation. In that context every chapter of this book will consider the benefits of this technology. Equally important it will note that implementation and ongoing management are critical. As a result the steps necessary to ensure success with the technology will be covered throughout the book.

The intended audience for this book includes engineers, building owners, utilities, contractors, industrial users and control manufacturers. These may be novices entering the energy industry today or generalists who want to get an update on the latest state of the art. The importance of building technologies like DDC to overall building operation makes reading this book useful for any of the above individuals. Further important new trends like Demand Side Management, Open or Standard Protocol, Indoor Air Quality and Environment Impact of Inefficient Operations heighten the importance of DDC. As this book unfolds the reader will see that DDC may offer the most effective management tool available to any facility operator. This technology is invaluable whether the manager is located in the building or across the country. It allows for optimization for day to day comfort and for information management to ensure long term records.

Information management is used to track progress and results of the energy program, to ensure preventative maintenance for extended equipment life and even to provide back up in litigation. The trend towards litigation to break leases, and to receive damages for adverse health effects from occupying a workplace has changed the dynamics of operating a building. Effective use of information and timely access to building data can also avoid cost from equipment failures and acts of nature. For example the author worked with a system recently that alarmed on severe freeze conditions and avoided equipment damage. Not only did savings result from equipment that did not have to be replaced, but litigation was avoided. This building owner had tenants that would have filed suit for "missed productivity and revenue" if the building had to remain unoccupied due to the freeze failure.

The reader will note that effective buildings are extremely desirable based upon the discussion above. This book focuses on DDC as an essential element to an effective building. The perspective that the book takes is definition of each phase of a system from planning through specification and implementation. It will outline the technology to ensure that the owner, consultant or contractor is able to make intelligent decisions. The author firmly believes that it is not necessary to understand every detail of a system. Yet it is critical to be able to converse intelligently about the technology and the process of implementation. Armed with adequate knowledge the reader will be able to ensure that the system secured meets their needs and provides expected results.

This book will address the intended audience by outlining the system process in a logical flow. It starts with preliminary chapters that clarify both reasons to install a system and ways to help justify the system to facility owners. This is critical because many individuals in the intended audience can see the benefits of these systems, but are unsure how to justify and sell the installation to decision makers. Next the book will consider overall strategic planning and conducting surveys to define the scope of DDC system required. This is essential because successful systems are well planned. Smooth and effective installations result from a clearly defined outline of equipment to be controlled, sequence of operations desired and steps to be taken in implementation.

This book also provides a description of DDC system components including architecture, communication and Central Operator Interfaces. Again this knowledge is important to make intelligent decision and thoroughly understand what the owner is buying. Key topics including financing and specifying will follow technology discussions. These steps occur closer to implementation and will be valuable for the reader who is armed with a clear concept of the work and technology required. Finally the book will provide successful case studies to outline for the reader examples of effective DDC installations.

With the broad coverage of system issues outlined above, this book may be expected to meet the goals of any reader who needs a clear understanding of Distributed DDC technology.

Chapter 2

Why Automate Buildings: Using Networked DDC As A Management Tool

By Dan Agne, Johnson Controls, Inc.

PREFACE

The buying decision to invest in Distributed Direct Digital Control systems is often quite complex, and may not always consider the true range of benefits offered by these products.

In this chapter, Dan Agne does an excellent job of defining today's corporate environment. He also describes both the obstacles and the opportunities that are associated with DDC systems. Traditionally control system benefits considered by building owners have been limited to energy cost reductions, yet there is much more that automation can offer. Mr. Agne covers the big picture benefits that can aide the building owner to effectively manage a facility. He also provides a conceptual framework for how systems work, and looks at more specific operational advantages to networked DDC.

Dan Agne is Market Manager for Facility Management Systems (FMS) and Direct Digital Control (DDC) with Johnson Controls, Inc. in Milwaukee, Wisconsin. This chapter was reprinted with permission from the Association of Energy Engineers. It was originally published in the proceedings of the World Energy Engineering Congress, Atlanta, Georgia in October of 1992.

The reader who perceives Distributed DDC as just a new generation of energy management system is encouraged rethink that perception. This chapter is important to expand the readers understanding of benefits available from existing systems. Equally critical is the process of developing a summary of the full DDC benefit, when seeking to secure corporate commitment for new systems. Ultimately these actions will raise both the technology and the energy manager in the esteem of top management.

❖

Forget about hardware for a moment. Let's assume the building controls and everything else are working properly. That would certainly please the folks responsible for occupant comfort and for keeping the utility bill in line. However, the value of a controls system can extend far beyond the boiler room. When properly configured and used, it can assist in vital management decisions that affect the entire corporation.

This can best be accomplished by a networked direct digital control (DDC) system. Today's sophisticated facility management systems are just as important as managers, of building control information, as they are as controllers of the actual building environment. In order to gain more support – as well as funding – for the facility management system, the perceived value of the system must be raised throughout the corporation.

As upper management becomes increasingly knowledgeable about technology and its possibilities, DDC users must fully understand how sophisticated facility management systems can help the corporation achieve its goals. This chapter explores the many advantages of networked DDC as they pertain to business management issues. Foremost among these is the control system's ability to organize enormous amounts of data into meaningful reports for a variety of audiences.

RECESSIONARY TRENDS

The late 1980s and early 1990s were not the best of times for American industry or DDC projects. The period was characterized by slow growth, no growth and recession. Business responded by cutting costs and undergoing drastic reorganizations. These took the form of mergers, plant closings, layoffs, delayed purchases and so forth.

Reduction of personnel put pressure on surviving departments to increase automation. Financial staff scrutinized operations more closely than ever before – and not just because of economic factors. Technological advances in DDC have made more scrutiny possible. Among DDC's benefits are tools for users to document procedures, to justify what they are doing and what they hope to do next budget year.

QUALITY EXPECTATIONS

The corporate-wide mission to reduce costs does not mean quality can be sacrificed. In fact, quality must be improved. Quality expectations and standards are increasing in at least three critical areas:
- Regulation
- Marketing and Competition
- Litigation

Regulation – In recent years, environmental and political pressures have imposed tougher requirements on companies that discharge waste of any kind. Many industries are also requiring higher quality standards. Two examples of such standards are U.S. Food and Drug Administration (FDA) validation for pharmaceutical manufacturers and Instituting and the Joint Commission for the Accreditation of Healthcare Organizations.

Marketing and Competition – Improved quality is also being driven by competitive factors. Yesterday's acceptable level of product quality is probably unacceptable today. More and more customers are requiring suppliers to attain International Standards Organization (ISO) certification for their manufacturing processes.

Litigation – The legal arena creates challenges of its own. Many activities that were conducted without a second thought in less complicated times now require careful documentation and justification. Instead of simply keeping things "normal", the trend is to keep detailed records of performance in case a company is ever asked to prove that everything was "normal" after all. This issue applies regardless of a company mission. Manufacturers are good examples, but building owners have seen an upsurge in litigation as well, regarding issues like indoor air quality.

A BIG INVESTMENT

The challenge of improving quality while holding down costs is particularly difficult for those in charge of operating a building. Typical payrolls for plant operations and maintenance can exceed 35% of the total department budget. These are significant costs, which means that many of the business issues discussed in this chapter are being felt at the operations and maintenance level.

As everyone knows, running a building more efficiently requires capital investment. When money is tight, it's not enough for the operations and maintenance function merely to justify its capital funding request. Cost-avoidance projects that look terrific on paper are frequently shelved in favor of more lucrative, though possibly riskier, revenue-producing projects. That's life.

In times like these, departments must compete harder among themselves to receive the money they need. This is where documentation of operations can come to the rescue.

FROM ENERGY MANAGEMENT TO NETWORKED DDC

Networked DDC systems trace their roots to the energy crises of the 1970s, when the skyrocketing price of imported oil prompted the federal government to urge Americans to use less energy. The control industry responded with energy management systems (EMS) for business that closely monitored energy usage.

These systems grew in sophistication and scope. An offshoot called building automation systems (BAS) emerged in the early 1980s, adding some historical data and trend logging, as well as fire and security systems to traditional energy management functions. Applications, however, continued to focus primarily on one objective – return on investment based on utility savings.

The emergence of direct digital control systems in the mid-1980s rapidly displaced analog closed-loop control as the temperature control scheme of choice for large mechanical equipment. DDC brought improvements in accuracy and reliability; however, these systems were force-fitted into existing architectures that didn't anticipate the presence of intelligent, standalone field devices. They were still arms-length interfaces to various building systems, with all major decisions made at a central computer.

Today's customers expect Building Automation Systems (BAS) to transcend the limitations of this interfacing by providing a seamless, integrated network. This is the promise that networked DDC delivers. The architecture of this network is not only fully distributed – providing independent controllers for systems and subsystems within a building – but all of the various components talk to each other in a common language.

The full power of networked DDC can be seen in these four areas:
- Instrumentation
- Decision Support
- Documentation
- Automation

Instrumentation – Intelligent, versatile instrumentation at the individual control level enables the facility management system to do double duty – control the building and allow data to be easily organized for management purposes.

Decision support – A highly organized data collection system allows facilities and operations staff to become more effective managers. They will also be better able to substantiate their decisions because a variety of reports can be obtained instantly.

Documentation – Hard-copy or soft-copy audit trails enhance the ability to substantiate proof of performance, as well as limiting future risk.

Automation – By pressing the right buttons, a networked DDC system can automatically document processes, generate reports and even issue work orders when serious problems are detected at any point in the network.

THREE LEVELS OF HIERARCHY

The primary focus of this book is to cover Distributed or networked DDC systems in extensive technical detail. Yet it is important prior to the technical discussions in the chapters that follow, to clearly delineate networked DDC benefits. These benefits are outlined here, because the building owner must understand them to justify a DDC investment, and to establish a program that makes full use of the technology.

To outline the benefits of networked DDC it is worthwhile to provide a conceptual framework for how these systems work. Networked DDC systems have three general levels of hierarchy or "Architecture"; Distributed Controllers, Building Wide or Island Host Level Control and Information Management. Each level serves an important purpose. The added value of networked DDC is that all three levels are interconnected, much like the personnel structure of a corporation.

Distributed controllers – This is the level where microprocessor-based controllers monitor sensors and control other devices to meet the needs of a specific application. (Think of standalone controllers as employees who are given a high degree of authority and responsibility, along with powerful tools to do their jobs).

Building wide or Island Host Level Control – This level coordinates building-wide control strategies. (Empowered employees need supervisors to coordinate their specialized activities and to provide global direction.)

Information management – At this level, data collected from various points on the DDC system is transformed into usable information. (The combined efforts of highly specialized employees and their supervisors allows top management to make informed decisions about selling the company's products or services most effectively.)

Let's look at each of these three levels in more detail…

DISTRIBUTED CONTROL

Distributed control is a complete control system inside a box – with all the necessary inputs, outputs and control processing logic. These controllers usually bring to mind traditional HVAC functions such as discharge air temperature, space temperature, humidity and fan control. However, distributed controllers can be put to much wider use, as we will discuss shortly.

First, it's important to understand the three major benefits of distributed controllers in an intelligent, integrated network:
- Repeatable success
- Individual control
- Employee productivity

Repeatable success – Multiple control strategies for a specific comfort zone must provide consistent and repeatable results – such as switching from occupied to unoccupied, then reverting back to the exact same comfort conditions for the next occupied mode.

Individual control – It is important to be able to control a specific zone, room or desk – whatever the appropriate level may be. For example, the system must allow individual employees to set their own comfort levels. These devices can be tied onto the BAS network, allowing data to be collected so that productivity can be measured.

Employee productivity – According to a 1992 study by the Rensselaer Polytechnic Institute, West bend Insurance Company in West Bend, Wisconsin, realized a 2% productivity increase in processing insurance claims when the company brought comfort control down to the cubicle level (through Personal Environments). Assuming a salary and benefits package of $45,000 a year, this 2% savings translates into an annual productivity gain of $900 per employee.

As noted earlier, networked DDC brings a Building Automation system far beyond traditional HVAC functions. By employing the same principle of a "complete controls system in a box", networked DDC also can manage lighting, fire and access control. Furthermore, these systems are fully integrated with HVAC. Expanded integrated control functions can also address environmentally motivated issues such as: indoor air quality sequences or CFC leak alarming. These issues will be covered under Chapter 3, "Integrating Automated Buildings".

BUILDING WIDE (HOST LEVEL) CONTROL

A higher level of integrated control is possible by tying together all distributed controllers via a communications network. This creates a reporting path that allows information from one controller to be passed to another.

Through coordinated control sequences, the entire Building Automation system can be monitored and its various functions optimized. All of this can take place behind the scenes, automatically.

INFORMATION MANAGEMENT

Information management is the highest level of control provided by networked DDC. Data from hundreds or thousands of I/O points throughout a Building Automation system can be accessed quickly to be easily organized to assist in management decision-making.

With the proper communications architecture, access to system information can take place at any one of several locations throughout the facility. The most obvious of these is a personal computer workstation. However, the same information can be provided at the Building wide control level and by the standalone controllers.

Information management is important for several reasons:
- Responsiveness to occupant problems
- Regulatory compliance and risk management
- Financial decision-making
- Public perception
- Quality assurance

Responsiveness to occupant problems – Automation can be used to speed the response to problems, as well as their resolution. A maintenance management feature issues and keeps track of work orders, thereby conserving people resources.

Once an automated information management system is established, a department's policies and procedures can be designed to maximize the use of automation, allowing problems to get solved even faster. Response times are now being benchmarked by some property management firms to ensure customer satisfaction. After all, happier tenants remain in the building longer and yield higher profits.

Regulatory compliance and risk management – Documentation is often required for regulatory compliance. This documentation may include testing, proof of performance, critical monitoring, incidence reporting, description of current systems and maintenance management. In addition to satisfying regulatory bodies, documentation is essential in managing and reducing risk.

Financial decision-making – Historical data can be used to identify cost-saving opportunities and provide justification data for budgeting purposes.

Public perception – Through detailed documentation, the results of energy reduction can be shared with the public.

Quality assurance – Networked DDC helps those in charge of quality assurance identify, analyze and improve operations relating to building comfort and security. The Building Automation system is able to monitor and evaluate the facility environment and other important functions of the quality assurance responsibility.

CLOSING THOUGHTS

This chapter has outlined the benefits of a Building Automation System employing networked DDC technology. These systems gather information about the operation of a facility, and make it available to buildings owners in an organized fashion. This information extends to all of the systems withing the building, and helps managers to make decisions about maintaining and staffing the building.

Information management is a significant feature of today's Building Automation Systems, that continues to increase in importance. Chapter 17, "Managing DDC Systems for Results" outlines techniques for putting DDC information to work to enhance systems results and improve the overall building operation.

Chapter 3

Integrating Automated Buildings: Expanded Benefits of making HVAC and other systems work together

EXECUTIVE SUMMARY

There may be something more difficult than getting your children to find harmony together! What is it? ...making the many computer based systems in your building integrate. In spite of this issue, it is unquestioned that Automation systems offer proven results in cost reduction and improved comfort. These systems have been so successful that the global market for systems products exceeds $1.5 billion per year.

This paper recognizes the significance of that global market, and the fact that the two most important trends are distributed processing and integration. In fact, distributed processing makes integration of various controllers a necessity within systems, and opens up the potential for coordinating the control of multiple complete systems. The purpose here is to aide the reader in understanding this integration process by removing some of the confusion around the topic. It is also to identify the best opportunity for integration benefits in the short term, Heating, Ventilating and Air Conditioning (HVAC) control. These benefits, which include cost savings, comfort and even life safety, will be discussed throughout this chapter.

It is important to clarify at the outset that the term integration is used with more that one meaning in the HVAC industry. The focus here is integrating the operation of multiple microprocessors controlling equipment throughout a building. This is critical to ensure that building owners achieve their ultimate goals of supporting the company's mission while optimizing the efficiency of the building. Another use of the term Integration in the HVAC industry has to do with how the microprocessors are installed, either factory mounted or field mounted. This will be discussed later in the chapter, however the reader should bear in mind that our focus in on operational integration of all the microprocessors in a given building.

The chapter will provide the system designer and decision maker with critical information necessary to understand the technology involved, and the basic requirements for integration. Bear in mind that the overall focus of this book is to describe DDC systems, therefore this chapter will do little more than introduce the key components of such systems. Readers who desire more than this brief exposure to DDC and Network Architectures may read beyond this chapter and then come back to revisit the issue of integration.

The focus of this chapter, as stated, will be implementing integration for Heating, Ventilating and Air Conditioning (HVAC) systems. A direct emphasis is on the role of integration in improving HVAC operation. Important topics and critical details relating to integration will be clarified through discussion of the following topics.

- Define Integration to ensure agreement on basic concepts and issues, and discuss Why Integrate.
- Describe system technology, architectures, and benefits of Integration.
- Introduction to implementing Discrete and Integrated system technology.

INTRODUCTION

Why Integration? Before we can ask the "Why?" question, many readers may still be looking for a good definition of the term. So, "what is integration?" may be the most critical issue to discuss first.

In simple terms integration is achieved when multiple microprocessor based systems work together to provide benefits for the building owner. The work required to achieve this is only justified if real benefits are actually available. This chapter will attempt to define integration, and document the process to achieve it, as well as some of the benefits of integration.

Throughout the past decade, system owners have voiced concerns about maximizing the capabilities of Building Automation or DDC control systems, especially when multiple control vendors are involved in a building or complex. In this case the owner has come to realize that each system is an "Island of Automation", and to accomplish some very simple functions requires integrating those islands.

Nevertheless, system owners want to achieve multiple "functions" with energy management, HVAC, security, life safety, etc., and to coordinate control. For example, HVAC equipment has commonly allowed after hours tenant use through a program override, or bypass. This function may be enhanced for the owner through integration with door access and vertical transport systems. The card access verifies that the person entering the space is authorized, and the elevator will not operate for access to higher floors unless a valid card is inserted in the reader. Secured stair well access from the lobby serves to further eliminate undesired access to building space.

Building owners and contractors also want simplified operator interface with all these systems through a single personal computer operator interface. What many have not realized is that achieving all of these things requires "control integration". Perhaps even more importantly, it requires very sophisticated computer skills and, very often, proprietary information, so it is very expensive to achieve.

Many terms have entered and left vogue over the years with regard to integration. One of the more visible concepts in this vein several years ago was "intelligent building", and this basic concept was, and still is, a good one. In fact the Intelligent Building Institute in Washington. D.C., serves as a focal point for information and research regarding this technology around the world. In fact the leaders in this technology are now in Japan, with several Pacific Rim and European countries making significant strides as well. A key factor in many countries is aggressive support from the government including funding and mandated building codes.

In this book, the author defines the term Intelligent Building as appropriate only if control integration has been achieved. An article published in Building Operating Management Magazine's December 1991 issue[1] identified key characteristics of these buildings. They included;
- the use digital control technology for all functional control systems,
- application flexible structure and data pathways to simplify modification and reduce costs by installing shared building communication and electrical wiring busses,
- active contribution to worker productivity through the above, and most importantly,
- integration of building systems.

The author adds a final requirement for the intelligent, or integrated building, that is to optimize building management by offering shared access to all building systems via one central interface for the owner.

Control integration is the true key to Intelligent Buildings because owners would have available to them both full system information and the leveraged power of multiple systems.

Unfortunately mass publication of the Intelligent Building concept in the late 1980s was ahead of the power curve for communication standards and technology implementation. The focal point for this effort at that time was a joint venture between American Telephone and Telegraph (AT&T) and United Technology. This venture proposed a share tenant services concept that revolved around telephone technology. The initial offering was a shared telephone system for tenants in large buildings, with opportunity to integrate other functions, HVAC, Fire, etc. over time.

---

[1] Intelligent Buildings Rebound; Geissler, Richard and Flax, Barry; Building Operating Management, Milwaukee, Wi., December, 1991

As we now know the marketing concept of full building functionality offered through telephone technology did not come to fruition, but the Intelligent Building concept took on a life of its own. Though the concept may still be ahead of the power curve with respect to enabling network and communication technology, Chapter 12 will discuss that topic in much greater detail. This chapter is intended rather to provide an introduction to what is required to accomplish the integrated HVAC functions, that the building owners want, regardless of what we call the building. The basic premise is that building intelligence is an evolving requirement of owners. Further that today's definition of intelligence must revolve around integration as a central theme, and requires new technology in this area to be achieved.

So technical issues aside, the reason to integrate remains constant. Building owners want to optimize system operations, and to achieve the best possible return on their system investments. Ultimately the benefit of integration is many faceted, and involves both cost and operational enhancements for the building owner. Operational enhancements include sharing information and control sequences between various functional systems, HVAC, Fire, etc., as well as Interface Integration. This means allowing Central Operator interface with those systems from a single piece of equipment. The purpose of this chapter is not to provide an all encompassing treatise on integration, but rather to provide the reader with an understanding of the basic issues and requirements.

DEFINITION

Integration, like many other terms in our business, is subject to more than a little confusion. As an author, it is my job to try to resolve that chaos by presenting answers to questions that cause the confusion. Whether I succeed, only you can tell, but this entire book is based on the premise that Distributed DDC systems offer significant value to building owners and operators. That value is unlikely to be pursued by owners if systems are too complex, therefore simplification is critical. In fact, the value offered by DDC systems has been demonstrated consistently, and further there are several other systems installed in a typical building that provide value as well.

Typically each building system is totally self-contained with a discrete purpose. Yet there is a combined or additional value that could be offered through enhanced integration features. Unfortunately, this has only been explored with large scale buildings and systems. This is because the investment necessary to leverage the power of multiple systems and offer new capabilities has only been cost justified in larger projects. Thus these greater benefits have been available to a very limited group of customers.

Given that integration offers value, the requirements to achieve it must be weighed against that value, to ensure that it is justified. As implied above, this is typically done based on cost. The installed cost of a DDC system has two components; labor and materials. This chapter discusses integration along with the cost to achieve it. It also attempts to clarify in general terms both the investment and the benefits that are available. So, we now understand that integration has a value and a cost, like all other aspects of DDC systems, but what is integration?

The answer to that question can be quite complex. Yet all humans seem to share the desire to simplify, therefore consider this. In the simplest of terms integration can be defined as "teamwork between computers." As with a quality team in the factory or a sports team on the field, we now accept that every aspect of a process or task can be improved through teamwork. Wouldn't it be great if you could get the same benefits in productivity and quality with your computer based building systems? That would be the ideal form of integration.

Integration in this sense would offer the building owner significant improvements in building operation, occupant comfort, and cost savings through energy management. This would all be possible by simply leveraging the power in each individual system with the combination of information and power available through multiple systems. The flaw in this concept of implementation is that computers are not people, they can not be motivated, they can only follow very specific instructions (programs) in an unwavering pre-determined order. Nonetheless, the benefits described are very possible.

A secondary issue is that the complexity of the issue increases with the distinction between new and retrofit applications. A new building may warrant awarding the control project to a single vendor who can meet all of the control requirements, including integration. There may be times when an owner, or a general contractor, feels that the way to meet cost targets, etc. is to competitively bid each sub-system; energy management or DDC, fire and so on. In these cases, as in retrofits, where the likelihood increases that multiple manufacturers systems will be installed, integration becomes much more difficult to achieve.

DISTRIBUTED DDC TECHNOLOGY AND ARCHITECTURE

Having outlined the issue and the meaning of the words we are using, it is critical for the reader to next understand DDC technology. The technology that is required to address these issues must be covered first, and the architecture that allows components to work together will be covered in the next section.

In this section the intent is to briefly describe the technology for the primary type of building system of our focus: HVAC. In addition, there will be a brief discussion of the other building systems that may be present. This will help to clarify the general function of these systems, along with the concept of building system interdependence.

An important precursor to discussing integration is to understand the concept of discrete functional systems. With few exceptions, the various systems (HVAC, Fire, etc.) installed in a typical building are proprietary and discrete. This means that they perform a discrete function, and they communicate and control via languages and programming structures that are held as the proprietary property of the systems manufacturer. The term discrete defines limits around their control capabilities. Bear in mind however, that this could include very sophisticated systems, such as building automation and HVAC control for a large central station facility with thousands of controllers. Even with the magnitude of such control however, it is rarely integrated with vertical transport or office automation.

Traditionally in the mainframe computer industry the cost and complexity of such systems has resulted in multiple system integration. This task was done by a specialized contractor called a "System Integrator" Neither the control industry systems themselves, nor the data that they contain, are considered valuable enough to stimulate the development of such specialized contractors. Therefore the "control contractor" is taking on that role. As a result the contractor puts on an "Integrator" hat, and must consider the systems technology in a typical building. The next step is then to determine if the systems are compatible. Perhaps the most important part of that compatibility discussion will be the architectures. To clarify this, a distinction will be drawn between integration with "functional system" and "discrete control" architectures. An example of Functional system integration might be control integration between the functions of HVAC and Fire/Life Safety. Integration within a discrete control system, on the other hand, might be a specific sequence implemented between the VAV terminal unit and air handler of an HVAC control system. Again, understanding these concepts requires at least a passing familiarity with automating the complete building.

The building...

Let's start with a general discussion of Building System Interdependence to identify the key applications. This is a key issue particularly for buildings with systems that provide multiple applications, that we are calling functional systems. Growing use of microprocessor based control for every function in a facility; HVAC, Fire, Vertical transport, etc. means that there is a tremendous duplication of processing power in buildings today. Unfortunately there is not a duplication of effort in providing services to the owner. Of greater concern to this discussion, the power of these systems is not leveraged to provide new coordinated control functions and corresponding benefits to the owner.

A programmer would say that the solution to integrating these building systems is simple, change the system software! Sounds reasonable but, for example purposes, just how many systems are likely to coexist in a typical building?

- Heating, Ventilating and Air Conditioning (HVAC)
- Fire, Life Safety and Smoke Evacuation
- Security and Access Control
- Lighting
- Energy Management and general purpose control
- Telecommunication
- Office automation
- Vertical movement
- Industrial Process

Integrating Automated Buildings:
Expanded Benefits of making HVAC and other systems work together

That's a lot of systems! but is it overwhelming? A good question, yet before we discuss the individual systems any further let's examine the building wide benefit more closely. We have already identified all the possible systems in a typical building, but remember the actual work performed by those systems is done by controllers.

For example purposes, consider a typical building of 150,000 square feet. How many controllers are there in that building for those nine systems? Obviously there are some systems that may not be present, and any number of variables that impact the number of controllers with those that are in the building. Variables aside however, there will undoubtedly be several hundred controllers. With some HVAC systems, for example floor by floor VAV, there may even be more than a thousand controllers. Bear in mind that these are controllers, not thermostats, sensors, motors, etc. A controller, in this context, is an intelligent microprocessor based device capable of monitoring data and executing control sequences, as well as, communicating with other devices over a building wide network. With HVAC or Energy Management functions the system technology generally applied is called "Distributed Direct Digital Control (DDC)".

Historically these building wide systems were not distributed DDC, rather they consisted of discrete control products that carried out specific functions from a central panel. A few of the major manufacturers were able to provide systems, often mini-computer based, that could integrate; HVAC, Lighting and general purpose control along with Fire, Smoke Evacuation, Security and Access Control. These systems, however, were often cost effective only in very large individual buildings or complexes. Smaller buildings have generally relied on totally discrete systems for all of the functions, and integration was not implemented. So there are unquestioned benefits to integration that have been proven over the years in larger buildings.

Smaller buildings could not apply integrated features because the discrete systems were self contained. This presented a problem that might be called an "Island (automation) Mentality". Each control system is an "Island of Automation" by itself, and there is no way to share either information or control.

Island mentality has been referred to by many terms in our industry; closed systems, proprietary protocols and the communication barrier to name a few. The meaning is always the same, regardless of the name. Systems are installed in the same building, often wired sharing the same conduit, but there is no way to share information and communication or to integrate control. If the lighting controller was notified of a break-in by the security system, it may be possible to scare away a thief by turning on all the lights. Unfortunately, this information resides in two separate islands of automation and can not be shared. These functional systems simply do not speak the same language. How many similar benefit examples can you think of for sharing island system data?

Again the Intelligent Building concept highlighted many building wide opportunities for system integration. This building wide integration is exciting, but to go any further discussing it challenges the scope of this paper. Rather than attempting to document the full breath of the opportunity therefore, this paper is intended to focus specifically on HVAC. As a result this section will now move to a brief discussion of individual functional systems. HVAC will be addressed first, and the other systems will also be covered briefly. The focus with these systems will be on opportunities to leverage internal system data and HVAC data to achieve integration benefits.

Under each topic heading four areas will be covered:
- intended function,
- technology,
- architecture and
- integration opportunities.

HEATING, VENTILATING AND AIR CONDITIONING (HVAC)

The topic of HVAC control is intentionally brief here because this technology is covered in extensive detail throughout the balance of this book.

Function

The function of HVAC control systems is obvious to most readers. These systems have evolved over the years from simple electrical and electro-mechanical to pnuematic, and eventually to microprocessor

Integrating Automated Buildings:
Expanded Benefits of making HVAC and other systems work together

based. The microprocessor based, more often called Distributed DDC, are state of the art today. There are two primary functions of any HVAC control; maintaining one or more conditions (temperature, pressure, etc.) and ensuring equipment safety. Historically, most HVAC systems have implemented discrete equipment for each of these purposes.

As noted earlier the use of the term integration with reference to factory mounted controls is a secondary issue to our topic of operational integration of those controller. However, a brief overview of the process that results in microprocessor based technology being installed in equipment is warranted. As noted, Original Equipment Manufacturers (OEMs) have always provided control devices with HVAC units for both temperature control and safety functions. Traditionally OEMs used discrete electro-mechanical, or perhaps pneumatic, controls to achieve individual functions. This was done for many reasons including simplicity, cost and reliability. As a result however, there were numerous control devices in an HVAC unit, and at best their operation was integrated via relay logic and electrical lockouts, etc. The approach to achieve Building Automation in this arena was then for a control manufacturer or contractor to overlay a building wide system in the field and disable or enable the OEM controls. Note that this was rarely done with control safeties, however the status of these was commonly monitored.

The current control trend is for OEMs to combine these various control and safety functions into a single microprocessor that is installed in the factory. The OEMs have also become more active in the field installation of building wide systems as well. This process was originally called "pre-integrating" controls because they were applied to the unit prior to the conventional approach of field application. However, this approach is now being referred simply as "Integration", and that may cause some confusion for the reader. In this chapter our focus will not be on the installation of controls, factory of field mounting, rather it will be on the more essential issue of "operational integration" which is critical to successful implementation of DDC control systems.

Regardless of the installation process, the controls at issue here are normally applied by the original equipment manufacturer (OEM) in the factory. Again building automation controls are usually overlaid in the field to replace or override the conditioning functions of the OEM mounted discrete controllers. Equipment safeties are usually not modified since this is generally unnecessary to meet the owners automation needs. This is also avoided because it can result in a voided OEM equipment warranty. It is common though to monitor safety conditions for reporting to the central operator interface. With the onset of DDC, and the corresponding computer power, it is becoming more common for manufacturers to combine conditioning control and equipment safety in one device.

Technology

Obviously this is an oversimplification of the HVAC equipment control function, which varies dramatically by application. This is being done because the focus here is on integration rather than documenting the breadth of application control functions.

This book will explore the current approaches to applying DDC technology in detail. However, in this section enough general overview will be provided to support a discussion of integration. DDC is applied at the load, or piece of equipment. This far exceeds the control sophistication available with HVAC equipment in the past, while adding the ability to communicate over a network with other controllers.

A critical aspect of HVAC control today is the use of high powered devices as Equipment Level Controllers (ELCs). ELCs provide custom or application specific control sequences. These controllers can then be used to integrate control with higher level (building host) and lower level (zone specific) devices for building wide functions.

Equipment Level Controllers are distributed DDC devices with onboard points: analog and digital inputs and outputs (I/O). These ELCs are microprocessor based, and capable of initiating application control sequences for conditioning and safeties as noted above. These control sequences allow implementation of sophisticated, custom or library sample, control strategies to meet any HVAC application. This is also important to simplify monitoring, alarming and remote communication with other system controllers.

Architecture

Architecture defines a Distributed System structure. An Architecture consists of a group of related devices with a common communication network. DDC "nodes" are intelligent devices that reside on those networks, and are capable of equipment control, networking and remote communication. The architecture and its networking software actually define Control integration capabilities for distributed systems. Such things as protocol, topology, transmission speed and allowable structure determine the compatibility of various building control devices with the architecture.

The control theory or strategy of the system is also important to integrating architectures. To illustrate this, consider a typical strategy: Temperature reset. One type of Reset requires that the air handler control make decisions based upon temperature data from distributed zone controllers. ELCs may then average values or reset to the high or low. Though reset is specific to the air handler and its associated zone equipment, it also requires building wide data. Such data would include outside air temperature, and whether the building is in unoccupied mode. The architecture must facilitate the flow of all this information via devices with varied levels of sophistication.

Benefits and Integration Opportunities

The results of distributing control are new benefits for your customers through precise control, technical capabilities, remote communication and energy cost reductions.

Precise control lends itself directly to occupant comfort and a host of related benefits. In talking with a major property development company, the author learned that their tenants will begin looking for new office space after the third complaint they must make concerning comfort. Comfort translates to productivity, the amount of time a retail customer will spend in a store to make purchases or the quality of a school's learning environment, to name just a few.

Technical capabilities refer to such features as being able to reduce building and equipment downtime through alarming and diagnostics. Other such benefits that come from intelligent building type features for

HVAC would include after hour tenant billing for space conditioning and the ability to enhance indoor air quality.

Remote communication complements all of the features noted above. It allows for a local system to dial out an alarm notification to a location that may respond. At the same time it is possible to call the system and conduct troubleshooting prior to a trip to that site by a service person.

The benefits of energy and cost savings remain viable reasons to invest in systems, and offer opportunity for enhancement through integration. Integrating HVAC and lighting systems to ensure reduction in use is important. At the same time coordinating access control or security systems with these systems would also be desirable particularly when occupants arrive late, and the building can remain in the unoccupied mode for a longer period. There is a direct correlation between these systems because occupancy determines the need for lighting and HVAC. At the same time lighting contributes heating load to the space, and increases the cooling load on HVAC system as well.

Many Operational Integration opportunities have already been discussed in this section, and more will follow. A second key aspect of integration however is monitoring and management of control systems. This is done via simple handheld devices and also with powerful Personal Computer (PC) based central operator interfaces. A final integration opportunity for discussion is the PC and its integration capability. One of the first major drivers for the "Open Protocol" movement of the mid-eighties was the demand by building owners to have one PC interface that could talk with multiple Energy Management Systems. As the author defines "Interface Integration" it includes both interface with multiple Building Automation Systems and with HVAC, Fire, Security, etc. This truly optimizes the owners systems, and increase productivity because they only need to learn one interface systems. The major manufacturers of control equipment can point to fully integrated systems of this type today, but the trend needs to be dramatically expanded.

Interface Integration should not be limited to building technologies however. The author has written for many years that the control industry must make use of "off the shelf technology" in development of equipment. This may be seen today in the growing use of Microsoft

Windows[2] as the basis for Central PC interface software. The true integration opportunity is to take this process a step further and make these systems compatible with off the shelf "Local Area Network" and even "Wide Area Network" products. Most building owners today have networks in their buildings. It is desirable for them to be able to run Interface Integration software on a server, and be able to access the program and manage their building systems from any workstation on the network. Again this type of integration opportunity offers significant enhancement to the value of the building owners systems.

FIRE, LIFE SAFETY AND SMOKE EVACUATION

The heading of this section refers to two separate systems: Fire and Life Safety and Smoke Evacuation. As appropriate, each will be covered briefly under the 4 headings:
- intended function,
- technology,
- architecture and
- integration opportunities.

Intended Function

Fire and Life Safety Systems enable sophisticated sequences in a building to help put out fires and protect occupants from exposure to fire and smoke. These systems also serve as communication links for use by fire fighting personnel. An important factor to note with regard to both Fire, Smoke Evacuation and to some degree Security systems that are discussed below, is that laws or regulations and codes are critical drivers that impact the technology. These also impact HVAC control, for the American Society of Heating, Refrigeration and Air Conditioning Engineers standard on ventilation is essential for control of air handling systems. In this case however, building codes and standards agencies are the primary driving force in defining control requirements and must be strictly adhered to. An example of such impact on these systems is the impact of the Americans with Disabilities Act. This piece of legislation has had major impact on every aspect of building design

---

[2] Microsoft Windows is a registered Trademark of Microsoft Corporation, Redmond, Washington

and operation. It has particular impact on systems designed to ensure life safety. In addition to features that aide in putting out fires, ancillary warning systems have been developed for hearing and sight impaired. Voice alarms that notify occupants of the fire and give instructions are another valuable feature in many systems. Enhancements in vertical transport systems are also critical to speed evacuation of those with physical impairments.

The core function of Fire and Life Safety systems is to aide fire fighters in putting out a fire as quickly as possible. This is done by:
- monitoring to detect the outbreak of fire via points including smoke detection, heat sensors and automatic sprinkler systems,
- warning to notify the local fire department, etc. and the building occupants that a fire is underway,
- communication functions for fire personnel to use during the course of fighting a fire. This feature is used in annunciating the outbreak a fire, giving evacuation instructions to occupants and for inter-building communication between fire fighters and occupants and
- integration with other building systems to aide in safe evacuation of occupants, slow the spread and speed the containment of fire.

In general Fire systems do not directly, or "automatically" control any non-fire related equipment in the building, though there is likely to be interaction between the fire panel and other building systems. Basically this interaction is either manual or automatic. The manual approach simply allows fire fighters to enable or disable equipment from a central control panel. The automatic control concept is more germane to a discussion of operational integration of microprocessor based systems. The most common integration feature in this area is smoke pressurization. The basic sequence has evolved over time to complement the systems primary mission, and aide in putting out the fires and protecting human life. In large buildings, particularly, research has shown that it is extremely difficult to fight a fire while occupants are evacuating en mass. Fire fighter could not reach affected floors, and occupants suffered from extended exposure to smoke. Again the impact of ADA is important here because previously handicapped occupants were in particular danger, and might be trapped in a building. For example,

Integrating Automated Buildings:
Expanded Benefits of making HVAC and other systems work together

improved elevator controls that allow vertical transport during a fire are essential to evacuate these occupants. Older elevator controls were sensitive to heat, and tended to stop on floors with fire.

Smoke pressurization is a strategy using supply fans to exhaust smoke and aid in putting out a fire. The sequence is enabled when a data transfer from the Fire system to the Building Automation initiates the smoke routine. Supply fans on the floor with the fire are shut down, return fans continue to run and this serves to exhaust smoke from the fire floor. The supply fans on floors above and below the fire will continue to run at full capacity, but return dampers are closed to maintained a higher space pressure. In addition on these floors, outdoor air dampers will be modulated to provide 100% fresh air.

The strategy above aides in putting out a fire, and also allows occupants to stay in the building longer. Where building wide communication is available from a Central Fire Panel using telecommunication in systems, occupants can be notified of status and whether they should move to another floor or evacuate. Again code requirements spurred by the ADA will raise the level of importance placed on these functions, because more control over egress is essential to the safety of disabled individuals.

Smoke evacuation systems are very different from the systems discussed above. True, they are initiated upon detection of smoke or fire, and in many cases are mandated by local building codes, however they are totally discrete. Rather than integrating functions of other systems, this is self contained equipment initiated by dedicated monitoring devices. Upon initiation the major component of the system, roof mounted variable speed exhaust fans, are enabled to run. Again, these systems are focused on protecting occupants from a major cause of injury and death, smoke inhalation. However, there is no integration and therefore we will not discuss these systems further under this section.

Technology

Fire and Life Safety Systems are highly sophisticated systems with the primary functions outlined above. The significant enabling technologies in a Fire and Life Safety system are:

- the sensors, manual pull stations and ancillary gear used to identify a fire condition,
- the communication capability of the central fire panels and
- the networked integration that is possible with other building systems.

There are several companies offering Automation systems that are approved by the National Fire Protection Association (NFPA). An automation system that meets NFPA 72 can serve as a Fire System, and this simplifies the integration process. This is generally only cost effective, however, in larger buildings. As noted Fire systems do not directly control any equipment. The primary functions as noted, have historically revolved around detection, annunciation and communication of alarms. These remain the hallmark functions of these systems.

In recent years there has been greater utilization of integrated functions, such as smoke pressurization, via the fire panel. Though the concept of manual and automatic integration functions was discussed above, it is reasonable to consider the technology involved as well.

Manual integration technology implemented through Fire systems directly involves fire fighting personnel. This technology allows fire fighters to initiate "start/stop" commands to any major equipment, such as supply and exhaust fans, fire pumps, emergency power systems and other equipment. Manual control of this equipment may be strategic to containing a fire or slowing it's progress in spreading. Again manual operation does not utilize integration power of other building systems and is less germane to this discussion.

Automatic integration functions are important in containing fire as well. In addition, these strategies may serve as preemptive approaches for both fire fighting and life safety. In essence the fire panel becomes a monitoring and communication system for use during the fire to coordinate and manage the containment efforts. In addition, these systems may also become the vehicle for enabling sequences in other building systems for integrated fire functions, such as smoke pressurization.

A well designed integrated system will detect a fire and respond with coordinated control sequences, and annunciation, well in advance of the fire fighters arrival. Ultimately this is critical to cost effective implementation of successful systems. Key technology required to enable such sequences falls in 3 categories: communication hardware and firmware. Hardware refers to the physical wiring bus connecting the systems and any interpreters or gateways that are required to translate commands from one system to another. Firmware is the control programming in the HVAC equipment that executes a sequence based upon a command from the fire system. To ensure successful implementation, the local controller must take charge of the sequence. In a fire situation, any number of variables could disable communication between these various systems and the central panel.

Architecture

As may be evident from the centralized nature of Fire systems, there architecture is quite simple. The primary fire panel is the focal point for all activity. In general this is a central processing type of system with a central master panel. The panels generally have inputs only that bring back information from all the monitoring devices installed throughout the building. Output capability is very limited, even with systems that allow manual integration. The primary output capability is for telephone communication to notify fire personnel of alarms.

The reader should not get the impression that networking is not applied with Fire systems however. Several manufacturers offer multiplexed communication between fire panels. These multiplexed systems enhance both communication speed and reaction time. One of the reasons for this is that they are addressable and take a zoning approach to fire detection by monitoring pull stations, sprinklers, smoke detectors, etc. Therefore critical areas can be pinpointed more quickly allowing greater effectiveness on the part of fire fighters. In addition, the systems that allow expanded functional level integration, with features like smoke pressurization, are capable of more sophisticated architectures.

Integration Opportunities with Fire and Life Safety

Integration with this application has already been discussed in several examples. There are significant enhancements possible through integration, and these take on a greater level of import when human life is involved. The ability to reduce human tragedy is a major integration opportunity. Sequences such as smoke pressurization are important in this regard. Other integration opportunities are important for building occupants with disabilities. Expanded integration with Close Circuit TV to identify where individuals may be trapped, and then to control vertical transport and automatic doors are new opportunities. Again Interface Integration with PC systems that can incorporate both closed circuit TV signals and monitoring/control of functional systems is an incredible opportunity. It would be possible for Firefighters to sit at a command console with large screen monitors and have access to both video and real time data. Using split screen technology the Firefighter could display both pictures and information on critical fire floors simultaneously. These and other integration enhancements may be implemented on wider scale in short term as a result of the Americans with Disabilities Act.

SECURITY AND ACCESS CONTROL

Again, the heading of this section refers to two separate, yet related, systems: Security and Access Control. As appropriate each of these systems will be covered briefly under the 4 headings:
- intended function,
- technology,
- architecture and
- integration opportunities.

Intended Function

The systems identified under this heading are related and yet there are any number of offerings in the marketplace that provide these functions through distinct products. The similarity of the technology however allows them to be discussed together except as appropriate.

Security systems are primarily intended to protect people, information, property, etc. through monitoring and notification of law enforcement for rapid intervention. These systems monitor any unauthorized attempts to enter a facility or, an area within a facility, and also serve as communication links to the outside world, because notification of law enforcement agencies and rapid response is critical to meeting the mission of these systems. The core functions of Security systems are:

- monitoring to detect any breach of security through doors, windows, etc. or tampering with specific areas, containers or display cases (ie: museum or retail jewelry store),
- notify law enforcement personnel if a breach of security occurs
- annunciating the breach for any occupants that may be in the building.

As with Fire apparatus, these systems do not directly, or "automatically" control any non-security related equipment in the building. There may be interaction between the security panel and other building systems. An important factor is that there may be varying levels of security by area within a building. This may require more sophisticated control sequences that integrate elevator and stairwell access even after initial entrance to the building is granted. Also there are impacts here from the Americans with Disabilities Act. These types of features will be discussed further under technology.

Access Control systems are different from those discussed above. These are automated vehicles for controlling entrance and egress from the building during off hours. Access control tends to be a discretionary product that building owners implement for various reasons. The function of these systems is highly complementary to building security because they enhance the owners control. As a result these functions are often combined, and acquired from the same vendor. Therefore they will be discussed together under the following headings.

Technology

Security and Access Control Systems are highly sophisticated systems with the primary functions outlined above. The significant enabling technologies with these systems are:
- hardware both distributed and centralized,
- the various sensors and ancillary gear used to identify a security breach,
- the card reading and individual tenant identification technology for access control and
- the networked integration that is possible with other building systems.

As noted Security systems do not directly control any equipment. The primary functions have historically revolved around detection, annunciation and communication of alarms. These remain the hallmark functions of these systems, but Integrated control function that are possible will be discussed below. Key technology factors include expanded use of distributed system architectures to limit single point of failure concerns. These enhancements also bring added demands because rapid response requires that hardware incorporate both high speed communication between network members, but also very fast control loop execution at the processor level. A wide variety of sensors are available to sense those changes in condition that the processor is monitoring including: motion detectors, heat sensors and x-ray scanners. At the same time network integration with other systems can bring added information from such sources as Close Circuit TV, card readers and elevator use logs/status.

Access control equipment again incorporates a subset of the features noted above. In addition there are some direct control requirements to enable door access through various mechanisms. A common integration function today is for these systems to be integrated with vertical transport. Various strategies are implemented with the goal of limiting access to the entire building simply because the front door is opened.

Architecture

As with fire panels security systems may be distributed or centralized. Architecture are somewhat simple, though again high speed data communication is critical for both internal system functions and integration with other systems. In particular access control architectures are simpler because the equipment is generally located close to the entrance and does not require complex communication buses. The panels generally have inputs only that bring back information from all the monitoring devices installed throughout the building. Output capability is very limited, even with systems that allow manual integration. The primary output capability is for telephone communication to notify law enforcement personnel of alarms.

Integration Opportunities with Security Systems

There are a number of integration opportunities with these systems, many of which have already been discussed. In general they fall into the same two categories: securing the building and controlling access. In the area of security, information is critical to effectively and safely dealing with a building break in. The interaction of systems can allow a Closed Circuit TV (CCTV) system or the motion detectors that are used for lighting control, etc. to provide clues to the felons whereabouts. As noted earlier, the author has worked on many systems in the past that turn on the lights after a breach is detected. Other functions, similar to those used with integrated fire systems, might involve monitoring door openings and elevator status with security systems. It would also be possible to use the same phone and public address media as the fire system to address felons that are still in the building.

Access control integration opportunities for information include dial up or card access data sharing to HVAC controllers. With this data it is possible to begin conditioning the tenants space prior to their arrival, and provide better occupant comfort. Integration with vertical transport can limit after hours access to floors other than those occupied by a specific tenant. Shared information on status of access system use could be essential to saving lives if a fire breaks out during a normally unoccupied. Fire personnel can determine quickly if anyone is in the building.

ENERGY MANAGEMENT, GENERAL PURPOSE AND LIGHTING CONTROL

The heading of this section refers to three separate systems: Energy Management, General Purpose and Lighting. Again the similarities allow these systems to be covered together with specific references as appropriate under the same four headings;
- intended function,
- technology and architecture and
- integration opportunities.

Intended Function

These somewhat distinct systems are linked here because of their historic interrelationship. Energy Management systems are typically related with a combination of building wide functions and specific control of general purpose and lighting equipment. Early computer based systems combined many of these functions, in some cases, including HVAC, fire and security as well.

Energy Management System is a term that has given way to Direct Digital Control (DDC) or Building Automation System. Yet many of the critical control functions implemented with any systems for either energy cost reduction or enhanced comfort are generally referred to under the category of energy management functions. Such functions will be covered in far greater detail later in the book under Building Wide control, but they include Time of Day programs, Power Demand Limiting and Optimization. General Purpose control refer to a subset of the Energy Management routines through it is generally focused on start/stop interface for a wide variety of equipment, usually simple electric equipment. The term general purpose is normally intended to distinguish control of HVAC and lighting from other equipment such as domestic water systems and ancillary pumps or fans. The equipment controlled may not be small, such control may be used with a 500 horsepower pump or a unit heater. The key is that control simply requires enable and disable functions generally implemented via electrical relays or contractors.

Lighting control is combined in this category because it is still common for on/off control of this type to be used for interior and exterior lighting. Lighting in the same area is grouped together on the same circuit breaker, and a contactor is interfaced to automatically switch that lighting on or off. A second and much more sophisticated type of control system takes direct control of the light load via power provided to the fixtures, and or electronic control of ballasts. This type of system is capable of much greater variability in the light output, measure in footcandles or lumens. Such systems are able to adjust for variations in ambient light available from windows with a proportional increase or decrease in measured light. These systems can also provide lumen maintenance, a strategy which ensures that a pre-programmed light level is always maintained. This is not an energy saving routine, however because lamps over time decrease in light output. To achieve lumen maintenance the system must increase power to the fixture, thus increase energy use. To simplify this discussion, our focus will treat lighting control as primarily general purpose on/off control noted above.

Technology and Architecture

The key technology driver in all these areas is distributed DDC. This means that primary technologies are both the dedicated controllers implemented at the application level and the distributed network. As a result there is a proliferation of distributed controllers available to provided the functions discussed above. These controllers may be HVAC, lighting, etc. however it is key that they are applied at the control level and maintain a complete sequence of operations. This sequence will outline both the requirement for the local apparatus, and control scenarios for implementing building wide routines.

Again the second critical technology is network integration. Technology and integration are discussed together because control integration between multiple devices may be the most significant requirement for all three systems. The key factor is to ensure that a hierarchy of interactive routines is implemented that allow local sequences that reflect building wide control scenarios. For example, a building wide routine is enabled at the Energy Management level for Demand limiting. Strategies must allow for a network wide call to implement shed routines that turn off electric apparatus and avoid a power demand peak. This strategy may involve decreasing lighting in non-essential areas, turning off pumps or sending a signal to HVAC units to increase a cooling setpoint. The local controller scenario is a critical element in these strategies because it may not be possible for this device to implement the command. For example, a command to increase a cooling setpoint in a zone that is experience high load conditions may result in a loss of control and long term unacceptable comfort. The controller must be able to compare current conditions to a set of parameters and determine whether or not to enable a routine. Ideally the controller should also be able to send a message on the bus confirming whether the routine enabled or not.

Integration Opportunities

The systems outlined in this section are totally dependent on integrated control. As noted earlier in this chapter, distributed systems must provide a measure of integration to match the functions that were provided by CPU based systems. Yet there are a number of new integration opportunities also. Many of these have been noted under discussion of other types of systems. Such features would include turning on lights when a security breach is detected. With expanded system integration it may also be possible to optimize lighting system use based on motion detectors from security systems. Rather than turning on lights throughout the building based on time of day, why not wait until the first tenant is detected entering the space. Ideally this would be done based on entrance to a buffer area which has already had the lighting energized. Further integration opportunities are limited only by the imagination of systems designers and users.

OTHER BUILDING SYSTEMS

The following systems will only be covered briefly. This section will not fully discuss the topic areas covered with the above systems, as they are new to the building automation integration. The reader should not mistake this to mean they are unimportant, rather it is more essential to focus on the key topic of building automation. The key topic areas for discussion will be intended function and integration opportunities with traditional DDC systems areas already discussed.

Telecommunication

Intended Function

The intended function of telecommunication systems is ever changing. These system were initially focused on voice communication. Yet the world has watched the dramatic growth in use of voice grade lines for many related functions. These include data communication via modem and fax along with Wide Area Networking and more.

Integration Opportunities

The integration opportunities for use of telecommunication are limitless and many have been discussed throughout this chapter. Some of these include; remote system communication and Fire integration for communication with occupants during a fire situation. There will be further discussion of remote communication later in this text to outline Interface Integration related opportunities for telecommunication. At this point it is only necessary to say that telecommunication is more a requirement than an opportunity with DDC systems.

Office Automation

Intended Function

This area is one of the faster growing in the building sector. Five to ten years ago the energy use from computers, fax machines, laser printers, etc. was negligible, if at all. Of course the large computer room was a significant user, however the decentralized computer based equipment was not an issue. Today this equipment accounts for 5 to 10 percent of the energy use in an average commercial building. The intended function is obvious, these are the technological devices that have become modern business realities in completing every critical office task.

Integration Opportunities

The reference above to energy consumption brings to light a direct and simple opportunity for integration. Energy Management based control of office equipment is a possibility. The reader may find this an unlikely suggestion at first consideration. You may be interested to learn however, that International Business Machines (IBM) Corporation is planning to introduce an energy management based system that will shut down main frame computers. This system will interrogate a mainframe to determine if it is running a batch operation, etc., and based upon the data will shut down the main frame completely or its peripherals that are not necessary at this time.

Vertical Movement

Intended Function

Again vertical movement, or elevator systems, are integrated with buildings across the size spectrum. Quite simply the function is to allow quick and easy access to all building floor. This is particularly critical when we consider the Americans with Disabilities Act and access for disabled individuals. There has been a significant technological leap with these systems as microprocessors have been integrated into the control of elevators. As we have noted in earlier sections the control has been improved, and these systems have become important to dealing with Fire and Security control for the building.

Integration Opportunities

These systems can be very energy intensive based upon horsepower requirements, thus offering energy saving opportunity for integration. It may be possible to shut down an elevator during peak periods where multiple elevators are available. There certainly have been a number of approaches taken by elevator manufacturers to address energy costs as well. The integration of elevators with Fire and Security has already been discussed extensively and is a real opportunity as well.

Industrial Process

Intended Function

Industrial Process control is not the focus of this text, and warrant a great deal more elaboration than is possible here. The function is automation of key aspects of any industrial process, heat, robotics, etc. Typically highly specialized and dedicated Programmable Logic Controllers (PLCs) will be dedicated to each critical aspect of a process. These devices focus on such important aspect of a manufacturing operations mission that they do not incorporate multiple process loops or building wide functions of any sort. A typical device might be dedicated to ensuring that a plastic injection mold is maintained at the precise temperature necessary to melt the media and inject it into a mold on a periodic basis. The requirement is for extremely fast update of key control points and execution of control sequences. Traditionally these devices did not communicate on bus of any type. As the reader may realize however, the concept of distributed control was born in this business and networking of controllers has been common for quite some time.

Integration Opportunities

There are a number of options for integration with other building systems. Operational Integration has been a critical short term need for this business. This is because industrial processes very often have specific requirements for temperature, humidity and other ambient conditions. In order to optimize the process and ensure the highest quality product, it is necessary to integrate these various functional operations. Further, a key need with Industrial Facility Managers in Interface Integration. Particularly where process and other functional control areas managers must be able to monitor and modify every factor that impacts product quality.

INTRODUCTION TO IMPLEMENTING INTEGRATION

This chapter has covered a wide range of extremely complex systems data. The functions of the individual systems are well established, and in spite of ongoing enhancement these technologies are not in a state of complete metamorphosis. The integration of these functional systems, both Operational and Interface, is a somewhat new and rapidly evolving area of technology. The intent of this chapter has been to give the reader a big picture view of the building and all it's systems. With this broad based knowledge the reader should be better able to continue through the text, and take a more detailed look at the various components that make up this technology. Prior to exploring that technology in detail however, there are several topics that equip the manager to understand what is required to develop a successful program.

In closing this chapter, it is warranted to briefly discuss some of the key requirements for ensuring that a DDC installation program considers integration. The three key topics that must be considered by a business that is investing in this technology are:
- monitoring the integration standards process,
- developing Intelligent Specifications and
- being technology aware.

Monitoring integration standards may sound like a good job for a University Professor or a researcher. Most building professionals don't spend time learning standards until they are finalized. The problem is that the technology itself, as well as the standards that regulate it change rapidly. This means that anyone who is purchasing, installing or managing and living with these systems should keep track of the process, particularly as you plan for the future. The leading Integration related standard process in the control industry today is being managed by the

American Society of Heating, Refrigeration and Air Conditioning Engineers (ASHRAE). This organization has developed many standard in a wide variety of areas. In this case a proposed standard, Building Automation and Control Network (BACNET), is under development by ASHRAE. ASHRAE develops these standards by mobilizing a highly qualified group of volunteers drawn from the industry. These volunteers are employed by manufacturers, engineering firms, end users, etc. and their companies allow them to contribute time to this process in the interest of long term industry benefit.

BACNET is not the only standard under development, but it is the most prominent. Participants in this business should track it, and know the timing for completion. Current projections are that it is not likely to be published in final form before 1995, and BACNET compatible controllers will not hit the market before 1996 at the earliest. Understanding the process and tracking it, however it important to ensure that the company who is investing in this technology gets the best possible solution to their control needs.

Developing Intelligent Specifications is directly related to monitoring industry standards. It starts much more simply though. The first step is to clearly analyze the users needs for control systems. It may sound simple, but there are many systems installed every year which did not include this step. Hiring an engineer to define the buildings needs and specify a system is not enough. The owner must clearly define "needs" It is easy to confuse needs with technology, but they are very different. An example of defining "Needs" is:
- comfort condition must be maintained at 70 degrees for heat and 75 degrees for cooling,
- energy management strategies be implemented to keep costs at or below $1.15 per square foot,
- the fire systems should meet local codes, but also provide for the optimum in life safety, etc.

The needs outlined above are not technology statements, they are requirements for the way the building is used. Once the owner understands and clearly defines what is needed writing a specification is much easier. It is actually recommended that this definition process be part of a "Strategic Plan" for the building that addresses both short and long term needs. It is then possible to make intelligent decisions about the investments that are made in the building. At the same time it is possible to choose a vendor who is aligned both technically, and even philosophically, with the owners long term view of the facility.

The final area to consider in implementing integration is knowledge. First and foremost the ultimate manager of this technology must be aware. Study this and other related texts to gain an understanding of the technology. Engage in intelligent specifying, but also remember that the industry is not static and monitor standards processes and other technology breakthroughs. The combination of these efforts along with the information you will find throughout the balance of this book will start you on the way to making truly informed decisions on automation investments.

Chapter 4

Energy Management 2001 - A View of DDC in the Future

By George Owens, P.E.

INTRODUCTION

In the first three chapters of this book topics have focused on the big picture issues facing building owners today. Current control technology was viewed as well as, the benefits that this technology offers to building owners. Chapter 3 outlined more traditional building management functions along with the broader benefits of integrated building control. Discussion of multi-functional systems we traditionally think of, such as Fire and security, were supplemented with new concepts like mainframe shutdown for energy savings.

It is valuable for the reader to consider all of these alternatives, because all humans have a tendency to get focused on the tasks and processes that are familiar. Unfortunately allowing this to happen can result in missed benefits from integrating the power of multiple systems. The author did an article several years ago about having a Building

---

George R. Owens is Director of Engineering with The Rouse Company of Columbia, Maryland. It is reprinted with permission from the Association of Energy Engineers and Mr. Owens. This chapter orignally appeared in the World Energy Engineering Congress Technical Proceedings, Atlanta, Georgia, October, 1991.

Automation System take an input from the security panel, and turn on all the lights if a building break-in occurs. This very simple hardwired feature can be dramatically expanded today through integrated communication and even coordinated control by these systems. Flexing our control application intellects to discover these types of benefits is essential to harvesting the greatest possible benefit from a control system.

One of the best examples of creativity in control application intellect to be published recently is George Owens' technical paper "Energy Management 2001." This paper is an excellent way to set the tone for a reader's perspective on the balance of this book. George does an excellent job of identifying trends and considering how those trends will impact users. Whether all George's predictions for 2001 come true is not the issue, rather the reader should accept that everyone in this industry must approach the future with his perspective. Probing and questioning the direct user benefits of breakthroughs in technology allows you to identify opportunities to improve the performance and therefore the value of systems.

The balance of this book will discuss the current state of the art in a very matter of fact way with some further prediction for the future. Use this as a baseline for approaching technology solutions to your problems, and always be open to new ways to solve those problems. This doesn't mean buy new equipment regularly, instead the suggestion is to optimize current products to the fullest and use new technology from every aspect of the industry to achieve greater increment gains in performance.

With that lead in, consider George's view of "Energy Management 2001."

❖

Since the first cave dwellers banked down the fire with ashes at night, we have strived (sometimes more, sometimes less) to reduce energy costs. Their motivation may have been higher than ours, with tyrannosaurus rex chomping around while they were out collecting firewood.

Even during the 1800s and early 1900s, when utility prices were decreasing (remember utility prices did not start rising until the 1970s), the control of energy was important. Automatic controls were developed in the late 1800s and a workable night thermostat was developed in the very early 1900s.

Not until the 1970s, with multiple energy crises, and increasing costs, did the ideal of conserving energy become really important to building owners. A profession was born (the Energy Manager), and equipment developed to reduce costs. Out of all this, the most important new technology was the advent of the computerized Energy Management System. These early systems provided centralized control, unattended, with electronic accuracy. However, these early systems were bulky, not user friendly, unreliable, and very expensive.

During the 1980s, technology changed the face of energy management. The personal computer (PC) and increasing computer literacy of the general population was responsible for improving the performance of Energy Management Systems in the area of increasing complexity and paradoxically, easing the use by the operator. However, the systems stubbornly refused to yield to major price reductions and even more complete user friendliness.

Now it is 2001 and, looking back at the history of Energy Management Systems in the 1990s, who could believe that you had to punch keys to get data in and out of the machine – how archaic!

Let's take a look at just some of the more mundane features of today's systems that, except for a select few brilliant and farsighted individuals (such as the author), were not even being considered just ten years ago.

ENERGY MANAGEMENT 2001
INTRODUCTION

This paper is written in an attempt to predict what an Energy Management System in the year 2001 might look like. The author has used a combination of fact, speculation, and a sprinkle of science fiction to take today's Energy Management System through ten years of evolution to the year 2001. Although I consulted with the Research and Development department of the major Energy Management System suppliers and several smaller companies' principles, the predictions contained herein are my own, based upon that research. The predictions are also based upon an owner's and a specifier of Energy Management System's desire (the author) to solve some of the shortcomings of today's offerings. I want to thank all who suffered through incessant questioning in my quest for the future. Thank you.

COMPUTING COST/PERFORMANCE

The single most important reason that the Energy Management System of today is even possible is due to the invention of the computer and the dramatic drop in price with a corresponding increase in performance.

A summary of the author's approximation of price and performance is tabulated below:

ENERGY MANAGEMENT SYSTEM

Computer Cost

Year	CPU Cost	Bytes	Cost/Megabyte	Comment
1970	$50,000	16000	$3,125,000	Mainframe
1975	$20,000	32000	$625,000	Minicomputer
1980	$10,000	64000	$156,250	IST PC's
1985	$5,000	512000	$9,766	IBMPC
1990	$1,000	100000	$1,000	Compatibles
1995	$100	4000000	$25	Micro
2001	$1	100000	$1	1 Chip PC

The earliest Energy Management Systems were installed in the early 70s and utilized small "mainframe" style computers that were costly, bulky and hard to program. The term "user friendly" had yet to be invented. The next generation (late 70s) graduated to a minicomputer or a vendor's proprietary CPU design.

Finally, someone got around to inventing the PC (Personal Computer) and the real revolution of computing power for the masses began. My first PC in 1979 consisted of an 8 bit CPU, 4k memory, 4 megahertz, tape recorder, and a monochrome monitor. When I expanded to 64k, a 5¼" disc drive and a printer, I thought I was on top of the world and at the pinnacle of technology development. The cost was $3,600. Today, about one half of that cost will buy a tenfold increase in performance. Sixteen and 32 bit CPU's at 25 to 35 megahertz utilizing one to four megabytes of RAM, hard drives, laser discs and color monitors are being offered by dozens of vendors.

In about five years, hi-performance CPU prices should drop to at least $100, and may approach $10. As these prices drop, you will see more and more computing power decentralized from the head end of the Energy Management System. Smart controllers will, then, be the norm not the new kid on the block.

Now take technology another five-year leap and imagine a $5.00 or even a $1.00 full function CPU taking up 25% of one large scale integration chip. At that price, the cost of the box is more than the computer itself. At that price, everything – motors, HVAC units, VAV boxes, relays, light fixtures, receptacles, appliances, and maybe even each and every light bulb will contain it's own uniquely programmed CPU for control.

With all of this greatly expanded computing power available at a very low cost, the next area that must evolve rapidly may also, in some ways, be the biggest obstacle to overcome.

SOFTWARE

Software development has not kept up with the quantum leaps of computer hardware evolution. Yes, I know that when compared to the systems of the 70s, the software of today has improved. Today we have computer graphics and many more software routines for controlling HVAC equipment. However, the basic control strategies that I wrote about and presented in a paper at the Association of Energy Energy World Energy Congress in 1980 are essentially the same today as then. These strategies are/were:

- Scheduled start/stop
- Optimized start/stop
- Temperature compensated duty cycle
- Temperature compensated demand shed
- Temperature reset:
    a) chilled water
    b) hot water
    c) condenser water
    d) supply temperature
- Night setback

The programs of today still primarily rely on the Energy Management System supplier's programmers to implement the above strategies, based upon their expertise after reviewing the building. It is rare (if it even exists) that major changes are made to the software, once installed, to improve the operations or to provide further energy cost reductions. The reason for this reluctance to make major changes is the number of hours and cost for the programmer to set up the system in the first place. Software cost (both factory and field) for a major Energy Management System may run from 20-30% of the total system installed cost.

The major breakthrough in both software cost reductions and improvements in the sophistication of control will be wrought by the computer itself and not the programmers. Artificial intelligence is a term that has been around for a few years but by 2001 will be firmly entrenched in our computer culture and Energy Management Systems. More specifically, expert systems will be developed, possibly utilizing some form of fuzzy logic, that will self-develop and continuously optimize control strategies. An expert is one that has a comprehensive set of rules in the base operating system. These rules would generally define how outputs should react to certain input signals. These signals might come from sensors, the operator or other outputs. But this is not enough. A self-optimization sub-routine would take over and modify the expert system as the computer itself learns how to uniquely control each building.

Even expert systems can only react to what is happening after it occurs. And, although these actions occur very soon after the even, that is often not the optimum situation. Imagine if the system could simulate fully the operation of the building and could accurately predict exactly what would happen prior to its occurrence. The level and accuracy of control and optimization of energy use would be phenomenal. By 2001, all of these scenarios may be commonplace.

COMMUNICATIONS

Again, going back to the earliest Energy Management Systems, all external devices were hard wired back to the computer. Then a distributed format evolved utilizing multiplexed signals over a common wire or even over the electrical distribution system (power line carrier).

The multiplex system did a lot to reduce the cost of wiring from the remote panels to the computer but did not reduce the cost of wiring input/output devices. Power line carrier systems have not taken over a large share of the market due to some earlier reliability problems. Both of these systems have a common drawback in that the response time goes down as the system gets larger. The system response time will probably be sped up in the next five years by a combination of putting more and more computing power out in the remote multiplex boxes and even in the input/output devices themselves. Also, the throughput of information can be increased by switching from the twisted pair wire signal carriers to coaxial cables or fiber optics. Still these improvements offer no solution to high installation cost.

By the year 2001, communications between sensors and multiplex boxes and the head and system will be by a combination of technologies:
- Traditional means such as twisted wire, coaxial, etc.
- Non-traditional methods such as:
    a) infrared
        - or -
    b) radio wave

OPERATOR/MACHINE INTERFACE

First there were pilot lights, then single line light emitting diode displays. Next came CRTs with text only in monochrome. Today's technology includes graphics and color with live data updated once a second. But again the programming of this data format takes a considerable amount of time, effort and cost.

In the future, due in a large part to the increase in computing power and expert system software, the building's complete set of plans will be loaded into the computer. On new buildings this will be a simple matter of popping in a disc from the CAD (Computer Aided Design) system that designed the building. For existing buildings the drawings will be scanned into a CAD system and then be utilized in the Energy Management System. The system will then start to automatically synthesize this information into formats that ease the operator's ability to interact with and direct the system.

Multi-media computers will combine video, CD-ROM, audio, text, pictures and sound to enhance communications and understanding of both the computer and operator. The system will fully customize this interface automatically for each and every operator, anticipating what the operator will desire to know and, then, provide a display/sound environment before being queried.

Supplanting today's operator's interfaces may be a concept called "virtual reality." This is currently being experimented with at major universities and research corporations. In virtual reality, you no longer are told nor observe what is happening – you become a part of it. By donning special headsets, gloves and even a full body suit, the operator will fully experience a building in operation. After receiving a complaint of a hot temperature, an operator would zoom immediately to the space to feel and measure the temperature. With a twist of the wrist, we would, then, be inside the VAV box looking at damper position, seeing readouts of air volume and temperatures. The thermostat would be readjusted while observing the whole system's operation. After the problem is solved, the operator then moseys over to the machine room and checks the operation of fans, boilers and chillers. All this without having to leave the control console.

I can envision the following scenario by 2001:

- An operator wants to add a sensor to a previously unmonitored room.
- The operator goes to the storeroom and picks up the $10 sensor.
- When the operator peels off the self-adhesive backing and sticks it on the wall, several things simultaneously occur:
  a) Power is supplied to the unit by a built-in solar cell with battery backup.
  b) The sensor starts broadcasting that it is now alive. This broadcast may be infrared, radio wave, or microwave.
  c) The computer recognizes the sensor, assigns a point number internally to the computer and within the sensor.
  d) The system will know the location by triangulation of the signal and its internal map of the building.
  e) The system will then start a self-optimization routine to discover the appropriate control strategies to utilize this new sensor.

- By the time the operator returns, the data for this sensor will be fully integrated with the man/machine interface. Information will be presented to the operator in the format that the operator prefers without ever once issuing a request.

CONCLUSION

The history of controls and Energy Management Systems indicates that the performance has increased more than tenfold each decade. By 2001, wireless communications, self-optimizing software and improved operator interfaces will far surpass any of our current expectations. The systems will be easier to use, provide better comfort, reduce energy costs and be less expensive to install. The term Energy Management System will no longer exist. All of this will be just "business as usual".

Chapter 5

Ensuring Successful DDC System Projects

PREFACE

This chapter is focused on success and return on investment for DDC Systems. In the author's experience there are two key criteria for ensuring results:
- good system design and implementation and
- effective System Management.

The importance of these two topics cannot be stressed enough. In fact, this chapter is nothing more than an introduction to DDC System Design and Operations. The balance of the book is primarily intended to provide the reader with enough information to make intelligent decisions in these areas. The technology discussions should aide the reader in becoming fluent with this equipment. In these specific areas there are additional chapters as well.

Regarding design and specification, Chapters 15 and 16 are intended to give you an understanding of this process. Unfortunately there are no hard and fast rules that can be learned because of the dramatic variations in customer requirements. These chapters however will provide you with a good foundation for establishing an effective system acquisition program.

In the area of operations, DDC System Management may be the most important chapter in this book. A thorough analysis of functions and tools for this process is included in Chapter 17.

Again to provide an introduction to these topics, this chapter includes information from two contributing authors. These two authors, a consultant writing on design and an end user describing system operation provide an excellent summary of concerns in system design and operation.

R. Krout of Heery Energy Consultants, in Atlanta, Georgia contributed a portion of his paper from the World Energy Engineering Congress on Design Guidelines for systems. In this section of the chapter the focus is system concept, design and installation. Though there are aspects of the section that may appear directed at larger projects, the suggestions apply to systems across the board. The topic of design is broached here as an introduction. Due to the crucial nature of system specification, two chapters have been devoted to these topics, Chapter 15 and 16.

R.F. Barsoum of Waterloo County Board of Education in Waterloo, Ontario, Canada provides an excellent overview of the operational requirements system success. In fact Chapter 17 goes into extended detail on this topic because of the crucial nature of this issue. In this introduction however, Mr. Barsoum does an excellent job of reviewing the "crucial links" of activity that are required to successfully operate DDC systems.

Design Guidelines for a Customized EMCS

By R. Krout

A customized engineering design is crucial to procuring an Energy Management Control System (EMCS) which is the correct system for the property and at the correct price. This paper outlines seven important steps in the EMCS design process and discusses how the design steps can be applied with great success.

GENERAL DESIGN PROCEDURE

In order to assist building owners and property managers in obtaining an EMCS which is the right system at the right price for the particular property, Heery Energy Consultants (Heery) utilizes the following seven steps in the design and procurement process:

1. Perform an Economic Analysis
2. Design to the Budget
3. Obtain Manufacturer Involvement
4. Obtain Operator Involvement
5. Develop Specification Appropriate for the Job
6. Competitively Bid with 2-Step Format
7. Perform Construction Monitoring

Perform an Economic Analysis

An economic analysis of potential hardware points and EMCS functions should be done as part of every EMCS design. The analysis

---

R. Krout, Heery Energy Consultants, Atlanta, Georgia. Design Guidelines For A Customized EMCS was reprinted with permission of the Association of Energy Engineers. This section was originally presented as a technical paper at the World Energy Engineering Congress, October, 1990 in Atlanta, Ga.

should include an estimate of the cost of the hardware points and the energy cost savings expected from EMCS control. Together, these provide a payback period for all points which may be used along with the Owner's financial criteria to develop an EMCS budget, or if the EMCS budget is already established the economic analysis is used to prioritize EMCS points for inclusion in the system. For example, if this analysis indicates that the two points required to monitor swimming pool temperature and control the pool heater have a 5 year payback as opposed to 3 years or less for all other hardware points, temperature control of the pool would receive a lower priority and may well be eliminated from the EMCS budget.

Design to the Budget

This step overlaps with the Economic Analysis. The EMCS budget may be established before design begins, therefore, the EMCS has to be designed to fall within the budget. Applicable installed cost per point data should be obtained, and the Economic Analysis should be used to design the optimum system for the budget. This may leave some HVAC equipment off the EMCS, but if it does not make economic sense to control it with the EMCS, it should be included.

Obtain Manufacturer Involvement

Heery's approach is to initially identify at least two good manufacturers, and their local representatives, of systems appropriate for the property, and solicit their involvement from the start. The manufacturers' representatives are a valuable source of information and experience, especially concerning product and cost information, operational specifics, and installation procedures. Utilizing this expertise is good design practice, and helps minimize addenda to the contract documents during the bidding phase and problems and change orders during the construction phase.

Obtain Operator Involvement

The capabilities and needs of the future EMCS operators is one of the biggest variables from job to job, and getting them involved early on is very important for many reasons. First of all, understanding their needs and including them in the design helps ensure the installation will be successfully used. Secondly, getting the operators to view one or more system demonstrations will help them get familiar with the basics and overcome any phobia they may have so they can begin to accept the idea that an EMCS will be replacing the electric and pneumatic controls they have become so used to. Finally, if the operators recognize their likes and dislikes for different systems, and if more than one EMCS bid is within the budget, the Owner may be able to choose the system which best meets his needs, which might not be the low bid system.

Develop Specification Appropriate for the Job

The specifications and drawings should be tailored to match the requirements of the job. A relatively brief performance type specification may be appropriate for a small job, whereas a detailed specification may be warranted for a large job. All specs should include a points list, which provides a listing of the equipment to be placed under EMCS control and the EMCS software functions which are required for the equipment. Figure 5-1 illustrates one example of a points list.

Competitively Bid With 2-Step Format

The value to the Owner of a competitive bid on an EMCS project cannot be overemphasized. Heery has found enormous cost savings resulting from competitive bidding as compared to direct quotes to the Owner from vendors, with no decrease in system quality whatsoever. In fact, because the engineering design defines the Owner's needs more precisely and optimizes the points included using the Economic Analysis, systems obtained through customized engineering design and competitive bidding are normally better.

Heery uses a 2-step bidding process. The first step requires the Bidder to submit a detailed Technical and Qualifications (T&Q) proposal along with a price proposal. The T&Q proposal normally includes product data, a system configuration diagram, installation schedule, training session agenda, previous project references, a sample of the

Figure 5-1 EMCS Points List

LOCATION: North Building SYSTEM: Interior Air Handlers POINT DESCRIPTION:	HARDWARE					ALARMS		SOFTWARE APPLICATION PROGRAMS
	OUTPUT FROM EMS		INPUT TO EMS				
	DIGITAL	ANALOG	DIGITAL	ANALOG	DIGITAL	ANALOG	
	On/Off · Open/Close	Pneumatic Transducer · Speed Control	Pressure Switch · KW · On/Off Status	Temperature °F · Relative Humidity · PSIg · Position · Speed	Equipment Status · Off by Safety	High Limit · Low Limit · Run Time · Speed	Scheduled Start/Stop · Optimum Start/Stop · Demand Limiting · Enthalpy Economizer · Run Time Total · Speed Control · CHW Valve Pos. by P · SA Temp. Control
West Interior AHU:							
Chilled Water Valve		1					X X
Supply Fan (2S)	1		1		1	X X X	X X X X X X
Supply Air				1 1	X	X X	
Mixed Air Plenum				1		X X	
Relief, Return, & Outside Air Dampers		1			1		X
Return Fan (2R)	1		1· 1		1 X	X X	X X X X X X
Return Air Plenum				1 1		X X	
East Interior AHU:							
Identical							
POINT TOTALS	2	2 2	1 1	3 1 1 1 2			= 15 x 2 = 30

programming language, and so forth. Requesting this submittal with the initial bid, rather than after the contract is awarded as is the traditional method with engineering projects, protects the Owner by proving ahead of time that the proposed EMCS meets the specification and that the Contractor is qualified to install it. This also expedites the submittal review process during the job.

The second step is the review of the T&Q proposal and the price proposal by the Engineer to determine if the proposed system meets the specifications and is within the budget. As part of this step, the Engineer may request clarifications and revisions to the T&Q proposal, and if these are required the Bidder is allowed to revise his price proposal. Proposals describing systems which obviously do not meet the spec or the budget may be dropped from further consideration. After the second step, a fairly good "apples to apples" comparison of the remaining bids can be made, and a recommendation on which system to select is provided to the Owner.

Perform Construction Monitoring

The size and complexity of the job and the Owner's staff's capabilities and experience determine the degree of construction monitoring the Engineer should provide. Heery's construction involvement in EMCS projects has ranged from only making a final inspection, report, and punchlist to making weekly inspections. A final inspection by the designer should never be omitted because, during the haste to install a system by the deadline, some important requirements of the specification are usually overlooked by the Contractor.

Smart Building Operations

By R. F. Barsoum

INTRODUCTION AND DEFINITIONS

There are numerous factors in successful Building Automation Systems operations. Many are addressed as if they are separate entities from the whole (eg. operator training); others are widely recognized as part of a universal means to achieving an end (eg. Management's involvement in energy conservation programs).

Consistent, prolonged, dependable, successful Building Automation Systems operations, in fact, depend on the above, plus some other requirements that are seldom addressed.

In the author's view, there are crucial elements that play nearly equal roles in achieving "successful" Building Automation systems operations. But first, the definitions:

Building Automation Systems (BAS) – are Direct Digital Control (DDC) or Energy Management Systems (EMS) that are first and foremost user-friendly, yet are powerful enough so that they may be manipulated to automate building systems operations.

Examples:
- Alarm limits that automatically change by either time of day, or because of a change in conditions (eg. to correspond to heating stages);

R. F. Barsoum of the Waterloo County Board of Education in Waterloo, Ontario, Canada. Smart Building Operators was reprinted with permission of the Association of Energy Engineers. This section was originally presented as a technical paper at the World Energy Engineering Congress, October, 1991 in Atlanta, Ga.

- Graphic Programming styles or Boolean logic format (if, then, else) coupled with prioritization levels, and "normal" programs. (eg, secondary water temperature control to track induction system air temperature, to minimize condensation during occupied hours, but are not limited while the fan system is off)
- Electrical demand forecasting based on the previous 15 minute sliding window period.

Operator(s) – are those that have energy management as their primary responsibility. Others may use the system, but someone within should be ultimately responsible for the electronic systems' behavior.

Most organizations have "engineers" whose sole responsibility is the hands on welfare of the mechanical and electrical systems. However, due to vast advances in BAS systems capabilities, some other individual – who would preferably report to the same authority as the "engineer" – should be responsible for the welfare of the electronic equipment.

Smart DDC or Building Automation System operator – Not the BAS. Not the operator. Both!

Successful BAS operations – Occur when:
a) The building occupants are comfortable;
b) The maximum savings (or cost avoidance) is attained without hampering comfort; and
c) The two above points are carried out automatically.

"Non-alarm alarms" – Those alarms generated by the BAS that are not alarms at all! (eg, an alarm every day at 8:30 because Mr. Hearn turns off his fan before he goes home).

Alarms generated by the BAS should be that – an alarm. (For the above example, a decision must be made to either bypass Mr. Hearn's controller, or the sensor wiring for the BAS.)

CRUCIAL CHAIN LINKS

As in a circular chain, there are seven crucial "chain links" that determine the building's actual energy cost.

1 – Operator Training

In most commercial buildings, other than specialized industrial and power generation plants, operator training has been inclined towards hands-on repair and maintenance. Technical tasks have been generally sub-contracted to a service firm. The service firm's staff, however, tend to have electronics diplomas rather than mechanical or electrical. The result: building staff who don't know what the BAS can do, and therefore rely on old and established procedures; and service firm staff who lack the proper background.

General – An operator should have a basic theoretical background, as opposed to practical. (In Ontario, there are college diploma courses available through night school.) It is the author's belief that less experience in hands-on building system maintenance may actually be an advantage, since the incumbent would not have preconceived ideas on "easier" ways to run a plant.

Specific – If at all possible – politically as well as practically – the operator should:

a) Attend design meetings (at a very basic level);
b) Play a key role in the commissioning phase;
c) Act as a resource, particularly while away (to alleviate situations where buildings run better while he/she is there); and
d) Be the individual responsible for formulating and/or maintaining the operations manual.

2 – Operations Manual

Of all the "links", this may be the most important, and usually the most frequently overlooked. If an operator feels comfortable enough to formulate and maintain this, the associated building should be half way to becoming a "successful" operation.

The "operations manual" should be the operator's first reference. 99% of the time – if set up properly – it will be his/her last. Thus, it must be detailed, organized, and easy to update. Furthermore, it is a means by which the operator will endeavor to resolve "non-alarm" alarms.

The manual should incorporate three sections:
- Quick reference
- Alarm response instructions
- Intention

Quick reference – Usually supplied as system graphics, and, unfortunately, usually limited to each specific system.

Whether by graphics or language, it is proposed that the operator be given as much opportunity as possible so scan as many inter-related systems as possible. (eg. Heat pump water loop with associated heat sink and source).

Alarm response instructions are rarely supplied.

Indexed by point group and point name, this will answer – particularly for those for whom energy management is a secondary responsibility – the questions below.

- What is wired to this point?
- How does it control and/or sense?
- Exact physical location, and point address.
- Why would it alarm?
- What to do in case of an alarm?, including a list of conditions that may require different responses.
- What other points should be checked if this point generates an alarm?

Intension – This also is usually never supplied.

Indexed by each individual point, this section should provide information as to which operations and/or energy management strategies rely on each individual point. Note also the discussion of point summary data in Chapter 15.

This section should also provide a fairly detailed outline of each strategy, pertinent engineering data, and reasoning behind the implementation – or intent.

3 – Coordinated Design

Too frequently an indication that the design engineering firm's right hand didn't talk to its left. Example: a heating system designed to run 24 hours on "design days", and an energy management system that will shut it down during the unoccupied cycle regardless of outside conditions.

The answer: "Deeper" involvement by the engineering firm.

It is rare to unheard of for an engineering firm to discuss with an owner the option of incurring additional costs to involve an EMS specialist. On the other hand, owners rarely evaluate design engineers by their building's energy performance. Both of these points may have political implications; yet if all the other links in the chain were fairly strong, the political weakness in this link would not be as difficult to overlook.

4 – Commissioning

Again, rarely involves the initial design engineers to any degree. At the least, the design specification should call on the contractor to step through each strategy with the owner (preferably the operator). At best, an engineer's representative would also attend this phase.

Commissioning of any piece of equipment should be confirmed with the operator subsequent to completion. Also, any repairs or modifications should include reconfirmation with the operator, and subsequent documentation. In this way, the operator assumes ownership, and problems caused by changes are easier to troubleshoot.

5 – Energy Monitoring

This is where some of the operator's training will pay off. His/her understanding will go a long way to alleviate costs associated with external bodies generating reports and recommendations, as well as determining the reasons for problems.

Aside from generating alarms when energy conservation is off track, the BAS should also generate reports to assist the operator in determining performance. If required, the operator should be able to set up supplementary spreadsheets. Chapter 17 on DDC System Management covers the concepts of energy monitoring in much greater detail.

6 – Preventative Maintenance

A subject discussed by many, and rightfully so.

The author believes that careful evaluation is required before maintenance management is integrated with a BAS – for many of the same reasons that Fire and Security systems should not. This does not mean that large BAS's don't do the above well, but if they are carefully analyzed, one would find that they are made of dedicated components that are then "tied together", and this may limit the value of integration.

Some BAS salespeople will tell you that they have an EMS that is capable of generating PM work orders based on equipment "run times."

What they don't tell you is that the BAS may not track work orders, or equipment, or people, or allow you several approval phases.

The author also believes that it is less important to integrate run times with PM task generation, than it is to ensure that the work is done properly. With that in mind, part of the PM program should include random double checks by a Supervisor to ensure "quality control".

Summary: For the sake of building systems operations:
a) Pay particular attention to equipment maintenance. Equipment in an energy efficient building usually has to work somewhat harder. Couple this with the necessity of keeping comfort complaints to a minimum, and PM becomes a crucial element.
b) Keep the PM and work order systems separate from the BAS.

7 – Quality Assurance

As opposed to quality control, where random checks are made on a specific task to ensure conformity to a standard, "QA" is more like a series of fishing expeditions.

If a consultant was brought in to review a department's operations, the first thing he/she does is "fish" – looks for loopholes. But any department can ensure a minimum number of loopholes exist by looking for them themselves.

In Ontario, health care institutions have been implementing this type of program over the last several years. Basically, it places the responsibility of supervision with the individual departments.

There are key points that should easily be adopted for BAS operations:
a) Identify the department's or individual's goals (eg. Air Quality). Using these as a guideline,
b) Set standards. These should be realistic and/or close to "industry standards". (eg. regardless of outdoor conditions, there should always be a minimum of 25% fresh air introduced through fan system F1.)
c) Survey. Gather data to assist in determining the facts (eg. On March 9th, return air was 75F, outside air at 55F, mixed air at 65F, therefore: 50% fresh air). This may require some time to accomplish, particularly if different conditions should be tested.

d) Compare. If the data shows a "fail" condition versus the standard set, there is a problem!

Now, one may think that this is simply troubleshooting. However, if the following is kept in mind, the benefits will become evident:

- The operator is now the "expert". How does his/her Supervisor find out how well he/she is doing? The QA report. If need be, the Supervisor may set a particular standard that he wishes "tested" and simply pass it on.
- If the employee recognizes that the Supervisor reads the reports, and must have them quarterly, the reports become a way of focusing each department or individual, on a regular basis, on their tasks.
- In large organizations, it is a way of finding out who does what!
- It is pro-active.
- It can be used to support budget requests, or justify monies spent. (eg. What type of response would you get if – in the name of Quality Assurance – you surveyed your building's tenants regarding comfort levels?)

SUMMARY

A BAS system is a high technology tool that is often in control of extremely important functions. Unfortunately, the benefits derived from up-front investments are usually too vague to measure, thus monies spent are usually only justified versus energy dollars "avoided".

However, the following should also be considered:

- "Intangible" energy savings. That is, those that could not be pre-calculated, and are a result of the operator's knowledge, care and involvement.
- Building equipment lifecycle
- The BAS's lifecycle (it is more difficult to justify a replacement BAS if the old one is already attaining can be saved).

Chapter 6

Developing a Strategic DDC Plan

The term "Strategic Planning" was in vogue for some time in the Energy Industry for some time. Like many other buzzwords however, it has fallen from grace, and is rarely used today. Yet the author still considers this an essential step to ensure the long term success of any energy program.

The focus of this book is Distributed DDC and as a result strategic planning with focus on this technology is very valuable. Yet the nature of this process is to address all possible and reasonable strategies, projects, etc. that will reduce operating cost and improve building operation. Improvements to building operation again may vary but may include improved comfort or establishing a profit center for after hours billing. As you may know after hours billing is used in office billings to charge tenants for energy used outside normal operating hours. Such projects are many and varied, and could include anything from lighting or mechanical retrofit to DDC. Examples and specifics here will focus on DDC though again the general process applies to any type of building project.

Developing a Strategic DDC Plan involves looking at the entire facility. In fact the planning process should include provision for ongoing analysis of the facility and its operation, energy cost and use. An effective program establishes a program for ongoing analysis and conducts periodic "audits" or studies to identify opportunities. The process should also involve regular monitoring of energy use and cost, as well as setting of performance goals for the facility. Such goals might be in energy use (ie: BTU) per square foot or per unit of corporate volume. Many food chains for example measure energy use per hamburger etc. Facilities or areas of a facility that fall short of these goals should be target for action. At the same time any facility program should incorporate energy efficient building design. This design must include a DDC system from startup to ensure the building operation is optimized.

Another area of planning that is beyond the scope of this book but should be considered is utility planning. There are two key aspects of utility planning: supply and incentives. Fuel supply is actually an alternative to buying energy resources from the utility. This approach has been common with gas nearly a decade. Sometimes called self-help gas programs, these have been the source of significant cost savings for many building owner and industrial users. With changes in electric regulation the same types of programs may result in further savings for building owners. This entire topic is beyond the scope of DDC, however it is important for users to explore many areas in pursuing cost reduction. Further DDC can be used to help manage and monitor these systems thus simplifying the Energy Manager's job.

As a result this chapter will address this in terms of energy management and DDC programs. Again many energy projects exist beyond DDC but the focus here will be on how this process leads to system implementation.

As a final note the reader should not look to this chapter as a "how to" for writing Strategic Plans. It does not instruct the reader on writing the plan or provide an outline for the essential sections for a plan. The author believes that a practical outline of the phases required to implement an Energy Management and DDC Program is of much greater value to the reader. Therefore this section will define the program requirements and the reader may establish a written action plan to address each area. Development of the program including essential actions and the chapter will briefly address each stage in the process.

The first stage of an energy management and DDC program is selecting an individual energy manager who can establish an energy management team to manage the energy usage in your building. Energy

Developing a Strategic DDC Plan

managers should have knowledge of the areas to be managed. These areas are technical or mechanical, financial, personnel, and management. The energy manager can then establish a team that should include the engineer or head of maintenance, the accountant, personnel manager, and the administrator. Other individuals will be included later in this report.

The energy manager must be responsible for the total energy management program and have the authority to take action. He or she must follow through with the whole program from surveying energy consumption, identifying energy conservation opportunities, planning for systems modifications, to implementing energy conservation techniques. A very effective technique that has been used in many organizations is to also appoint a "champion" in management. This ensures that the energy manager always has direct access and ultimately improves the success possible from the program.

The second stage of the energy management program is surveying your building to identify building characteristics, equipment, and energy usage. This stage will be covered in more detail under Chapter 7, but a brief outline will be provided here for introduction. The energy survey is useful in identifying personnel, documents, and other sources of information explaining equipment operations, heating and cooling systems, fuel used, quantity used, cost breakdowns per fuel source, etc., that will be used to determine the energy savings resulting from implementation of energy projects, including the focus of this book, Distributed DDC Systems.

There are four main areas of the building to be surveyed.

1. *Energy Consumption: Record monthly energy usage and cost* from utility bills. Separate electricity, natural gas, and other fuels so that each fuel source can be viewed individually. It is best to record consumption on a monthly rather than a yearly basis. To establish a data base, record the energy consumed for the past 36 months. Many types of Forms are available. Examples are shown in Chapter 7.

2. *Identify the major energy using systems* within your building. These high energy users are mechanical/electrical equipment, heating and cooling systems, lights, ventilation equipment, etc.

It is useful to get accurate data on each piece of equipment pertaining to size, capacities, and hours of operation.
3. *Identify the materials used in the construction of the building.* This can be done by referring to building blueprints, or by walk-through inspection.
4. *Record the size of your building* in square feet, and break down in square feet the major use area of the building.

When deciding who, among the energy management staff or team, is to do each part of the survey, remember to have the most qualified personnel in each specific area supply the information. This will not only give your survey credibility, it will also make the best use of time and effort spent on each area. The energy manager should coordinate and participate in all aspects of the study, but internal allies can be earned by involving others within your organization. The accountant should survey energy usage. The engineer should survey energy using systems. The administrator, engineer, or building consultant should survey building construction materials, building size, and area usage information. As a final note, it may be worthwhile to consider doing a consultant depending and the complexity of the facility or physical plant.

The third stage of the energy management program is identifying energy projects and determining energy and operating cost reductions, as well as return on investment. Bear in mind that DDC system benefits include comfort, cost reduction, and in some cases establishing profit centers for features such as after hours billing. After the data is collected from the survey the analysis of the total operation can be done to identify cost effective mechanical and system retrofit projects.

Examine the large energy users of your building and under-utilized areas that can be scheduled off. These are the areas where the potential for cost savings is the greatest. Determine if heating, cooling, and ventilation systems are operating efficiently, if lighting and work areas are at optimum levels, if electrical motors are operating both effectively and efficiently, etc. It may be useful to consult a utility representative. This will help with the analysis of your utility bills and also identify rebate and other Demand Side Management programs. Determine the amount of energy used, its unit cost, and time of day it is used.

Operational techniques and maintenance procedures of equipment must be evaluated. The energy manager should be knowledgeable in these areas in the evaluation or contract with a consultant.

Develop a benefit summary to determine energy and dollar savings of proposed energy projects, as well as income from projects like after hours billing. The survey will result in several types of projects. *No-cost* energy projects can be done by in-house personnel with materials on hand. Low-cost energy conservation techniques can be handled within the existing operational and maintenance budget. The benefits from a DDC system retrofit generally require that such project be done to ensure effective implementation of a system. Major modifications include projects that usually require large financial outlays over a number of years.

To Summarize:
1. Review building characteristics, physical plant, food preparation equipment, etc., to determine energy cost and energy savings opportunity.
2. Determine the dollar savings and energy savings of each energy project.
3. Divide the estimated dollar savings into the cost of the ECO to determine the payback period. Use the payback period to prioritize the various ECO's.

The fourth stage of the strategic energy management program is to set up an action plan. Historic energy data is developed to determine base year data prior to implementation of the energy management program. Future energy data will show the effects of implemented energy projects. Record future energy consumption data to monitor the effect of conservation techniques. All energy sources should be converted to BTUs to achieve a common measuring unit of energy usage. Remember to include energy usage and unit cost, because over a period of years unit costs will rise, while usage falls, which may cause total costs to remain unchanged. Therefore, using total costs as an indicator of strategic energy program effectiveness is unrealistic without unit cost information. It may also be useful to graph key relationships to avoid cumbersome numbers, while monitoring energy consumption trends. A final note is that the energy manager should carefully monitor changes in scope of use. When more users, more space or more energy intensive program (ie: computers) are added to the building, the baseline must be adjusted.

The fifth stage of the energy management program is implementation of the energy projects. Begin implementation with the no/low-cost techniques so that unnecessary energy consumption is eliminated. This will assist the energy manager in gaining credibility in the energy program. It will also ensure the DDC systems and other projects are not inhibited by faulty or non-function equipment.

In the past, motivational techniques have been successfully used by energy managers in encouraging total staff involvement in the energy program. Such programs included suggestion boxes, prizes given for the best suggestions, and recognition of employee participation. Consider again that this process must be tied to the company's mission. If a property manager wants happy tenants, show how comfort is critical to achieving this. In this view, the team approach is essential in program effectiveness, and all staff members should feel part of the big team. Training sessions are helpful in educating personnel about DDC systems and why implementation is necessary. This can even be tied to long-term corporate health and a concerted effort to control cost without workforce reductions.

As the author will emphasize throughout this book it is critical to follow any DDC project with ongoing management. This process is critical to ensure that the system achieve the results projected. In fact this is so essential to DDC projects that an entire chapter in this book is dedicated to DDC system management. Chapter 17 focuses on the process of DDC Program Management, and the reader is encouraged to read this carefully. The reader is further encouraged to establish some form of program to implement this type of management for any new or existing systems.

Another critical aspect of this chapter which could not be covered in the detail desired is the DDC survey. This is an extremely important step for any system. Further, accuracy in this phase can ensure better cost estimation, smoother start up and transition and fast results. The survey provides the basis for a clear and concise scope of work to ensure that the contractor and the owner know what is expected. All of this is possible because a well done study will provide details necessary to avoid last minute scrambling for essential data, such as schedules, and repeated call backs to address areas of confusion.

Chapter 7

Conducting a Building DDC Survey

The topic of conducting a survey or energy audit was discussed along with the Strategic Plan. In general, there should be a comprehensive building survey and energy audit or engineering study conducted during the planning phase. This chapter will expand on the discussion of that survey from Chapter 6 and focus on data required to implement the DDC system.

The term "walk through" will be used somewhat here to address the practical site survey of building equipment. The walk through process will be considered, though, in the context of an overall energy survey. Energy survey or "energy audit" is the term that identifies procedures typically used to develop the report which identifies available energy related projects. A list of the basic types of information required for a detailed energy audit is shown in Figure 7-1. The list is far from complete, but does serve to indicate the different types of concerns which must be addressed.

Developing the information involves two distinct types of procedures: The review of documents (energy consumption data, operating logs, plans, etc.), and a physical inspection of buildings and equipment, including interviews with an observation of appropriate personnel.

The review of documentation usually occurs first, to provide an overview of what is involved in general.

Identity and Use
    Name
    Use

Physical Data
    Floor area
    Number of stories
    Roof area
    Roof construction
    Wall construction
    Window type
    Glazing
    Shading employed
    Floor construction
    Exterior building dimensions

Operating Schedules
(weekdays, Saturdays, Sundays,
and holidays)
    People
    Lights
    HVAC
    Process
    Custodial
    Etc.

Energy Sources
    Electricity
    Gas
    Oil
    Steam
    Purchased chilled water

Energy Cost Data
    Electricity demand &
        consumption
    Gas
    Oil
    Steam
    Purchased chilled water
    Rate schedules

Historical Monthly Energy Demand,
Consumption Data for Past 2 or 3
Yrs.
    Electricity
    Gas
    Oil
    Steam
    Purchased chilled water
    Coal

Cogeneration or Backup Power
Sources
    Type
    Capacity

Electrical Characteristics
    Voltage
    Power factor

Type of Heating
    Space

Type of Cooling
    Space, Process

HVAC System
    System number
    Area served
    Critical/noncritical
    Type of air handling
    Type of water-side system
    Type of control and existing
        control device
    Outside air
        Minimum required
        Maximum available
    Measured running amps

Domestic and Service Water
    Size
    Rated input
    Temperature setting
    Usage
    Heat recovery application

Other Energy Consuming Devices
    Item
    Energy demand
    Operating requirements
    Existing control
    Critical/noncritical functions
    Measured running amps

Pumps
    Service
    Capacity
    Critical/noncritical
    Measured running amps

Telephone System
    Existing capability
    Spare capacity

Existing older generation control
systems
    Type
    Capability
    Reuseable sensing and control
        components

Figure 7-1 Typical Data Gathered in the Walk-Through

Building staff should provide the energy manager or consultant with information and materials required for the development of an effective energy management report. Materials which are needed include original plans and specifications, as-built drawings (if they exist), operating and maintenance manuals and logs, and such other information and materials as the consultant may request.

Once a study of documents and related tasks has been completed, a walk-through survey can be conducted. Through this procedure, energy managers can identify the overall nature and specific elements of major systems, and the way in which they interact with one another to consume energy. In addition, they can identify the condition of equipment, and other factors that may be used to suggest the applicability of various DDC strategies.

The participation and cooperation of building operations and management staff is critical to development of DDC strategies. Since limited hard data is available in many cases, soft data (from staff) will have to be developed. As such, the Energy Manager should direct the consult or energy survey team to depend largely on data provided by building operations and maintenance staffs. During the physical aspect, the walk through team should be accompanied by building personnel to provide data on operation and maintenance and on the feasibility of the potential DDC function identified for the system.

GENERAL AND ARCHITECTURAL DATA

It is of value for the reader to see some examples of data gathering instruments for the walk through. Form 7-1 provides an example of the types of general and architectural data that should be gathered for the building. It is important to carefully define the scope of the walk through and its intended goals prior to beginning. This is because a survey is time consuming and typically occurs prior to formal approval of project funding, as a result resources required should be carefully assessed to avoid wasted time and money. At the same time, however, it is often useful to have general architecture data, particularly for calculating energy savings from such strategies as night setup/back and optimization.

DDC Architectural Survey

I. General Information   Surveyed by: _____

                                                             Survey Date: _____

IDENTITY:

Company/Organization _____

    Address _____

    Primary Building Use

    _____

    Person(s) in charge of building

    _____

PHYSICAL DATA:

    Length, Feet and Orientation _____ Ft.   N.W.E.S.   Date of Construction: ___/___/___

    Width, Feet and Orientation _____ Ft.   N.W.E.S.   Date of Prints: ___/___/___

    Number of floors: _____   Architect: _____

    Height from floor to floor _____

    Height from floor to ceiling _____

Floor area, gross conditioned square feet _____

_____

    Floor area, special use:

        _____ Ft^2 Office

        _____ Ft^2 Sales

        _____ Ft^2 Warehouse

        _____ Ft^2 Other (describe)

_____

_____

HVAC Contractor

_____

Form 7-1-1

WINDOWS:

　　Type:_____ Fixed sash; _____ double hung: _____ casement

　　Glazing:

Exposure	No.	[1] Type	Gross Area	% Glass/Exterior wall area
N	_____	_____	_____	_____
S	_____	_____	_____	_____
E	_____	_____	_____	_____
W	_____	_____	_____	_____

[1] Type: Single, double, insulating, reflective, etc.

　　Glass shading employed outside (check one):

　　　Fins_____ Overhead_____ None_____ Other_____

　　Glass shading employed inside (check one):

　　　Shades_____ Blinds_____ Drapes, open mesh_____

　　　Drapes opaque_____ None_____ Other_____

DOORS:

　　Door types and numbers:

　　　1-single; 2-vestibule; 3-revolving

　　　　North – No._____ Type_____　　West – No._____ Type_____

　　　　East – No._____ Type_____　　South – No._____ Type_____

WALLS:

Orientation	Gross Wall Area	Net Wall Area
North	_____	_____
South	_____	_____
West	_____	_____
East	_____	_____

Detail: _____

Form 7-1-2

Exterior opaque wall construction: Circle Type:

1-frame; 2-curtain wall; 3-solid masonry; 4-brick and masonry; 5-masonry cavity

Exterior opaque wall insulation: Material: _____

Thickness:_____ U Value _____

FLOOR:

Floor Construction: Circle Type: 1-slab on grade; 2-over heated space;

3-over unheated space; 4-wood; 5-concrete; 6-other

ROOF:

Roof Construction: Circle Type: 1-masonry; 2-wood; 3-metal; 4-flat;

5-sloped; 6-pitched; 7-light; 8-dark

Room/Ceiling Insulation: Type:_____ Thickness:_____ Detail: _____

"U" Value: _____   _____

                                                            _____

PHYSICAL DATA SUMMARY SHEET

		Cond.			
GROSS ROOF AREA	_____ SQ. FT. _____	[1]HEATED VOLUME	_____ CU. FT.
GROSS FLOOR AREA	_____ SQ. FT.	[2]NET ROOF AREA	_____ SQ. FT.
AVG. CEILING HEIGHT	_____ FT.	[3]NET WALL AREA	_____ SQ. FT.
SINGLE PANE GLAZING	_____ SQ. FT. _____	[3]GLAZING AREA	_____ SQ. FT.
DOUBLE PANE GLAZING	_____ SQ. FT. _____	R_r _____	
SKYLIGHTS	_____ SQ. FT. _____	R_w _____	
EXTERIOR DOORS	_____ NO. _____		

1) Heat Volume = Avg. ceiling height x gross floor, per floor
2) Net Roof = Gross roof − skylights
3) Net wall = Perimeter x wall height − (glazing + door area)
4) Glazing = (single pane x 1) + (double x .5) + (doors x 10)

Form 7-1-3

OCCUPANCY AND USE DATA

A second critical area of concern for the walk through is to verify the actual hours of operation for the building. Further, it is important to verify whether any apparatus is currently in place to control equipment on an occupied/unoccupied schedule. This will again be critical for both calculating cost savings and for ensuring a smooth system start up. Callbacks and remote communication programming time to modify scheduling direct reduces the contractors profit on the job and inconveniences the owner. At the same time take careful note to actual and desired temperature setting throughout the building. Form 7-2 shows the type of scheduling data to gather in the walk through.

BUILDING OCCUPANCY AND USE

	Weekdays		Saturdays		Sundays, Holidays	
	No. of Occupants	Period of Occupancy	No. of Occupants	Period of Occupancy	No. of Occupants	Period of Occupancy
Administrative						
Laundry processing						
Cafeteria						
Stores						
Laboratories						
Production Areas						
Others (Describe)						
_____						
_____						
_____						

TIME CLOCKS
Load   Day(s)   On   Off

Notes: _____

Form 7-2-1

MAINTAINED INDOOR ENVIRONMENTAL CONDITIONS

	Occupied Hours				Unoccupied Hours			
	Temp (°F)		Rel. Humidity (%)		Temp (°F)		Rel. Humidity (%)	
	Summer	Winter	Summer	Winter	Summer	Winter	Summer	Winter
Administrative Areas								
Service Areas								
Store Areas								
_____								
_____								
_____								

Form 7-2-2

EQUIPMENT SURVEY

The next and perhaps one of the most critical phases to any DDC walk through is the equipment survey. It is essential to clearly identify all equipment including its configuration, current control and physical characteristics. Again this is important for several reasons. It is critical to know all building systems so that the job is designed properly and adequate system sizing is provided to meet building needs. Further the programming phase of the installation relies heavily on clear definition of the sequence of operations required as well as the specific number of monitoring, control and alarming points required. Finally, an accurate analysis of energy savings cannot be done without accurate equipment data. Form 7-3 shows an outline of survey data for Heating, Ventilating and Air Conditioning (HVAC) equipment as well as domestic hot water, ancillary and some process type equipment.

SYSTEMS AND EQUIPMENT SURVEY
HVAC SYSTEMS:

a. AIR HANDLING SYSTEMS (check as appropriate):

    Perimeter system designation:

Single zone _____	Multizone _____
Fan coil _____	Induction _____
Variable air volume _____	Dual duct _____
Terminal reheat _____	Self-contained _____
Heat pump _____	

    Interior system designation:

Fan coil _____	Variable air volume _____
Single zone _____	Other (describe) _____

    Principle of operation:

    Perimeter:

        Heating-cooling-off _____

        Air volume variation _____

        Air mixing control _____

        Temperature variation _____

    Interior:

        Heating-cooling-off _____

        Air volume variation _____

        Temperature variation _____

b. VENTILATION:

    1. Minimum Outdoor Air Required by Code:

    Percent Total cfm _____

    cfm/Person _____

_____

_____

Form 7-3-1

2. Air Handling Systems:

	Area				
	1	2	3	4	5
Unit or Systems[1]					
hp Totals					
Supply					
Fan					
Control Method[2]					
cfm (ft3/Minute)					
Total Supply					
Measured Present					
Outdoor Air					
(% Total cfm)[3]					
Outdoor Air Leakage					
Dampers (Closed)					
Hours System "ON"					
Weekdays					
Saturdays					
Sundays					
Holidays					

1. Type (rooftop, multizone, fan coil, central system).
2. No outdoor air, wild economizer, integrated economizer, fixed air.
   If economizer is used:
       Changeover Control: ☐ Dry Bulb   ☐ Enthalpy   Set Point _____ F.
       Mixed Air Control Set Point _____ F.
3. Measure 3 temperatures with O.A. Damper in minimum position.

$$\% \text{ Outdoor air} = \frac{T_R - T_M}{T_R - T_O}$$

(T_R = Return Air Temp., T_M = Mixed Air Temp., T_O = Outdoor Temp.)

c. **HEATING**

    Source of Heating Energy:

        Hot Water _____ Steam _____ Electric resistance _____

        Other _____

    Heating Plant:

        Boiler No. _____ Rating _____ MBH or kW

        _____

        _____

Conducting a Building DDC Survey

Boiler Type:

    Firetube _____ Watertube _____ Elec. resist. _____

    Electrode _____ Other _____

Fuel Used _____ Standby _____

Hot Water Supply _____ °F, Return _____ °F

Steam Pressure _____ psi

Pumps No. _____ Total hp _____

Room Heating Units:

    Type: Baseboard _____ Convectors _____ Fin Tube _____

    Ceiling or Wall Panels _____ Unit Heaters _____ Other _____

    Extent and condition of metering equipment _____

    _____

    Night Setback: Yes_____ No _____

    Morning Warm-up Thermostat Used _____

d. COOLING:

    Type system: ☐ Packaged rooftop   ☐ Packaged multizone   ☐ Central system

                         ☐ Chilled water   ☐ Direct expansion

Form 7-3-3

Cooling Plant – Compressors and Chillers:

No.	Type (Centrifugal, reciprocating, absorption)	Capacity (Tons)	Motor horsepower	Area Served
_____	_____	_____	_____	_____
_____	_____	_____	_____	_____
_____	_____	_____	_____	_____
_____	_____	_____	_____	_____
_____	_____	_____	_____	_____
_____	_____	_____	_____	_____
_____	_____	_____	_____	_____
_____	_____	_____	_____	_____
_____	_____	_____	_____	_____
_____	_____	_____	_____	_____
_____	_____	_____	_____	_____
_____	_____	_____	_____	_____
_____	_____	_____	_____	_____
_____	_____	_____	_____	_____
_____	_____	_____	_____	_____

Heat dissipation device:
    Evaporative condenser _____
    Air cooled condenser _____
    Cooling tower _____
Chilled water pumps _____ Total hp _____ kW
Condenser water pumps _____ Total hp _____ kW
Evaporator blowers _____ cfm _____ kW input
Condenser fan _____ kW input _____
Cooling tower fan _____ hp _____ kW input
Cooling night shutdown _____
Cooling night setup _____ F
Control System:
    Electric _____
    Electronic _____
    Pneumatic _____

Form 7-3-4

Misc. fans:

Location	Horsepower	Type	Function (Supply, Exhaust, etc.)	Method of Operation (Time Clock; Manual etc.)

Form 7-3-5

SELF CONTAINED UNITS:

   Type: Room air conditioner _____ Thru-the-wall unit _____

   No. of Units_____ Basic module served _____

   Total Capacity (Tons) _____

   Horse Power/unit_____

   Extent and condition of metering equipment _____

   _____

   _____

   Equipment condition and maintenance _____

   _____

   _____

   _____

ENERGY CONSERVATION DEVICES:

   Type:

      Condenser water used for heating _____

      Demand Limiters _____

      Energy Storage _____

      Heat recovery wheels_____

      Enthalpy control of supply-return-exhaust damper _____

      Recuperators _____

      Condensate used for heating domestic hot water _____

      Others _____

      _____

      _____

      _____

      _____

      _____

Form 7-3-6

DOMESTIC HOT WATER:

  Size _____ Rated input _____

  Energy Source: Gas _____, Oil _____, Electric _____, Other _____

  Aquastat setting _____ °F

General:

  Types of usage – Location of usage, estimated GPD, distance from heater. _____
  _____
  _____
  _____
  _____
  _____
  _____

  Water temperature – check several locations (reduced temperature during nonoccupancy)? ___
  _____
  _____
  _____
  _____
  _____
  _____

  Type and condition of heat exchange equipment. _____
  _____
  _____
  _____
  _____
  _____
  _____

Form 7-3-7

DISTRIBUTION SYSTEMS:

Condition of Steam Systems (leakage, insulation, etc.) _____
_____
_____

Condition of Hot and Chilled Water systems (leakage, insulation, etc.) _____
_____
_____

Condition of Refrigerant Piping (leakage, insulation, etc.) _____
_____
_____

Condition of Air Distribution Systems (leakage, insulation, etc.) _____
_____
_____

OTHER EQUIPMENT (Elevators, Escalators, Data Processing, Kitchen, Laundry, etc.)

Equip. Description	Quantity	Size/Capacity in Btu, kW, hp, etc.
_____	_____	_____
_____	_____	_____
_____	_____	_____
_____	_____	_____
_____	_____	_____
_____	_____	_____
_____	_____	_____
_____	_____	_____
_____	_____	_____
_____	_____	_____
_____	_____	_____

Form 7-3-8

DEPARTMENTAL DATA

FOOD PROCESSING – KITCHENS:

Insulation of inbuilt refrigerators. Door locations and seals. _____

_____

_____

Anti-cycling controls on compressors? _____

_____

Shutdown of steam/electric equipment when not in use? _____

_____

_____

Exhaust arrangement – hoods, air quantities, velocities? _____

_____

Use and condition of hot/cold food cards. _____

_____

Excessive food serving counters for number of meals served? _____

_____

_____

Variations in number of meals/day. _____

_____

LAUNDRY EQUIPMENT:

Age and condition of major equipment _____

_____

Maintenance procedures _____

_____

Evidences of condensate loss _____

_____

Equipment loading practices _____

_____

Hours of equipment operation _____

_____

Form 7-3-9

ELECTRICAL SURVEY

The final area of concern for the walk through is an electrical and lighting survey. This is particularly important if any type of lighting strategy is to be implemented. Such strategies include simple on/off scheduling and can be expanded to provide light level control with and without outdoor ambient light adjustment. Any electrical loads that are to be controlled such as motors and pumps will also require specific electrical data to interface on/off control. Form 7-4 shows general data required to document the building electrical systems as well as lighting loads. Note with all these loads it is again important to determine current schedules and control equipment.

Panel _____ Location _____ AMPS _____

Volts _____ Phase _____ Wire _____

LOAD	WATTS	CIRCUIT		WATTS	LOAD
		1	2		
		3	4		
		5	6		
		7	8		
		9	10		
		11	12		
		13	14		
		15	16		
		17	18		
		19	20		
		21	22		
		23	24		
		25	26		
		27	28		
		29	30		
		31	32		
		33	34		
		35	36		
		37	38		
		39	40		
		41	42		

Form 7-4-1

LIGHTING

Area Served	Type (Fluorescent, Incandescent etc.)	Density (Watts/Ft.^{2})	Method of Control (breaker panel, wall switches, control switching)
Interior			
Adm. Off.	_____	_____	_____
Storage Rm	_____	_____	_____
Corridors	_____	_____	_____
Lobbies	_____	_____	_____
Computer Rm	_____	_____	_____
Kitchens	_____	_____	_____
Toilets, etc.	_____	_____	_____
_____	_____	_____	_____

	Type	Total kW
Exterior		
Parking lot	_____	_____
Decoration	_____	_____
	_____	_____
	_____	_____

Equipment condition and maintenance schedules _____

_____
_____
_____
_____
_____
_____
_____
_____

Form 7-4-2

Schedules

Lighting – Interior	Area Served	Weekdays	Saturdays	Sundays or Holidays

Form 7-4-3

Lighting – Exterior	Area Served	Weekdays	Saturdays	Sundays or Holidays
_____	_____	_____	_____	_____
_____	_____	_____	_____	_____
_____	_____	_____	_____	_____
_____	_____	_____	_____	_____

Domestic hot water heat _____

Elevators

_____
_____
_____

Escalators

_____
_____
_____

Other (describe)

_____
_____
_____
_____
_____
_____
_____
_____
_____
_____
_____
_____

ENERGY HISTORY

Facility staff or the local utility/fuel supplier should provide energy use records for the building. This data is extremely critical for review along with the building survey data and energy savings calculations. Energy bills are the documentation of consumption, waste or opportunity in a facility. Much as a consumer can compare automobiles based on miles per gallon, a manager can compare buildings based on standardized energy consumption factors.

Building comparisons are done with an "Energy Trend Analysis" and can be developed by an energy manager with access to accurate utility history for the building. The Energy Trend Analysis measures energy consumption in terms of British Thermal Units (BTUs) per conditioned square foot of space per year. Very often weather data is also factored in with degree days for the local area to allow buildings in various climactic zones to be compared. An example of the form used for gathering and computing this data is provided in Figure 7-2. There are several organizations such as the local State Energy Office and the Building Operators and Managers Association that publish both energy per square foot and dollars per square foot data for various building types. Owners can therefore compare their buildings to national averages or to other buildings that they operate.

The final note on energy consumption data is to ensure that cost savings calculations compare with actual bills. This allows the Energy Manager to ensure that projected savings fall within reason.

Conducting a Building DDC Survey

Building _____

Gross Conditioned
Square Feet _____

MONTH 1	HEATING DEGREE DAYS 2	COOLING DEGREE DAYS 3	PURCHASED ELECTRICITY								BTU's 12
			kWh COST 4	DEMAND CHARGE 5	POWER FACTOR ADJ. 6	FUEL ADJ. 7	TOTAL COST 8	BILLED DEMAND (kW) 9	ACTUAL DEMAND (kW) 10	kWh USED 11	
JAN.											
FEB.											
MAR.											
APR.											
MAY											
JUNE											
JULY											
AUG.											
SEPT.											
OCT.											
NOV.											
DEC.											
YEAR TOTAL											

Btu Conversion Factors: Electricity, kWh—3413; Purchased Steam, 10^3lb—1,000,000; Natural Gas, 10^3ft^3—1,030,000; Oil (No. 2) gallons—138,700; Oil (No. 6) gallons—149,700

Figure 7-2 Energy Trend Analysis

MONTH 1	PURCHASED STEAM						OIL, HEAVY			OIL, LIGHT		
	STEAM COMMODITY COST 13	STEAM DEMAND CHARGE 14	TOTAL STEAM COST 15	DEMAND ACTUAL (lbs/hr) 16	DEMAND BILLED (lbs/hr) 17	STEAM USED 10^3 LBS 18	BTU's 19	TOTAL COST 20	GALLONS USED 21	BTU's 22	TOTAL COST 23	GALLONS USED 24	BTU's 25
JAN.													
FEB.													
MAR.													
APR.													
MAY													
JUNE													
JULY													
AUG.													
SEPT.													
OCT.													
NOV.													
DEC.													
YEAR TOTAL													

Figure 7-2 Energy Trend Analysis

MONTH 1	NATURAL GAS					COAL				TOTALS			
	COMMODITY COST 26	DEMAND CHARGE 27	TOTAL COST 28	10² FT² USED 29	BTU's 30	TOTAL COST 31	TONS USED 32	BTU's 33	TOTAL COST 34	COST FT² 35	TOTAL BTU's USED 36	EUI 37
JAN.												
FEB.												
MAR.												
APR.												
MAY												
JUNE												
JULY												
AUG.												
SEPT.												
OCT.												
NOV.												
DEC.												
YEAR TOTAL												

Figure 7-2 Energy Trend Analysis

The final area of concern here is DDC system savings. This is an important area for the Energy Manager to consider the aide of a consultant. The method used to determine energy savings and the impact on other systems of any given change is also a matter of concern. Approximations, such as 10% savings for any system, should not be relied upon.

There are a number of computer based models that may be applied by a qualified consultant. In addition, there have been some simplified approaches published by vendors and manufacturers. These can be extremely effective for allowing the Energy Manager to estimate savings from specific functions such as start/stop of electrical equipment and night setback/up. In addition, both seminars and published material are available from the Association of Energy Engineers (AEE) and the American Society of Heating, Refrigeration and Air Conditioning Engineers (ASHRAE).

In general, software packages automate calculations that are available from AEE and ASHRAE. These programs are designed to simulate a building energy systems using a variety of techniques. The more complex versions simulate a building's energy systems using refined input such as hour-by-hour weather data. Simpler systems offer technical shortcuts to arrive at variables for use in savings calculations. To evaluate the impact of a DDC system, or other retrofit project, these programs compare building consumption with and without the specific functions to be applied. Form 7-5 provides a list of the typical DDC system functions that are applied with various types of equipment. Cost savings are then calculated based on current energy costs, and perhaps some escalation factor for future price increases.

As noted earlier it is critical to ensure that the savings do not combine savings from multiple related projects, and thus double count. Finally, the Energy Manager must add all the cost savings available. The total system cost will be divided by the cost savings to arrive at a system payback. It is also important to document other related benefits such as those discussed throughout this book. With the completion of this survey it is now possible to proceed with designing the DDC systems and scheduling installation.

Calculation / Strategy Checklist

LOAD
_____ Heat   _____ Cool

SCHEDULING
Lighting
- _____ Interior
- _____ Event #1 _____
- _____ Event #2 _____
- _____ Other _____
  _____

- _____ Exterior
- _____ Parking Lot
- _____ Sign
- _____ Canopy
- _____ other _____

Ancillary
- _____ Pumps _____ _____
- _____ DHW _____ _____
-        General
- _____ Motors _____ _____

Night Set
- _____ Night Set Back
- _____ Night Set Up
- _____ Shut Down (cool)

Optimal Start / Stop

____	____	Fans	Heat ____		Cool ____	
____	____	Heat	Gas ____		Elect. ____	
____	____	Cooling	DX ____		CHW ____	
____	____	Chillers				
____	____	Boilers				
____	____	Pumps, Ancillary Watts				

Reset
- _____ Hot Water
- _____ Supply Air
- _____ Condenser Water
- _____ Chilled Water

Improved Control
- _____ Heat (____%)
- _____ Cool (____%)

Ventilation Reduction
- _____ Heat
- _____ Cool

Demand Limit
- _____ Lighting
- _____ Other _____
  _____

Duty Cycle
_____

Investment Analysis
- _____ System Cost
- _____ Investment
- _____ Tax Credit
- _____ Depreciation
- _____ Energy Savings (Escalation ___%)

Form 7-5

Chapter 8

State of the Art Distributed Direct Digital Control

Through the first seven chapters in this text the author has provided a general overview of key system management issues. These included general issues like why install systems, and discussing more specific issues like integration. The intent thus far has been to provide the reader with a general understanding of the management aspects of automation, or the big picture.

With this chapter the focus will shift to a more detailed overview of the system technology that makes up Distributed Direct Digital Control (DDC). The next seven chapters of this book will focus on the most significant components of a DDC System.

System Architecture
Hardware and DDC Platforms
Building Wide Functionality
Equipment Level Control
Zone Control
Central Operator Interface
Networking and Communication

An entire chapter will be devoted to each of these components. In each chapter the intended function of the component will be clarified. Key topics will include: typical applications and functionality, software design, and specific technology issues for that component such as expert systems style adaptive algorithms or OEM preintegration.

This chapter will introduce each component with a simple overview that is expanded in the following chapter. Obviously it would not be possible to cover the full breadth of DDC technology in one chapter, but even to address the topics as brief segments will require a knowledge base of the reader. For example, the author will assume that the reader feel comfortable with the material covered so far. Also we will assume that he or she already has had some exposure to with Heating Ventilating and Air Conditioning (HVAC) equipment, and basic control concepts.

In this chapter we will provide an extremely brief historical perspective on the control industry. Also it is essential to discuss the current DDC market and to begin to outline DDC system technology. DDC technology includes communication and networking requirements for these systems which will be discussed as a separate topic here and in a later chapter. Finally it is critical to look at key drivers for future growth in the industry. This will be done as part of the technology discussions, yet there must also be some general conclusions drawn about these trends and overall system requirements for the future.

The topic of Future Systems is critical because it defines the outer limits of the requirements that can be set for today. In a fast paced industry, such as DDC control, specifiers, owners and installers must anticipate change to achieve the greatest potential from their systems. Topics like communication and protocol were barely on the periphery of our thinking about systems only a few years ago, and now they are integral aspects of any systems discussion. This outline of the types of systems available along with a general overview of the features provided by these systems now and in the future, will provide the reader with a valuable perspective for discussing the technology in general, and for evaluating new offerings.

HISTORICAL PERSPECTIVE

Conventional control systems over the years have evolved from pneumatic and simplistic electro-mechanical systems to the highly sophisticated microprocessor based control products of today. The focus of this section is decidedly not to provide a detailed history of conventional equipment. The primary intent of this chapter is to clarify the types of control systems that are applied today and will be in the future. Therefore an understanding of the features provided by Distributed Direct Digital Control (DDC) systems is important. Some benefit may be derived from reading texts outlining the evolutionary progression that has led to Distributed control products, and the reader is encouraged to do further research. In this chapter a cursory discussion is provided covering two of the important control system types common for conventional systems: Load Controller and Central Processing Unit Systems. These systems were distinguished from one another by a number of characteristics, including a significant difference in the level of automation provided as will be discussed.

Load Controllers

The key topics to discuss under Load controllers are:
- hardware,
- control applications and
- architecture.

Load controller hardware was quite simple by comparison with today's systems. This was based on the relatively simple control applications that the devices were targeted for, and also based on the control functionality required. In general the hardware was self contained and included a microprocessor based controller with a real time clock and calendar and onboard inputs and outputs. Inputs were usually limited in number and used to sense various analog conditions such as outdoor air temperature or electrical demand, or digital conditions such as an equipment alarm or power failure.

Control applications and functionality were critical in defining the scope of load controller products. The applications for these devices were generally small to medium sized buildings. Market conditions that accompanied rapid growth with these systems were dramatically impacted by the energy crisis of the mid 70s through early 80s. As a result the control functions were oriented specifically to energy management and the functions included strategies like start/stop for electrical loads such as lighting, etc., temperature control included day/night setpoints and routines to save electric costs such as demand limiting and duty cycling. For the reader who is unfamiliar with these routines, they are discussed further in Chapter 9.

Outputs from Load Controllers were almost exclusively digital for simple start/stop operation. As time progressed a variation of the Digital Output became available with some devices and is called Pulse Width Modulation. This was in fact a Digital output that could be pulse to gradually change a percentage of output with an ancillary device and approximate analog control.

The final load controller topic is Architecture. Actually this term is not appropriate for the conventional load controller technology. These devices employed internal communication structures to access information from inputs, etc. and share it with control algorithms: this was not architecture by the usual definition of the term which refers to an organized process for multiple controllers to share information, coordinate control, etc. These devices, on the other hand, were generally "stand alone", meaning that there was no means of communication between controllers. Each controller provided discrete operation and was only able to access data from and initiate control with inputs and outputs that were physically connected to the controller.

As time progressed it did become more common for these devices to allow remote communication over telephone lines, though this could barely be called an architecture. Rather the function was simple as it allowed direct interface via a remote Personal Computer style workstation using RS-232 technology and standard voice grade telephone lines. In fact this is the same type of interface used today with most distributed systems, yet there was no on site network, and communication was only with an individual load controller. Remote functions were generally limited and not user friendly. Often the software employed to achieve this interface was borrowed from the computer industry and commands for monitoring and programming were cryptic at best. Again the term Architecture by definition means a communication structure for a network of intelligent controllers that are also capable of remote communication.

Central Processing Unit (CPU) based systems

Again the key topics for discussion include:
- hardware,
- control applications and
- architecture.

CPU based systems employed much more sophisticated hardware than Load controller of the same vintage. The hardware typically consisted of a mini-computer based CPU with a simple network allowing data and command access to a large number of dumb input and output panels. The basic hardware components were therefore a large mini-computer based terminal/workstation with operator interface, processing and data storage combined and Field panels for input/output. These panels often included banks of input terminals for both analog and digital information. The most intelligent of these were capable of analog to digital conversion to interpret the electric signal from a sensor and convert it to engineering units prior to being polled by the CPU for fresh data. The output capacity of these panels was analog or digital. In addition to contact closures, one of the most common output panels would include multiple electronic to pneumatic transducers. This allowed conversion of a pneumatic signal to electronic data the computer can understand and vice versa. The result was converting an electronic command from the CPU to a change in pressure on to the output device, i.e.: actuator.

An important issue to discuss initially is the use of the term "Host" to refer to these systems. It became common over the years to refer to CPU based products as Host systems. The use of the term is quite different than when it is used with today's systems. The use of the terms "Host" or "Host-based" varies significant as it is defined for use with Distributed DDC later in this book. The reader should note that the key distinction here is that these systems were totally dependent on one main computer based system for all monitoring and control.

The use of the term Host with distributed systems refers to a level of architecture that requires higher order processing. In fact the "Host level" of the architecture mirrors many of the control functions found in earlier CPU based systems. However, the essential control functions for all equipment are carried out by distributed smart controllers with operation that is completely discrete from the Host level. Even a failure at the Host level of a distributed systems will not result in failure of all equipment and total building downtime. Yet the hardware employed with CPU systems was relatively simple, and some control applications that the devices were targeted for, were also, so a CPU problem caused centralized failure and would bring a building to a complete standstill. This was and is one of the major drivers for Distributed DDC. For an illustration of the various components and functions residing in a CPU based host system refer to Figure 8-1.

Control applications and functionality were critical in defining the expanded scope possible with CPU based systems vs. load controller products. The applications for CPU systems were typically focused on the large building or campus style buildings. The same Market conditions that accompanied rapid growth with Load Controllers spurred sale of CPU based systems. Yet there were many other functions available with these systems that stimulated their application. For example this level of system was able to provide much more sophisticated control, and handle much more complex equipment, In addition these systems also integrated other functions such as Fire and Life Safety, Security and Access Control for building occupants.

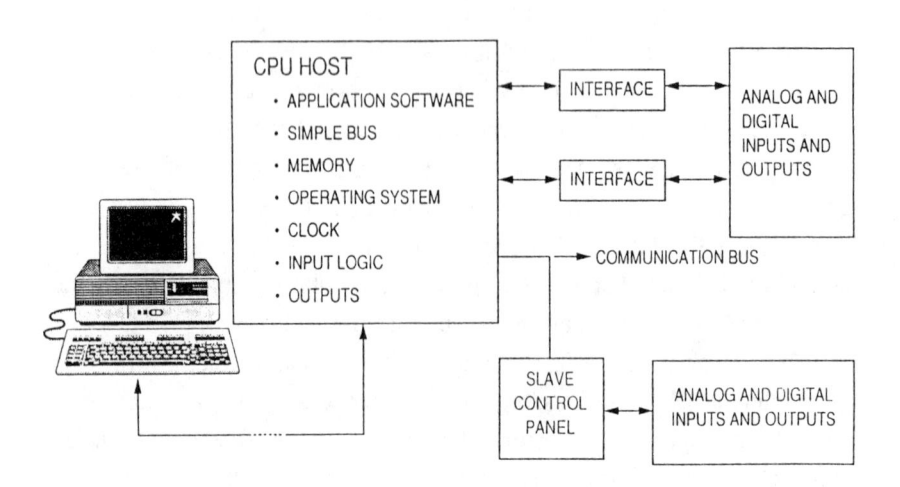

Figure 8-1 CPU Based Host System

Perhaps the primary energy management related control applications were the same as with Load controllers. These strategies again were: start/stop for electrical loads such as lighting, etc., temperature control include day/night setpoint and routines for saving electrical cost such as demand limiting and duty cycling. The added power of the CPU however also introduced new functions such as: Optimal start/stop for Heating, Ventilation and Air Conditioning Systems, Optimization for Chillers and Boilers, Event Initiated Programs and more, Again, for the reader who is unfamiliar with these routines, they are discussed further in Chapter 9.

A final important note on outputs. CPU capability that varied from Load Controllers in the availability of true analog output. Of course the high cost of electric actuators and valves made limited broad application of such devices. Again the analog outputs were used most often to interface with electronic to pneumatic (E to P) devices. True analog outputs were also an improvement however, because of the enhanced control action that was possible with a true analog device.

The final CPU topic is Architecture. This term is much more appropriate with this controller, because a network was required for communication between the various system panels. The network however, was quite simple because the field panels could only speak when spoken to, and simply responded with a data dump of all information available upon request from the CPU.

Again as noted there was similar communication functionality to load controllers for internal communication structures to access information from inputs, etc. and share it with control algorithms, etc. The added complexity that required a simple network was due to the fact that these inputs and outputs were not in a single panel, but scattered through out a building, or even campus, in multiple panels. This fact also impacted the speed of access to data and executing of command which implies a hierarchy of data priorities and structures, as well as data integrity for information being transmitted. These issues will be covered in greater detail under Chapter 12 on Networking.

CPU remote communication used Personal Computer style workstations, RS-232 technology and standard voice grade telephone lines were a critical requirement for many CPU based systems. However, these were often installed in buildings large enough to have on site staff with enough expertise to manage the system, and therefore it was not always applied. In addition this communication was generally via a Personal computer interface rather than modem interface to a panel. Even where full communication interface was not possible however, at least dial out for after hours alarms was very common.

Evolution of building control systems from Load Controller and Central Processing Unit (CPU) based systems involves a wide range of technological issues. For purposes of this text, the primary evolution of note with control systems is the movement to intelligent microprocessor based devices. These devices are capable of both control and communication, and often hundreds or even thousands of controllers are combined to implement a Distributed Direct Digital Control (DDC) Systems.

The term DDC developed early on to refer to application of control directly at the equipment contrary to the CPU approach that simply applied dumb Inputs and outputs. Yet the term DDC did not necessarily mean local micro-processor based control. The advent of technology that allowed installing a computer in every piece of equipment resulted in borrowing the term "Distributed" from the computer industry. This term simply means that the jobs normally done by one CPU, and multiple field panels, are divided up among many smaller microprocessors to make the systems more reliable and allow them to respond to local needs more quickly. Yet to understand some of the dramatic impacts of this technology, and the speed of its evolution it is necessary to view control in the context of the DDC market.

DDC MARKET

The Distributed Direct Digital Control market is expected to reach approximately $1.5 billion by 1997.[1] This market has evolved in recent years through some interesting stages. Significant growth occurred through the late 70s as a result of the energy shocks of 1973 and 1977. Control systems had been continually evolving yet a number of factors combined to make computer based control technology viable. The downward cost trend of electronics in general made control systems more affordable, while at the same time the interest in energy savings, and the incentives and tax credits available also stimulated the market. All these factors resulted in increased demand for technology that would aide building owners to save energy, and implementation of these systems led to development of new features. A new name was also given to these products, Energy Management and Control Systems (EMCS).

The primary systems approach at that time was use of minicomputers, or central processing units (CPUs) as discussed above for Building Automation. Systems also applied new energy saving features for optimizing equipment operation, offsetting electrical demand and scheduling equipment to be shut down when not necessary. The next step in the evolution was the rediscovery of Direct Digital Control. Rediscovery is an appropriate term because this technology was used in industrial process control, and even for some building applications as long ago as the 1950s. But what is Direct Digital Control (DDC)?

The term DDC has become very common in our industry, but it still warrants a slight digression to define the term. The American Society of Heating Refrigeration and Air Conditioning Engineers (ASHRAE) defines DDC as a closed loop control process implemented by a digital computer. Closed loop control, the reader may know, simply means that a condition is controlled by sensing the status of that condition, taking control action to ensure that the condition remains in the desired range and then monitoring that condition to evaluate whether the control action was successful.

A simple example is proportional zone temperature control. The zone temperature is sensed and compared to a setpoint. If the temperature is not at setpoint a control action is taken to add heat or cooling to the zone, and then the temperature is sensed again. A requirement to accomplish this by digital computer does not impose any particular complexity, because in the 1990s even the simplest residential thermostat is likely to have electronic or digital circuitry. The author therefore expands that definition to require that the digital control be provided by a microprocessor based device that implements a sophisticated control loop, or sequence, and is capable of communication over a local area control network (LACN). Chapter 12 is dedicated to a discussion of LACN technology.

Sophisticated control sequences in this case will expand on the simple proportional example above and include Integral and Derivative computations. In that example the function of the integral would be to calculate the distance that the space temperature is from setpoint and adjust control action to avoid wide swings around the setpoint. These swings are often caused by overshoot of the setpoint and delays in responding to a call for conditioning which are common with proportional control. The derivative feature in very specialized, and is used with highly dynamic applications such as pressure control. Derivative measure the speed of change in the controlled condition, and adjusts action of the control algorithm accordingly. Rapid changes in a condition require that the control is adjusted to react quickly.

The result of a Proportional, Integral and Derivative (PID) control loop is a condition that is accurately maintained at desired levels with very little deviation. Bear in mind that the control sequence or algorithm that is appropriate varies by application, temperature control, pressure, start/stop etc. There are a number of other sequences like PID that integrate the control of multiple pieces of controlled equipment to provide seamless operation. Combining this highly accurate control with networking is an essential feature of Distributed DDC for building control. In fact as noted above, this text emphasizes these systems as Local Area Network Building Automation Systems, or what author refers to as Local Area Control Networks (LACN).

These technical progressions were spurred to a great degree by ongoing demand for computerized Energy Management Control Systems (EMCS). Yet by the mid 1980s there was a glut of oil, and an absence of a national energy policy leading to a decline in the demand for EMCS. Interestingly however, the implementation of these systems over the prior ten years led to an awareness of the many benefits available through computerized control. Building owners saw real energy cost reductions which continued to be attractive, as well as the benefits of better control.

Benefits of better control include longer equipment life, more satisfied tenants and a host of new features available due to expanded building information. The communication features available through these systems allowed for managers to track their buildings more effectively and respond quickly and intelligently to problems. Intelligent response is a function of system communication and the ability to remotely diagnose a problem via telecommunication and send a service person to the site with a good idea of the task at hand.

These DDC features that go beyond energy management are extremely desirable particularly in such applications as office buildings. In talking with property managers, the author has learned that it is not uncommon for tenants to begin looking for another office space after the third time they must complain of space comfort problems. The DDC market therefore continues to be driven by savings, energy but more importantly dollars, and further by enhanced control and information.

The future for this market will continue to be impacted by many factors. Future Technology Surveys, Inc. recently updated their market research on this industry. This research involved gathering information from experts in the controls industry and notes drivers for growth of this market in coming years[1]. The author has observed the market for over 15 years and has added some new trends as well to this list of growth factors below.

- Higher energy cost and ongoing energy crises, along with issues like Demand Side Management.
- Lower cost systems and the cost effective replacement of conventional controls.
- The need for quality operation of buildings.
- A natural evolution to DDC technology combined with these controls being driven into the market place through pre-integration by OEMs.
- The growing trend toward Performance Contracting overcoming first cost issues and making this technology more accessible.
- The need to address Indoor Air Quality and related issues.
- Growing application of Intelligent Buildings worldwide, and the control integration that they entail makes DDC integral to facility technology

These key factors indicate an ongoing demand for DDC systems. Further the discussion of important current trends and characteristics of future systems unquestionably targets distributed DDC as an essential requirement for this technology. Both market size and growth are also discussed in that survey, and again it projects that the market for Energy Management or DDC control products will exceed 1.5 billion dollars by 1997 in North America alone. The market however will continue to be driven by advances in technology. The author has long espoused use of off the shelf technology with DDC, and this is becoming the norm today. What the reader may catch is the tone of excitement this brings to our industry because DDC can share in technology advancements whether they occur with; microprocessors, software, electronics, controls or communications

In summary the DDC market continues to grow based upon the ongoing importance of energy savings. There has actually been a resurgence of interest in this area due to the efforts of Utilities to avoid building new power plants. This is actually both an environmental and an energy related driver for automation. There are also expanded benefits available to the building owner through automation technology that include the implementation of an Intelligent Building. This is a strong selling point to occupants and tenants for many buildings.

DDC TECHNOLOGY

The advent of Distributed DDC technology requires that technology be discussed in the context of a control architecture. An architecture will be defined with more complexity later in this book, but for this discussion it may be viewed as a group of control related devices and a common communication network. The concept of a Local Area Network (LAN) Building Automation System (BAS) was discussed briefly above, but a definition of the LAN is still necessary. One comment prior to that definition however, is that this text will use the terms Building Control System (BCS) and Building Automation System interchangeably. Both of these terms refer to the focus of this book which is equipment that provides building wide control, and integrates that control via many distributed devices. The term automation does sometimes imply office automation, telecommunication and other systems, but bear in mind that these are not the focus of this book. Distributed control devices rely by necessity on networking as will be discussed later in this chapter and throughout this book.

A computer Local Area Network (LAN) as defined by Thomas W. Madron "covers a limited geographic area...," where every "node on the network can communicate with every other node, and... requires no central processor."[2]. A node is any device on the network, and Distributed DDC system nodes are intelligent microprocessor based controllers. The control industry adds an order of magnitude more complexity to computer LANs through requirements for control and multiple levels of architecture within the system. These concepts will be the focus of this chapter to set the stage for more detailed discussions of both control and DDC networking.

The overall control architecture will be discussed first in terms of the individual components, and in the next section communication and networking will be viewed. To consider the control architecture there is merit in briefly looking at a simple control application. Figure 8-2 shows a simple Distributed DDC System controlling HVAC equipment and ancillary loads such as lighting. Accomplishing this requires that distributed controllers be applied at each piece of equipment to provide application specific control. These controllers must be able to implement closed loop control via monitoring of status conditions and execution of control actions. At the same time it is also necessary to integrate overall building control provided by these devices for such functions as optimizing the building wake up or start time. These types of features will be discussed in more detail under Chapter 9, as they are implemented with distributed systems.

This chapter will introduce the reader to DDC technology and begin to provide a basic understanding of the control architecture necessary to implement a distributed system. This Control technology will be discussed in much greater detail in the next four chapters, and throughout the balance of this book. These systems are made possible through application of powerful microprocessor based controllers with distributed intelligence capable of equipment control, networking and remote communication. Figure 8-3 focuses on the architecture required for distributed systems, and the levels applied.

There are four levels of architecture defined: sensor/actuator, distributed controller, Building Wide/Island Host and Central Operator Interface. Of these levels the most important layers to our this discus-

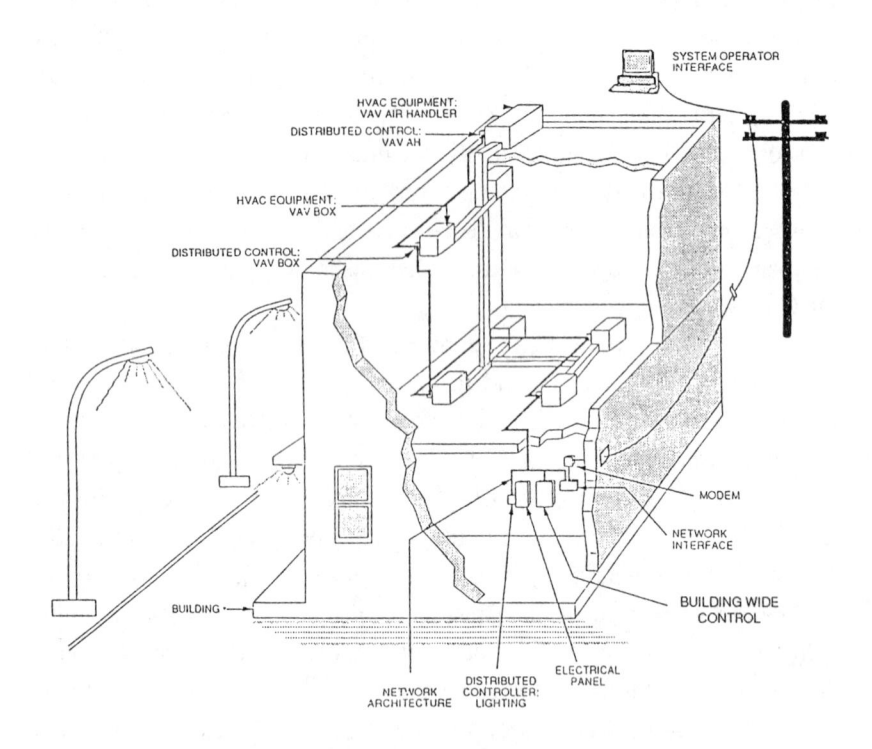

Figure 8-2 Distributed Direct Digital Control (DDC) Application

sion are the distributed controllers and the Island Host. The sensor and actuator technology at this point remains hardwired and uses electrical signals rather than networking technology. Industry standard signals like 4 to 20 milliamps that are wired to a controller are most common. Controllers must then use analog to digital, or A to D, conversion technology to convert the signal to an engineering unit for use with the control sequence. Though there have been a limited number of products introduced to the market that use communication networks, sometimes called "sensor buses", this is not a generally applied technology to date.

At the upper level of Architecture the Central Operator Interface (Central OI) is a critical system component, yet in many cases it is ancillary to control. For this reason it is not necessary to discuss the

CONTROL OPERATOR INTERFACE

CENTRAL OI

BUILDING WIDE/ ISLAND HOST

DISTRIBUTED CONTROLLERS

LOWER LEVEL ARCHITECTURE
 – HARDWIRED ON COMMUNICATION BUS

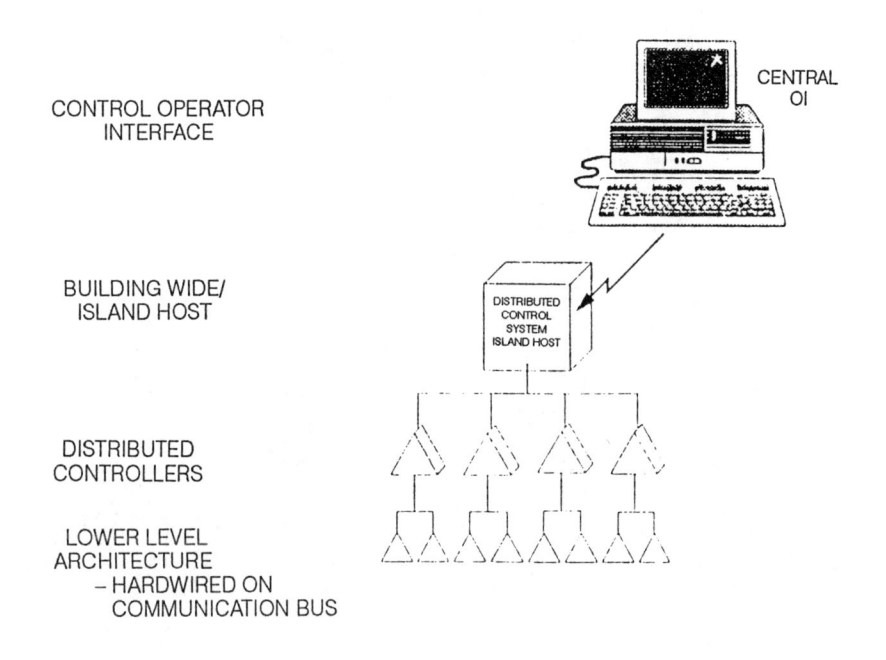

Figure 8-3 Levels of Architecture

Central OI, however the reader should note that a Central has been integral to larger DDC systems for many years, particularly due to the on site management functions provided. These and other features of this architecture level will be discussed in greater detail under Chapter 12 later in this book.

An interesting note regarding DDC systems is that there may be sub levels within the architecture. The Industry accepted common implementation for Distributed systems is to have both a Zone level controller, tailored to simple applications like the VAV terminal unit, and an Equipment level, or general purpose controller. These two controllers will have different requirements and designs to accommodate the given application, yet they both fit within the level of a distributed controller. This discussion introduces, and the next 4 chapters focus in great detail on control specific components. These system components include:

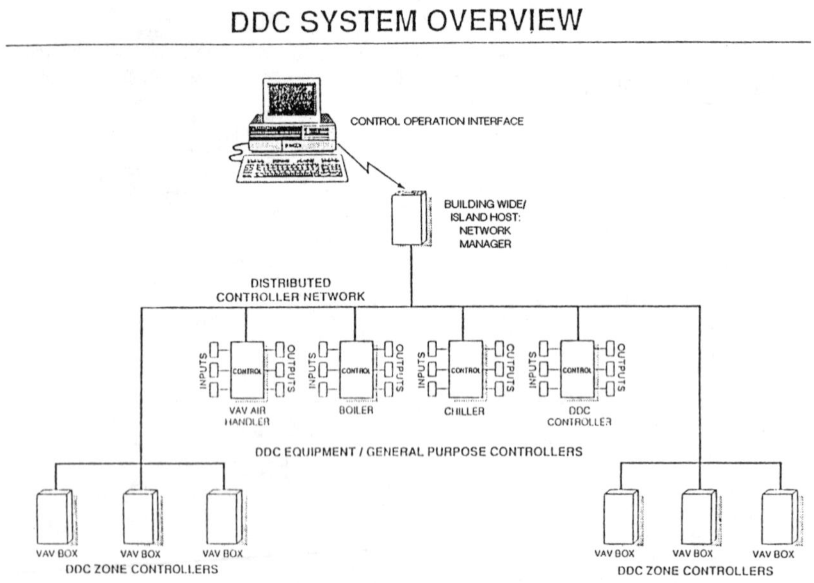

DDC SYSTEM OVERVIEW

Figure 8-4 Typical Distributed DDC Architecture

- Building Wide/Island Host,
- Distributed DDC Equipment Controller and
- Distributed DDC Zone Controller.

Figure 8-4 indicates a more typical distributed DDC system architecture with the Island Host and two levels of distributed controllers. The discussions of each of these levels of control will consider specific hardware and sequence of operations for that particular component. It is also important to cover the concept of building wide, or integrated, control. Finally this technology is extremely dynamic, so a general view of trends will be provided for each major component of the systems, as well as overall trends for the entire system. Though the reader has only been introduced to the components of a typical system, this chapter will digress for a moment to the future. It is important to consider DDC as a fast paced evolving technology. As a result the reader should be ready to carefully evaluate any offering regarding the state of its technology. Therefore while considering systems in general, and prior to more detailed discussion of specifics it is worthwhile to address the future.

THE FUTURE OF DDC SYSTEMS

The primary intent of this book is to clearly define our current state of the art in DDC technology. Yet this technology is so fast paced and changes are so rapid that it warrants a brief discussion of the future for these systems. Some discussion of future trends will also be found elsewhere in the text as well. Yet the intent here is to consider future systems from a big picture system perspective.

This topic of Future Systems affords a tremendous opportunity for crystal ball prediction. Rather than becoming sidetracked however it is absolutely critical to maintain focus on the basic topic of this book Distributed Direct Digital Control. One of the undisputed characteristics of future systems will be distributed networking controllers with substantial control flexibility. The communication capabilities of these devices will be equally important to the control sequences. The basic architecture discussed with regard to current DDC technology is setting the stage for the future. It is the authors belief that the face of the control industry will not vary greatly from the general approach outlined into the next century through the mid to late 1990s. Without question networking controllers will continue to expand their penetration into all types of equipment, and at the same time network communication capability will continue to expand. In fact these future systems may be viewed in those two categories: Control and Communication.

Future control capabilities of distributed systems include a number of new technologies. Among those noted in the recent Future Technology, Inc. market research are Modular expandable systems made up of smaller faster and more distributed controllers that are applied at each control process or local unit[4]. These controllers will require adaptive PID or expert system types of control capabilities, and networked communication is an absolute essential. Adaptive DDC is a concept that allows control algorithms to "learn" over time and provide better control. This is an expert system type of approach that has been applied with Industrial process and other types of control equipment, and is now becoming viable with Building Automation. A number of expanded system enhancements are also noted in the above survey such as expert equipment diagnostics.

The control characteristics, however are only one aspect of this topic. Delivery of these controls and the method of implementation is also critical. The concept of pre-integration of controls at the factory is also important. This is because the control sequence must address more than control functions for the equipment in which it is mounted. More and more the success of future control systems will require that the functions required for an entire building be integrated. This means that even controllers that are pre-integrated with a specific piece of equipment like a VAV air handler must be able to implement more functions for the entire building, like demand limiting or after hour access and billing. These control functions will be discussed in greater detail in the next four chapters, but integral to all these functions will be the communication and networking technologies that are implemented to share this information and ensure its integrity.

Communication characteristics and technologies of future systems will be explored throughout this text as they apply to Distributed DDC. At the heart of this topic is one of the most exciting, and undoubtedly one of the most challenging, future trends is standard communication. The reader may question why standard communication is even an issue for Distributed DDC. Quite simply a number of factors have merged to make this an extremely volatile and critical question for the control industry. First the trend toward distributed DDC necessitates communication, both networked and remote communication with systems via telephone lines. Second the fact that these controls are distributed to the industry through a wide variety of channels, and end up in buildings through equipment, site installation, etc. is also critical. Ultimately as noted, the building owner will want all of the equipment in the building to provide a seamless quality operation. Standard communication is one way to achieve this goal.

This topic is discussed in greater detail under Chapter 12. Yet there are a number of issues in addressing the control industry needs for communication. This industry has only passing familiarity with data communication through experience with proprietary control system development. Also because it is ancillary to our control products, there is some real question as to the appropriate level of internal research and development that should be funded. At the same time investment is hampered because users have stated that they are willing to pay only a modest premium for standard. communication[5]. Yet networking is essential to support distributed control, and it has been verified that the industry wants this standard for networking. Prior to availability of any standard though, there will be transitional products. Transition refers to the attempt by users and others in the industry to find a short term solution while the standard is being developed. This demand has led to a number of related trends and products in the industry that will be discussed in later chapters. These transitional efforts are likely to change the concept and implementation of future systems. Standard communication protocols for systems networking are not likely to evolve until the late 1990s. In the interim these transitional approaches will be implemented to provide short term solutions to the communication issues that are discussed here and later in Chapter 12.

In addition to communication standards there are some other specific communication trends that will evolve with future systems. These include:
- faster data transmission speeds over the network and via telephone modem,
- fiber optic and other options for transmission media,
- voice communication for monitoring and alarming,
- expanded network data sharing capabilities and
- eventually multi-functional integrated networks for building control, security and other systems to truly implement the Intelligent Building concept, and
- expanded pre-integration of communicating controllers in all types of building equipment.

There are also a number of ancillary products that will likely be integrated with control systems, such as PC based Local Area Networks for data access throughout a building and an entire organization. These future systems may also be expected to be characterized as "easy to use".

Ease of use is extremely important point because the expanding power of microprocessors and software has allowed for "user friendly" control products. Control industry users are being exposed to a wide variety of PC and computer related products that shape their expectations of the technology available. This industry must meet those expectations with control products that allow an average user to acquire sophisticated building operation without complex programming.

A PC based Central Operator Interface (Central OI) is likely to be integral to any future system. The Central OI will be the window through which the user is able to view the system. It will also serve as the focal point for ease of use through a high technology simplistic user interface. The industry has seen a growing trend toward availability of these products as two distinct packages. In the past Central OI packages were provided that allowed programming, monitoring and a host of utility functions for interface with control systems. The author believes that the ongoing trend toward easy to use systems will result in specialized packages designed for the intended user. One of these Central OI tools will be focused to the contractor or system integrator that must design, install and start up the control system. A second tool will be specialized with features for the building owner, who must maintain the system over time. There will be expanded discussion of this topic in the Central Operator Interface Chapter 13.

This chapter has provided a brief, yet comprehensive, overview of the DDC Market, Digital Distribution Technology and Future Systems. The intent was to provide the reader with an introduction to DDC technology, and set the stage for further discussion throughout the balance of this book.

FOOTNOTES

1. "Survey on Energy Management Systems - Survey # 208", Future Technology Surveys, Inc., Lilburn, Ga., P 26
2. Thomas W. Madron, Local Area Networks, The Second Generation, John Wiley & Sons, Inc., New York, N.Y., 1988, P. 3
3. "Survey on Energy Management Systems - Survey # 208", Future Technology Surveys, Inc., Lilburn, Ga., P. 9
4. "Survey on Energy Management Systems - Survey # 208", Future Technology Surveys, Inc., Lilburn, Ga., P. 29
5. Alan Aramson, P.E., Honeywell, Inc., Report On Intelligent Buildings Institute Focus Group Survey of Control Industry Users, World Energy Engineering Congress Proceedings, Atlanta, Ga., October 1990

Chapter 9

Building Wide Sequence of Operations

INTRODUCTION

This section begins the detailed coverage of Distributed Direct Digital Control (DDC) systems technology. In older generation systems it was possible to view system sequence of operations or "sequences" based on the device in which they were handled. This is no longer the case, as we discussed in Chapter 8 the sequences carried out by the system are becoming more and more distributed. This means that it is no long possible to say the CPU carries out building wide tasks and temperature control is done by an air handler control. Think of controllers as employees sharing a common goal for their company, and carrying out a series of complex tasks, or sequences, in cooperation.

With a distributed approach the sequences provided by a control system can actually be equated to those carried out by a corporate organization. At the day to day operations level, there are micro-level tasks, such as temperature control, that are specific to a particular area of the building and piece of equipment. Just as work direction for an employee team is best provided in the work area, these tasks are more appropriately located in a local level controller at the equipment.

Just as an employee must have some access to the corporate vision, policies, etc. to do a job in the way an employer expects, a local controller must have access to higher level data. Actually the examples are quite simple for controllers. For example, the controller must know, and synchronize, time of day and calendar data to coordinate building control. More examples will be covered throughout this chapter.

This chapter is designed to provide the reader with a big picture view of the control system. It will also define some of the initial concepts that will be essential to intelligent evaluation, installation and management of control systems. Again do not think in terms of Building-Wide sequences as occurring in any particular device. With the trend toward more intelligent controllers that have "Peer" networking privileges for communication, these sequences can be distributed to any controller in the system.

As will be discussed in the network chapter this trend can be expected to continue until every sequence resides in the most logical controller for that sequence to be carried out. First though it is essential to understand what these sequences are and how they fit into the overall process of building management.

The topic Building Wide Sequences was selected for discussion first under this DDC technology section because it is responsible for overall coordination of all automation In many cases the greatest cost reduction is available as this level because it ensures consistency in day to day operation through such data as time and calendar dates. Also control sequences that are initiated based upon data like electrical consumption must be coordinated on a building wide basis, because there is generally only one meter for the building. That meter must be interfaced to the system and a controller must monitor the meter, compare current status to desired levels of consumption and take action accordingly. Key topics to this discussion that will be discussed in this chapter are listed below.

      Application overview
      Introduction to DDC
      Hardware
      System Programming Techniques
      DDC Control Sequences
      Building Wide Coordination
      Building Wide System trends

APPLICATION OVERVIEW

The term application is used extensively in the DDC or control industry to refer to the equipment being controlled. Applications should be distinguished from functions. This book considers a DDC System to comprise Building Wide automation of all the functions in a facility. Chapter 3 on Integration provided a description of the various functional systems, such as Heating, Ventilating and Air Conditioning (HVAC), Fire and Life Safety, etc. Depending upon the needs of the customer, the size of the building and other factors, the functions may be integrated into one DDC system or provide through discrete systems.

Within the context of DDC, this book will focus on functional systems, and note that each function is made up of a large number of applications. These applications provide some critical condition to the occupants of a building, and a sequence of operations must be established to maintain that condition. For example, it has been demonstrated repeatedly that temperature in a work area impacts worker productivity. The equipment that maintains temperature for that area is an application. A sequence of operations must be implemented to control that equipment and ensure that the desire temperature condition is consistently maintained.

The primary intent of this chapter will be to address DDC functions related to cost savings sequences for applications of the HVAC function. These are the most complex functions in the building, and so the most time will be devoted to sequences for these applications. Some discussion of control sequences for the Fire and Security functions will be provided, yet these are generally fairly simple in concept as we discussed in the Integration Chapter.

To clarify the term application further, it refers to specific types of equipment within an HVAC or Fire systems. For example, a Variable Air Volume Air Handler would be one application within the HVAC function of a DDC systems. A Sprinkler system would be one application of a Fire and Life Safety systems. Focusing on HVAC, the first step in designing any DDC system is to identify the types of applications to be addressed.

The building survey provided forms and an overview of the process required to capture this information. The purpose of the walk through is to define exactly what equipment resides in the facility, and how it is used, schedules, temperatures maintained, Fire & Life Safety requirements, etc. With this data the functional systems required to meet the customers needs may be identified. The Energy Manger or team can take this information and target functional systems and sequences to be implemented. The opportunities are significant and the complexity of sequences available requires extensive systems knowledge. The capabilities of a DDC system go far beyond what will be discussed in this chapter because customized sequences may be developed. These customized sequences may be for any application from industrial processes to a unique type of HVAC equipment.

For the most part sequences discussed in this chapter will be focused on building comfort and cost savings through equipment optimization and energy conservation. These functions are directly almost exclusively to HVAC, yet many of the programming techniques discussed below are applicable to other functional systems. For example, scheduled start/stop functions are applicable to lighting, building access control systems and monitoring for break of security.

It is also important to note that the expertise of a DDC professional is extremely valuable in defining sequences that are applicable to a facility. Temperature setback for a computer room would not be acceptable, for example. Advice of a professional can be instrumental in effective implementation and smooth start up to avoid problems associated with blanket application of these sequences. Before discussing either examples or sequences further, it is desirable to provide some added definition. Up to this point this book has been written with the assumption that the reader has very limited background in DDC.

The balance of the book will provide more detail and rely more heavily on a working knowledge of the both controls and building technology, particularly HVAC. It is not possible to provide a primer on HVAC in this book, the reader who needs more background in this area is encouraged to do additional research and study. The Association of Energy Engineers and the American Society of Heating, Refrigeration and Air Conditioning Engineers are associations that have a wealth of information available on this topic. It is reasonable, however to provide some general background and definition on DDC for the reader. Chapter 8 provided information on CPU and Load Controller systems. Yet the internal sequences of those devices were only discussed to a limited degree. As this chapter and the next two will be more focused on applications and sequences a brief description and discussion of DDC is appropriate.

INTRODUCTION TO DISTRIBUTED DIRECT DIGITAL CONTROL (DDC)

The best place to begin this introduction may be with a definition. One of the most credible organizations in our industry is the American Society of Heating, Refrigeration and Air Conditioning Engineers (ASHRAE). This organization has published a definition for DDC. DDC is a term that has been in use since the 1950s, and the concept of Distributed is a later development, actually an enhancement, to the technology. Therefore a definition will be provided for each, and then further discussion will be provide to identify the current state of DDC control and briefly how it evolved.

ASHRAE defines DDC as closed-loop, modulating control of an output device by means a a digital computer. The author expands that definition to include the use of a sophisticated sequence of operations that is used by the digital computer to implement the DDC control. Combination of the sophisticated sequence and the computer or microprocessor based control provides DDC.

Distributed DDC relies on expanded use of the DDC computer. ASHRAE defines Distributed Processing as the practice of dividing computational tasks among a number of discrete computers, each with its own program. Microcomputer controllers, for example, can be programmed to collect, process and store sensor data and perform control actions at remote locations while still communicating with each other or a central computer.

Combining Distributed Processing and Direct Digital Control results in systems that provide sophisticated control. This control is implemented on a building wide basis via independent smart controllers that carry out control strategies for individual pieces of equipment and for the entire building. The result is Distributed DDC, a technology which relies on Local Area Control Networks to implement a full building system.

Distributed DDC offers many benefits. It is state of the art technology, but beyond that the power of these microprocessor based controllers working both individually and in coordination is significant. The individual controllers can be application specific and therefore provide an expanded sequence of operations for controlled equipment. This means a great deal with HVAC, because the control requirements can be quite complex, and also it is then possible to do equipment safety and information oriented routines. Equipment safeties might be targeted at alarming when a refrigeration compressor reaches high pressure, thus avoiding a costly failure and loss in productivity due to downtime.

The benefits and drivers for DDC were discussed in earlier chapters. The intent here is not to repeat that data, but to very simply describe DDC. As the term and the definitions above imply, the key to this technology is applying intelligent control at the equipment control level. This author has written for over a decade that it is critical for that control to implement sophisticated strategies capable of improving the output of the equipment, i.e.: temperature control, while optimizing efficiency. Control is a primary function of DDC but only part of the requirement, communication is the second key component of a DDC controller. Without this communication the overall functionality of the systems will not be achieved, nor will the essential benefit of information management be provided.

It should be clear at this point that implementing Distributed DDC required a combination of hardware and software. There are specific hardware requirements to support the DDC control strategies held in software. These hardware issues tend to be specific to the controllers carrying out the application related control tasks. Hardware will be discussed in more detail under the next two chapters, in particular under Equipment Level Control. Powerful microprocessor based devices are essential to DDC, and key to implementing software strategies are two factors: execution time and scan rate for inputs and outputs.

Execution time is a function of the processor along with the number and complexity of control loops. More effective control typically requires faster execution time so that closed loop control corrections occur rapidly enough to ensure desired conditions are maintained. For execution of the control sequence to be effective the information used must be current. This means that the controller must scan inputs and outputs quickly, as often as once a second. To be effective, DDC hardware must ensure control data is refreshed and sequences are executed as quickly as is appropriate to the controlled condition. Duct static pressure loops require execution with fresh data as often as once a second, while an outside air reset loop could execute every 30 seconds or more. More on hardware later, first let's consider the heart of these systems, the DDC software itself.

An entire section of this chapter is devoted to the various styles of software programming that are applied in this industry. What is more important here is how the implementation of this software makes DDC effective. Ultimately DDC software produces results because closed loop control is both implemented directly at the equipment level and coordinated throughout the entire building. At the end of Chapter 8 future trends for DDC were viewed in terms of control and communication. In keeping with the premise that there is not DDC without effectively merging these capabilities, this section will define DDC by both control and communication.

Control is the obvious key component and requirement for any Distributed DDC device. Again the focus of this Chapter will be HVAC control, and therefore examples are drawn from that arena. What makes DDC unique and different from its predecessors is that the many control loops within a given piece of equipment are coordinated. Prior to DDC is was common to have 2 to 10 discrete control devices in an Air Handler to control the various sequences, supply air temperature, static pressure, mixed air, equipment safeties for flow, freeze, etc. The result was a piece of equipment that was difficult to maintain, degraded in its performance over time and was likely inefficient as well. DDC coordinates those loops by having discrete control sequences for each, but also by having additional sequences overlaid within the controller to monitor and tune their interaction.

In terms of the critical feature which define DDC control, sophisticated strategies have been targeted as essential. Again focusing on HVAC a brief discussion of these strategies is important. In fact a point of clarification is that DDC software consists of three levels:

1. DDC Control sequences
2. Sequence components and
3. programming language.

DDC Control sequence is the primary focus of this chapter and these will be discussed in more detail later. In essence these are pre-programmed sequences of operations that are implemented with the majority of systems. Examples would be start / stop control or Demand limiting. Think of these as higher level requirements for any systems. In addition to these basic overall sequences however, a system must accomplish a variety of control tasks. These tasks including more equipment specific requirements like temperature control.

Sequence components are the essential building blocks for all DDC control sequences including temperature control. These functions are in most cases one level removed from most users who interface with a system. Yet without these components, sophisticated control would not be possible.

To use HVAC equipment as the focus again, consider temperature control. DDC was defined as closed loop control, meaning that a condition is sensed and compared to a desire value, setpoint,. If the condition varies from the setpoint action is taken. Closed loop control means that the condition is then sensed again to determine if further action is required to meet the desired setpoint.

Control sequence components are required to ensure that closed loop control results in optimal temperature conditioning to the desired setpoint. For temperature, pressure, humidity and other conditions, this is done via three sequence components:

Proportional (P)
Integral and (I)
Derivative. (D)

These components are combined as necessary for the application to provide the optimum control required. In essence the best possible control is achieved when all three components are combined. More than likely, the reader has heard control systems touted as offering PID control. To simplify discussion of these components, they will be covered as they are applied.

PROPORTIONAL CONTROL

This control component has traditionally been applied in simple control devices such as thermostats. Proportional control senses the controlled condition or "variable" and takes an action that moves the controlled output proportional to the difference, or "deviation", between the controlled value and the setpoint.

In essence the position of the controlled device (valve, damper, etc.) will vary proportionally to the value of condition (temperature) measured. Consider a cooling sequence that must modulate a valve on a chilled water coil to provide conditioned air to the space. An increase in space temperature will result in a valve opening proportionally to provide cooling.

There are two relevant parameters for Proportional Control: set point and throttling range. Setpoint defines the desired condition for control, and throttling range defines the band of temperature around the setpoint that the sequence is expected to control. Accurately setting the throttling range, often called proportional band, for a sequence is essential to ensuring that stable consistent control is provided. If the proportional band is too wide control of the output device may be smooth but temperature control may not meet occupant needs. Tighter throttling ranges, on the other hand, can result in hunting. This means the control device is turned on and off, or modulated full open to full close too frequent, again with potentially unsatisfactory temperature control.

The combination of factors contributing to possible temperature control problems with proportional control has clear results. Proportional controls must be carefully maintained and tuned, even so the nature of this control is that there is a deviation between setpoint and the controlled condition, called offset. Control offset is often further complicated by overshoot or droop. Droop is the idea the proportional control tends to react somewhat slower to a change in temperature thus resulting in the condition dropping below setpoint before control action is taken. At the same time this control generally overshoots the setpoint. If you look at a trend of temperature over several hours therefore, the trend will look somewhat like a sine wave. With the setpoint in the middle, temperature points above and below will depict a pattern of droop below setpoint before action is taken and overshoot above setpoint before corrective action is made. These conditions lead to the use of Proportional Plus Integral control.

PROPORTIONAL PLUS INTEGRAL CONTROL (P + I)

This sequence effectively uses two components Proportional and Integral to enhance control. The proportional band for the PI loop will be set wide enough to ensure smooth control action. The primary function of Integral is to continuously recalculate the distance from setpoint, or offset. With this information the sequence can then issue corrective action and avoid droop and overshoot. It becomes obvious that execution time is important in these loops to ensure adequate reaction time.

Reaction time in a P + I loop allows for incremental changes in the position of controlled outputs thus ensuring tighter control around the setpoint. The integral component of the sequence calculates the offset and determines required action each time the sequence executes, or based upon an integral time interval. This time interval must allow for scan time on the controller. This ensure that integral action is calculated only after enough time has passed to sense the result of the previous correction. Most systems will initialize a control loop with default integral values that take into account the performance of the controller. Nonetheless some tuning of these values may be necessary depending on the application.

A key concern with the P + I loop is "integral wind up". This occurs when a control sequence is executing to maintain a setpoint, but the control action is ineffective because the equipment is shut down. For example a P + I cooling sequence may sense supply air temperature and try to open the chilled water valve. If the unit is shut down, or in unoccupied mode however, setpoint will not be achieved. After several hours in occupied mode, the unit is restarted and the integral component creates a problem. This component has been sensing no effect on setpoint for hours and therefore it will modulate the chilled water valve wide open and it may take some time for the loop to stabilize. As a result more sophisticated P + I loops disable the control loop during unoccupied mode, etc. through several techniques. The simplest is to have the cooling loop monitor fan status, and to disable the loop if the fan is off.

Integral control is extremely effective for eliminating offset and achieving steady state temperature control. There are a number of control conditions however, that change must more rapidly than temperature. This introduces the need for one final control component, Derivative.

PROPORTIONAL PLUS INTEGRAL PLUS DERIVATIVE (PID) CONTROL

The PI portion of this sequence operate as described above, and the Derivative is added to monitor the speed at which the controlled condition is changing. This is extremely effective for rapidly changing conditions like duct static pressure. It is very useful in correcting major problems with droop or overshoot. Derivative becomes a third value that is mathematically applied to the Proportional and Integral values. Ultimately the Integral and Derivative temper the Proportional reaction resulting in a tuned response to a dynamic environment.

It is important to note that conditions like temperature do not benefit greatly from Derivative. This is because an element of inertia is required to overcome temperature conditions. The mass of the controlled environment combined with the various control equipment masses results in slower changes in condition that are more easily anticipated. Pressure in the supply duct of a Variable Air Volume Air Handler however, can change dramatically in a fraction of a second, thus requiring rapid correction to avoid droop or overshoot.

DDC therefore is defined essentially by the complexity, and therefore accuracy that can be achieved through these control components. These components when used effectively in a control sequence for a given application result in optimized equipment control. Further human beings that interact with the controlled equipment are provided with optimized control conditions. A relatively new enhancement to these control components that should be addressed is adaptive ID.

Artificial Intelligence and Expert Systems are terms that have crept into our vocabulary over the last decade. As a rule concepts like these become cost effective with high end computer industry applications first and then migrate to industrial process control and eventually to building technology. A key reason for this is that the cost to initially implement these technologies is great. Once the technology is developed and use reaches some economy of scale, though it's possible to apply it more broadly, i.e.: to the control industry. Adaptive PID is a technology that implements these types of technologies in the control industry.

Adaptive DDC recognizes that the sequence components discussed above required tuning for optimization. Control requirement will vary based upon load conditions and optimal operation could require tuning proportional, integral and derivative parameters on a seasonal basis at least. As noted these loops typically start will general defaults, and in many cases no one ever modifies these values. As a result control is good, but never optimized. The cost of optimization would require ongoing system maintenance and in most cases this is an added expense the building owner will forgo.

Adaptive DDC would provide for optimized control without added maintenance and tuning tasks. In essence these sequences would be self-learning. Based on the existing inputs and outputs available with the system, the Adaptive portion of the sequence would automatically modify PID parameters to optimize control response. This would result in a self-learning control loop that adapts to ever changing control conditions. Adaptive DDC type feature were first introduced to the industry with Optimal start Stop (OSS). This feature is discussed later, but in essence it learns from day to day how early to start equipment if set-point is to be reached at a specific time.

This has been a simple introduction to DDC as a technology and a buzzword. Given that introduction the reader must now consider applying DDC to an entire building. Again the approach is to consider first the high level tasks that must be carried out for the entire building, and then specific requirements for individual pieces of equipment. As a result we will transition from DDC in general the Building Wide DDC Sequences, and look first at requirements for hardware.

HARDWARE

Hardware for Building Wide control varies based upon the implementation of Architecture. The Architecture define both the topology, acceptable wiring layout for the system, but more importantly it defines privilege. Network privilege is quite simply the authority a controller has to perform tasks on the network. ASHRAE's BACNET refers to this as a device "class". Why is this important to Building Wide Hardware? Because there may be no discrete hardware at all for this function. With a peer to peer control network, each device has the same authority and sequences are divided amount the controller. Primarily the reader should think of these sequences as required but not get confused by looking for a specific device to do the job. Simply ensure that the system you are evaluating provides the sequences that are required for the building. Still a brief discussion of hardware is worthwhile.

In general Hardware for Building Wide Control varies greatly. Again from an Architecture point of view, this often called the Island Host level as described in Chapter 8. With Hardware at this level, approaches have been taken which involve discrete custom microprocessors, or even Personal Computers, dedicated to complete all building wide sequences, and distributing sequences among peer controllers.

This discussion will focus on the carrying out the tasks through dedicated controllers, whether master/slave or peer because it is most appropriate to a true distributed system. At this point in the discussion it is not critical to understand more about Master/Slave and Peer Controllers. These topics will be covered somewhat under networking, and are simply styles of implementing communication on the bus. The Master/Slave approach is the most widely implemented networking technique applied for both computer and controller networks. The Peer approach is relatively new to our industry and introduces more power and to some degree integrity to controller networks. This is because each device has full access to the communication bus and can access data from any controller. As a result partial power failures or other interruptions in DDC control do not affect an entire building because the peer controllers continue to operate.

This custom microprocessor based approach to Building Wide, or Island Host level control, must be viewed primarily as a vehicle for system level functions. System level functions include: managing the network, integrating building wide control among all the distributed controllers and serving as a central focus of all communication with the network. These concepts will be expanded significantly under sequence of operations, yet they are mentioned here to indicate the hardware complexity required at the Building Wide Level. In spite of the large number of variations on manufacturer approaches to development of this hardware, the author will describe a somewhat typical hardware example. In fact this will consist of two examples which will be called implementations: Master/Slave and Peer, which is often called "Peer-to-Peer".

The Building Wide Master/Slave Implementation is typically focused on providing communication and networking functions, and therefore often does not have on board inputs and outputs (I/O). In some cases these devices will have a limited number of I/O, but the trend is certainly toward use of system wide data in the form of pseudo points.

A pseudo point is a piece of information that physically resides elsewhere in the system but is held in a "data field" at the Host or other distributed controllers. Perhaps the best example of this is outside air, which is needed by all controllers. The physical point will be connected to any particular distributed controller and then shared via the network with all other controllers. These data sharing functions will be discussed as part of control integration under sequence of operations.

Peer Implementation varies from the description above in that each Distributed DDC peer is fully capable and responsible of carrying out both local level control and Building Wide functions. That is a significant task and one example of how increased processing power has become cost effective for the control industry. Some Peer level control devices are implementing 32 bit microprocessors that are capable of high control execution and extensive communication functions. Back to the topic at hand however, this implementation differ from Master/Slave because it will have onboard inputs and outputs. The control related aspects of Peer controllers will be covered more under Chapter 10. This discussion focused on building wide functions which include control sequences and communication.

In either implementation at the Building Wide level hardware becomes the focal point for communication. It will provide direct outside access through a modem and telephone lines, or interacts with another peer dedicated for this purpose, and allows access to all system data. The Building Wide job also requires some overall network management functions, though in many cases these systems are self policing. Self policing in this context means that each network member, or controller, has responsibility for ensuring communication and sequences that are automatically enabled when a failure occurs. This is true whether these are Master/Slave or Peer, but the specifics of this discussion go beyond the scope of this book.

The picture that we are painting of Building Wide hardware therefore is a powerful microprocessor based system under both implementations. This system is focused primarily on the two levels of system communication: networking and remote interaction. The Host or Peer must also coordinate building wide control functions and availability of system wide data sharing, which will be discussed in a moment.

Building Wide functions are typically implemented in one piece of hardware for Master Slave systems or share among Peers. However there may be other functional systems required such as life safety, fire and security systems. Again we are focusing in on the most complex Building Wide System, HVAC, but the level of System Level Integration varies from one Building to another.

This chapter will establish the premise that functional systems in the building, HVAC, Fire, etc., are provided via separate Building Wide Controllers or Island Hosts. In fact there may be a control requirement to integrate control between these various functional areas, yet this book will focus on building control. For the most part it will be assumed that the Central Operator Interface is only concerned with monitoring and access of these other functional systems. In summary therefore, this hardware overview defines of a communication master through two different implementations. Hardware to accomplish these task varies between the implementation, yet the most important thing here is the functions themselves and not the implementation. As a result this discussion will move to software.

The balance of this chapter will focus on the actual Building wide coordination and sequences that must be implemented. Prior to describing these techniques and sequences however, it is important to understand the various approaches for programming. These software programming approaches vary greatly and can dramatically impact the cost to program, commission and maintain a system. Understanding these techniques will also better equip the reader to compare between systems.

SYSTEM PROGRAMMING APPROACHES

There are a number of approaches for programming control systems. In this section it is not possible to describe every approach in detail. The intent is rather to provide a general understanding of programming required for DDC, and its role within system development. Programming, commissioning and installing a DDC system will vary based on the programming approach or control software type. In spite of the variety of approaches most systems on the market employ one of three general programming types. These types will each be discussed briefly and include:

- Data Field Entry
- Code Entry and
- Graphic-Icon programming.

"Data Field Entry" programming, is used with Setpoint Entry and Expanded Parameter systems. It is typically tabular in fashion with prompting and fairly straight forward. Code Entry for "Equation Driven" systems is often called the blank slate approach, and has a very different appearance. "Graphic-Icon" based systems are the newest form of programming, and offer tremendous potential to reduce the time required to engineer systems.

The Graphic approach may be used with either of the above systems, though Mouse driven software and Icons replace much of the text entry required with both types. In Graphic Icon packages, the programmer selects from graphic segments, characters or "icons" to program equipment sequences. These graphic representations are generally converted to machine level code with a utility that runs in the background.

Regardless of the programming type, taking a controller from carton to start up requires commissioning and initialization. This discussion will focus on initialization and the software tools that are required to customize a controller sequence for the intended application. For example, the programmer must allow for adding or deleting a control point. Basic system programming will obviously differ by system type.

Data Field Entry is generally menu driven or mouse point and click operated, and directs a user through a series of screens for parameter selection. Actually many of the "Data Field Entry" systems are hybrids for they allow a significant number of complex parameter choices, which modify the controller operation. These hybrids in many respects offer the best of both worlds, because a controller can be customized to an application without a concern for the viability of the control loop.

The data entry process in this case is more complex yet it does not depart from the basic entry field style which requires a response to a prompt. Such systems also employ error trapping which limits responses to a particular prompt to answers within a specific range. Graphic-Icon based packages are similar except that the programmer selects an Icon which represents a sequence or control strategy. With Data Field Entry the icons are fixed as they are in the tabular formats, but it is possible to speed and simplify the process by selecting the appropriate icons during programming.

The most sophisticated systems on the market allow Graphic programming via control diagrams that depict the application. For example, you program by building a diagram and developing simplified control strategies. The system then automatically converts those Graphics into machine level code, which makes them easy to use. Another interesting benefit to this approach is that the programming sequence can be printed out for a detailed submittal package.

As noted "Code Entry" system commissioning is often called the "blank slate" technique and varies dramatically from a menu approach. In this scenario the user has a blank screen and a language available.

Equation languages offer IF, THEN, ELSE type logic, along with a mix of arithmetic, Boolean and relational operations. Arithmetic features include algebraic expressions for: addition, subtraction, multiplication and division. Many systems also offer complex mathematical operations such as: sine, cosine, square root, and exponents. Relational features compare one element to another and determine if they are: "greater than", "less than" or "equal to" one another, along with several variations. More complex code may include such sophisticated operators as calculations for proportional, integral or derivative control.

These languages must also allow the user to define the physical devices which are connected to input and output channels. Equation language features combine to require that the user have a high level understanding of the process to be controlled, and its interaction with all other controlled equipment. At the same time they certainly provide that knowledgeable user with expanded control of the ultimate sequence that is implemented.

Code Entry systems vary in the amount of control programming that the user must generate. For example, some systems will provide a library of strategies for the programmer to draw upon or link together, and incorporate into an individual program. These libraries may reside in the controller, but this is uncommon due to device memory constraints. More common is to have the Central Operator Interface or PC Commissioning Tool maintain a wide variety of control loops that the user may select and download. This is preferable to the writing programs or equations to accomplish each specific task because it saves engineering design time and is more reliable.

Again considering the Graphic-Icon based programming it is possible to key control loops to a graphic character. Control packages which allow the user to write custom loops and assign icons of various types are highly desirable to the sophisticated commissioner. These packages are somewhat new but the best ones offer all the enhanced flexibility and power of a pure programmable equation type device. Yet they are much easier for the average user to apply and maintain.

One final note in the area of programming has to do with the point data itself and how the system identifies that information. In the past system programming always required information to be reference by a cryptic point identifier, such as "Analog Input" or more typically, AI. This made it even more difficult to program because a programmer could not refer to a control point as "chilled water valve". Instead they had to know where the wires from that valve were terminated to either program or later to get status. The latest trend is call object orientation. An object is a piece of information referred to by its "English language" name such as "chilled water valve". Regardless of the process for programming or monitoring, this allows for the simplest possible interface over the life of the system.

DDC CONTROL SEQUENCES

This section is the heart of the chapter. To the reader it may seem that a large number of topics have been covered prior to reaching the fundamental discussion of sequences. Yet it was essential to set the stage for this discussion by ensuring that the reader commands a reasonable understanding of DDC technology. DDC technology is a very complex technology and, as you are becoming aware, requires a broad base of knowledge in many disciplines. The reader must have a working knowledge of the equipment to be controlled, as well as a base of understanding in control theory, computer technology, electronics and electrical systems. Much of this knowledge must be derived outside the pages of this book. However it has been the authors intent to at least touch on the basic concepts prior to launching into control sequences.

The topic of DDC Control Sequences is covered here under the Building Wide level of control. There will be two categories of sequences discussed though: Building Wide sequences and Equipment Related Building sequences. In both cases implementation of the sequences requires Building Wide coordination.

Building Wide sequences are unique in that they are generally required by every application in the facility. In addition these sequences often require some coordination or synchronization at a total facility level. As a result, these sequences are often managed by a Master Controller or charged to an individual Peer. The Master controller can issue a command to override an equipment level controller sequence, and thus implement these building wide sequence. With Peers on the other hand, one controller with typically be charged with monitoring for the condition that initiates the sequences. There may be a backup controller charged with this function as well in case of power failure, etc. The other Peer controllers however must request current data on each scan of the network to determine necessary action. In either case the Building Wide information is uniform and affects nearly all controllers.

Equipment Related Building Sequences vary from those above in that they are specific to one controller and its associated application. These sequences generally require Building Wide information for implementation, but they do not result in command action beyond the discrete controller. It is important to note that in a typical building there may be hundreds of controllers on, for example, air handlers. Each of these controllers would likely be implementing such sequences as supply air reset. This sequence requires Building Wide data like Outdoor Air

Temperature, yet the control action taken by Air Handler Controller A would not affect Air handler B. As a result these Equipment Related Building Sequences are appropriate for discussion here.

The DDC Control Sequences to be covered here, by category, include those below.

Building Wide Sequences
- Real Time and Calendar Access
- Start/Stop and Scheduling Functions
- Optimal Start/Stop
- Metering
- Demand Limiting
- Alarming - Local Notification and Dial Out
- Monitoring
- Trend and History Logs

Equipment Related Building Sequences
- Night Cycle - Setup/Setback/Shut down
- Temperature and Humidity Based Economizers
- Warm up and Purge
- Reset
- Optimization
- Duty Cycling
- Temperature Compensated Duty Cycling
- Diagnostics
- Data Sharing
- Status and Safety Alarms

BUILDING WIDE SEQUENCES

Real Time and Calendar Access

It is critical for the system to maintain a Real Time Clock and Calendar for all time related sequences. Typically the size of the controller will determine whether a clock is resident. The Real Time Clock may be a Quartz clock or firmware that controllers use for line synchronization of the electrical power to the device to derive time. There will be more discussion of that issue under Equipment and Zone Level controllers.

Regardless of how many controllers contain a clock, however the Building Wide Sequence requires time synchronization. This means that a designated clock in the system, typically the most accurate, is chosen and all controllers synchronize to that clock. This Building Wide Sequence therefore ensure that all schedules controlled by the system operate as programmed.

The Calendar is a second key requirement for Building Wide Sequences. Typically DDC Systems will provide a Calendar for 20 to 40 years including leap years. This ensures that the Building will be controlled as designed into the indefinite future. The Calendar is essential for scheduling including day to day operations and holidays. A component of the calendar function that will be covered at this level or as part of scheduling is Daylight Savings Time. This is a big picture calendar function that is essential for smooth Building Wide operation.

Start/Stop and Scheduling Sequences

Building Wide Real Time and Calendar capability allows for one of the simplest and most effect Building Wide Sequences, Scheduling. Time schedule operations consists of starting and stopping a systems based on time and type of day. Type of day refers to weekdays, Saturdays, Sundays, holidays and any other day which has a unique schedule of operations. Time schedule operation provides the greatest potential for cost avoidance and energy savings. This is because a large number of energy consuming pieces of equipment may be operating during unoccupied hours.

Scheduling Sequences are the simplest to install, maintain and operate. They fall into two categories, start stop and occupied/unoccupied sequences. Occupied/unoccupied sequences will be covered in more detail under Night Cycle which is an Equipment Related Sequence. Quite simply the schedule is used to determine whether different conditions should be maintained based on the time of day.

Start Stop sequences are self explanatory. These are most applicable for electrical loads such as: lights, pumps, etc. The sequence simply starts and stops the equipment based on a time and calendar schedule. Typically an electrical relay, contactor or motor starter is interfaced to a digital output on some controller to accomplish this sequence.

Optimal Start/Stop

This is again one of the most effective and widely applied Building Wide DDC sequences. As the sequence is described it may appear more relevant to Equipment Related function. However, this sequence is required by all controllers of temperature related equipment throughout the building, and is provided in essentially the same form. As a result it truly meets the definition of a Building Wide Sequence.

Mechanical systems serving areas that are not occupied 24-hours-a-day can be shut down during the unoccupied hours. Traditionally, the systems are restarted prior to occupancy to cool down or heat up the space in time for the first persons to arrive. Normally this function is performed on a fixed schedule independent of weather or space conditions. The optimized start/stop feature of the system automatically starts and stops the system to minimize the energy required to provide the desired environmental conditions during occupied hours. The function automatically evaluates the thermal inertial of the structure, the capacity of the system to either increase or reduce temperatures in the facility start-up and shut-down times, and weather conditions to accurately determine the minimum hours of operation of the HVAC system to satisfy the thermal requirements of the buiding (see diagram).

The diagram on the following page approximates graphically how Optimal Start Stop works. As noted, for both cooling or heating the system monitors outside air and space temperature. The lead-time on start-up is adjusted based upon the actual conditions. Points on the line represent varying amounts of time and are recalculated everyday. Further the controller learns from each days actual response fine-tunes its performance over time. Ultimately the least amount of energy will be consumed to achieve occupied conditions at the time of occupancy.

Metering

This function is to accept input signals in the form of binary contact closures and analog values from building and/or equipment energy meters to be installed in new buildings. The Building wide Controller maintains a continuously updated cumulative meter reading in memory, calculates, using stored meter constants, a weighted 5-minute average consumption value in engineering units, stores these values for 15 or 30 minutes, based on the utility rate schedule, in memory. These "sliding window" averages are held in a building energy file. A building energy file should be established for each meter installed, and the file should contain a file record for each elapsed calendar day. The file record should contain, in addition to file identifying information, the 15- or 30-minute weighted average consumption in fields representing each period in the day. This information can be utilized to examine building operating strategies and to verify the energy savings of the many different system sequences.

Typical metering functions include:

1. Building electrical consumption (KWH) and demand (KWD) for each building, project, or facility covered.
2. Cooling electrical consumption (KWH) and demand (KWD) for each electric driven refrigeration compressor with capacity greater than 40 tons.
3. Cooling steam consumption (pounds-hr) for each absorption refrigeration machine with capacity greater than 50 tons.
4. Cooling load (ton-hours) for each chilled water system with capacity greater than 50 tons. Wherever possible, located water

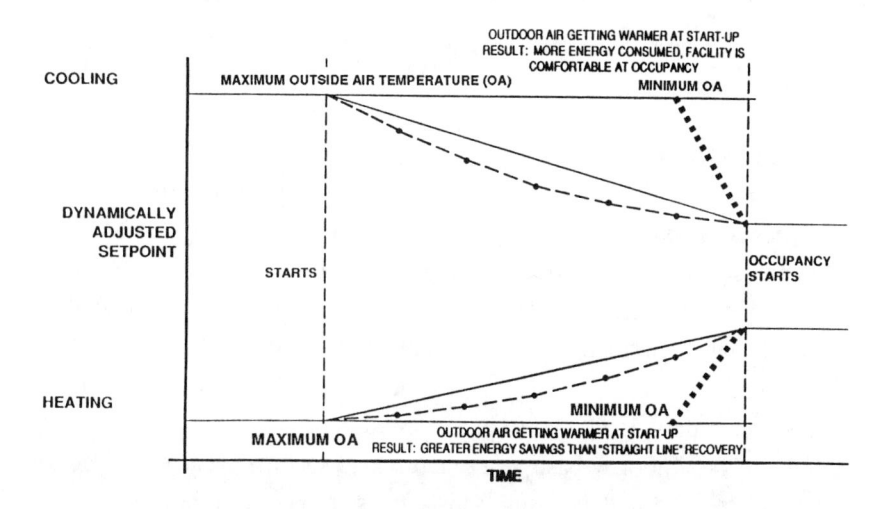

Optimal Start-Stop

flow meter and supply and return water temperature sensors so as to allow metering of both cooling load and heating load from or to the building.
5. Heating load (Btu-hours) for each heating water system with capacity greater than 500,000 Btu per hour. Whenever possible, locate water flow meter and supply and return water temperature sensors so as to allow metering of both cooling load and heating load from or to the building. Utilize water metering for hot water heating systems fired by oil, gas, coal, or refuse, and for systems exchanging heat from steam, geothermal, or solar sources.
6. Heating steam consumption (pounds-hr) for each heating system with capacity greater than 500,000 Btu per hour utilizing steam directly in convectors, radiators, unit heaters, and AHU or H&V coils.
7. Heating fuel oil consumption (gallon-hour) for each fuel oil fired steam boiler or direct fired furnace with capacity greater than 1,000,000 Btu per hour input capacity.
8. Heating fuel gas consumption (Btu-hour) for each gas fired steam boiler or direct fired furnace with capacity greater than 1,000,000 Btu per hour input capacity.
9. Solar domestic water system (Btu-hour) for each solar heated domestic water heater with collector area greater than 1000 square feet.

Demand Limiting

The demand control sequence stops electrical loads to prevent a predetermined maximum electrical demand from being exceeded. Many complex schemes are used to accomplish this function. Generally speaking, most of these monitor the meter, described above, and the base electrical demand continuously. The sequence predicts whether or not the preset limit will be exceeded. If it will be, certain scheduled electrical loads (secondary loads) are shut off. Additional secondary loads are turned off on a priority basis if the initial load shet action does not reduce the predicted demand enough to satisfy the sequence requirements.

Building Wide Sequence of Operations

When demand control is applied to water chillers, special considerations are necessary.

Water chillers are generally equipped with a manually adjustable control system which operates to control the chilled water supply or return temperature that the chiller operates against. An interface between the DDC system and this control current allows reset of the temperature setting in a load shedding situation, thereby reducing electric demand without shutting down the chiller completely. The method of accomplishing this function depends on the specific water chiller involved. In general, however, when the chiller is selected for load shedding, an analog reset signal is transmitted to the interface which then increases the chilled water temperature setpoint by a like amount. Chiller power demand is monitored by the chiller power meter and the chilled water temperature is reset upward until chiller power consumption is reduced to an operator setable minimum KWD setpoint. *Extreme caution must be exercised with application of this sequence.* Incorrect interface and control may cause the centrifugal refrigeration machine to operate in a surge condition, and may cause the reciprocating refrigeration machine to suffer from overheating of the hermetic motor or poor oil return, ultimately causing considerable damage to the equipment. Thus, the intent of this sequence is to leave intact the manufacturer's safety and operating controls and to reduce the power consumption through operation of the centrifugal vanes, or the reciprocating unloaders by resetting the chilled water setpoint. Reciprocating machines, like centrifugal machines, should be provided with the manufacturer's recommended recycle timers to limit number of starts per hour.

Demand control also can be integrated with cogeneration operations. When electrical demand approaches a peak, this sequence starts the engine or turbine generators which feed electrical power into the building where they are located, or drive specific items of equipment such as well water pumps, thus reducing base electrical demand. Extreme caution must be utilized in using this sequence to ensure proper control of the cogeneration system and thermal equipment.

Monitoring

Monitoring is a very simple yet essential Building Wide Feature. This feature will be discussed is greater detail under the Central Operator Interface and System Management chapters. As a result this section will not devote a great deal of time to the topic. Essentially monitoring allows for both on-site and remote access to all system data. This is a Building Wide Sequence because monitoring requires access to all data including point status, program parameters, etc.

Alarming - Local Notification and Dial Out

Many items of mechanical equipment are provided with various types of alarms for equipment protection. These types of alarms, such as high and low water for boilers, gas pressure alarms, and various temperature and pressure alarms. These temperature and pressure alarms apply on refrigeration machines, and are typical of the types of functions that can be monitored. Monitoring of such alarms provides the operator with information regarding the failure of equipment or the development of potential problems with the system operation.

This alarm data can be monitored locally through an operator interface or with a bell or light annunciation. With installation of a modem, phone lien and telephone interface they can also be dialed out to another location.

Trend and History Logs

These are tabular listings of data that has been monitored by the system. They are typically analog points like temperature, but they might also be status points to indicate when piece of equipment start and stop. The user must specify a point and a time interval for trending along with a duration to trend the point. This is a simple feature yet a very valuable diagnostic tool for Building Wide system issues.

Equipment Related Building Sequences

The sequences that are discussed here are for the most part equipment specific rather than Building Wide per se. However the sequences are universal or Building Wide in the sense that they may be used in tens or even thousands of controllers within a given facility. In addition, many of the sequences rely on data from outside the DDC controller or Building Wide data. This data may include Outdoor Air temperature, building Kilowatt demand, etc. As a result it is important to list these sequences here and allow the reader to gain a greater understanding of DDC control.

Closed Loop Control

The most critical equipment related sequence is closed loop control. This sequence is used to control temperature, pressure, humidity and other condition which are at the heart of maintain building operations. Closed loop control has already been discussed in this chapter, but for clarification purposes this is essential to DDC. The vast majority of DDC sequences involve either simple start stop control or some type of closed loop control. In particular, Equipment Related Building Sequences will rely on both simple and sophisticated PID type sequences to achieve the goals establish for the DDC system.

Night Cycle - Setup/Setback/Shut down

The cost and energy required to maintain space temperature during the unoccupied hours can be reduced by changing setpoints. During the heating mode space temperature setpoints can be reduced from 70 degrees to around 55 degrees. These new setpoints are still maintained by the DDC system, and therefore equipment will run to ensure that no danger is posed to the facility or equipment due to freezing conditions. In the cooling mode it is possible to raise set points from 75 degrees to 85 degrees, or even to completely shut down equipment. These sequences obviously apply to unoccupied, or perhaps non-critical areas. As the reader may guess, these functions rely on the Real time, Calendar and Scheduling capabilities of DDC Systems, along with temperature control sequences.

Warm up and Purge

These two functions are related to night cycle and are critical to maintaining comfort in a building. Bear in mind that comfort is a major driver for DDC systems, and directly impacts the mission of many businesses that install this equipment.

Warm up is the mode that HVAC equipment enters when the building begins to transition from Night Cycle to occupied mode. The thermal load imposed by outside air, used for ventilation, may constitute a significant percentage of heating and cooling requirements. At the same time is may dramatically increase the time required to meet comfort requirements in the building. The warm up cycle sequence controls outside air dampers when the introduction of outside air would impose a thermal load, and the area is unoccupied. This sequence would also apply during warm up or cool down cycles prior to occupancy of the area.

Purge uses the outside air dampers in a completely different way. The purge cycle opens the dampers and flushes the building with fresh air. This is typically used in the cooling mode to pre-cool the building or for indoor air quality. Indoor Air Quality applications involve flushing the building to clear carbon dioxide or other undesirable materials.

TEMPERATURE AND HUMIDITY BASED ECONOMIZERS

The utilization of an all-outside-air economizer cycle can be a cost-effective energy conservation measure, depending on the climactic conditions and the type of mechanical system. Where applicable, the cycle utilizes outside air to satisfy all or a portion of the building's cooling requirements when the temperature or the total heat content of the outside air is less than that of the return air from the space. Outside air is introduced through the mechanical system and relieved during this cycle instead of the normal recirculation system.

A dry bulb (db) economizer cycle is regulated by an outdoor air temperature sensor which discontinues compressor operation when outdoor temperature falls below the design supply temperature, normally 50F to 60F. Cooler outside air is then drawn into the system and used to reduce space temperature.

An enthalpy economizer cycle is similar to a dry bulb economizer cycle, except it measures total heat content (enthalpy) of air. Outdoor air is used for cooling when its enthalpy is less than the enthalpy of return air.

Choosing between a dry bulb or an enthalpy economizer cycle involves a decision which should be based on economics. If periods of high humidity seldom occur in your area, the additional expense of an enthalpy economizer will not be cost-justified.

Temperature Reset

Reset of temperature for air and water media in a variety of types of equipment is an extremely effective way to conserve energy and reduce cost. This section will review approaches for using reset with various pieces of equipment. The basic strategy is to monitor a control point, discharge air temperature, and several variables that impact that point such as return air temperature or outside air temperature. As these variables indicate a diminishing load, temperature can be reset to a value that will consume less energy to maintain.

Hot/Cold Deck Temperature Reset

HVAC systems such as dual duct and multizone utilize a parallel arrangement of heating and cooling surfaces (hot and cold deck surfaces) to provide simultaneous heating and cooling. Both heated and cooled air streams are mixed to satisfy thermal requirements in a given space. Without optimization controls, these systems are extremely wasteful, because the temperature settings of the hot and cold decks are fixed, usually at 90F and 55F, respectively. With temperature reset, the system selects the individual areas with the greatest heating and cooling requirements and adjusts hot and cold deck temperature accordingly, minimizing the inefficiency of the system by reducing the difference between hot and cold deck temperatures. Return air humidity may also be monitored for high limit.

Discharge Air Temperature Reset

The discharge air temperature reset function adjusts the cooling coil discharge temperature upward until the zone with greatest demand for cooling has closed its reheat coil valve. Return air humidity may be monitored for high limit.

Chilled Water Reset

The energy required to generate chilled water in a reciprocating or centrifugal electric-driven refrigeration machine is a function of a number of parameters including the temperature of the chilled water leaving the machine. Because the refrigerant suction temperature is a direct function of the leaving water temperature, the higher the two temperatures, the lower the energy input per ton of refrigeration. As a result, since chilled water temperatures are selected for peak design times, most chilled water temperatures can be elevated during most operating hours, unless strict humidity control is required. Depending on the operating hours, size of the equipment, and configuration of the system, energy can be saved by resetting the chilled water temperature, allowing it to rise to satisfy the greatest cooling requirements. Generally, this determination is made by the position of the chilled water valves on the various cooling systems. The positions of the control devices supplying the various cooling coils are monitored and the chilled water temperature is elevated until at least one control device is in the maximum position. Other control schemes may be necessary to satisfy different system configurations.

Outside Air Schedule Reset

Hot water heating systems are designed to supply system heating requirements at outdoor design temperatures. Depending on the specific system design, the hot water supply temperature can be reduced as heating requirements are reduced, usually in response to increased outdoor ambient temperatures. Where applicable, the capability to reduce the temperature of the supply water as a function of outdoor temperature will effect operating savings. To accomplish this function, the temperature controller for the hot water supply is reset on a predetermined schedule as a function of outdoor temperature.

OPTIMIZATION

A sequence that builds on the basic concepts Optimal Start Stop and temperature reset is called optimization. Optimization is used with a variety of pieces of equipment, particularly energy intensive central station applications of chiller and boilers.

Air Distribution Optimization

The air distribution optimization function involves control of zone dampers to stop the flow of conditioned air to nonessential areas during unoccupied periods. Dampers which in most cases will have to be installed, are positioned automatically by an optimum start/stop program. Supply fan dampers and associated exhaust fans are positioned by static pressure sensors.

Chiller Plant Optimization

Chiller efficiency usually decreases with chiller load below 50%. When a multiple chiller installation is involved, therefore, it generally is desirable to load each chiller as much as possible. In other words, it generally would be better to have two chillers loaded equally, as opposed to having one loaded to 100% and another to 30%. To achieve optimization, it is necessary to control loading.

Chiller loading can be controlled by resetting the supply and return water temperatures sensed at the unit. The temperature reset scheme should be selected in light of the specific installation involved, since there are many different variations. Some multiple chillers are connected in series, others in parallel, etc. Chillers with constant water flow can be controlled by return water temperature. Variable water flow must be controlled by supply water temperature.

A multiple chiller loading optimization function is one that controls temperature resetting in a manner to minimize energy consumption without affecting comfort conditions in any space served by the system. This normally is accomplished by measuring flow through the evaporator and condenser of each chiller and the reclaim coils. Differential temperature across evaporator sections, condenser sections, and reclaim sections also are monitored. This enables the computer to calculate the total energy being produced by each machine. Total electrical usage is monitored continuously thus enabling instantaneous efficiency calculations, When the temperatures are monitored the minimum energy requirement for the system can be calculated. The computer then will cause the most efficient chiller or combination to be operated.

Boiler Plant Optimization

Several techniques are available to accomplish the boiler plant optimization function, depending upon factors unique to the facility involved.

In one approach, electronic probes can be used to continually monitor the amount of O_2 (oxygen) that passes through the boiler plant stack.

Data collected permits the DDC to determine heat losses associated with that stack. If the O_2 level is outside optimum limits, the DDC can initiate control to regulate air flow through the boiler and stack at constant firing rates, to bring the O_2 level within proper operating limits and retain maximum combustion efficiency.

Another method of boiler plant optimization involves decreasing the fuel flow to the boiler as load on the boiler decreases. During winter, higher outdoor air temperatures reduce building heat requirements, so reducing the amount of heat the boiler has to produce. Boiler outlet water temperature can be reduced by mixing the outlet water with boiler inlet water, to create a mixture determined by the outside air temperature and the boiler water temperature difference. This unloads the boiler and decreases fuel consumption. As the boiler unloads, flue gas temperature decreases, but still remains high enough to not precipitate corrosive substances inside the boiler stack. It is recommended that temperature sensors be placed at the top of the stack to warn if flue gas temperatures become too low. Depending upon the sulfur content of the fuel being burned, the flue gas temperature should be maintained above 25F to 300F. Also, automatic flue dampers should close when burners are off to prevent cooling of boilers and loss of heat.

Secondary Loop Chilled/Hot Water Optimization

There usually are many secondary loops serving similar coils, each tied into the primary loop main distribution system. The primary loop feeds the secondary loop with supply water from a central source and conveys the return water back to the central cooling or heating machinery, as shown in Figure 9-1.

As can be seen, the control valve adjusts coil capacity to meet load by modulating coil flow, or by changing the temperature of water to the coil. Whenever the valve lowers coil output, it indicates that full capacity of the supply water is not needed. Accordingly, secondary loop optimization logic is based on the premise that if no coils need full supply water capacity, supply water temperature can be adjusted to save energy, by raising chilled water temperature and lowering hot water temperature. This action also saves energy by reducing pipe heat gains and losses.

Building Wide Sequence of Operations

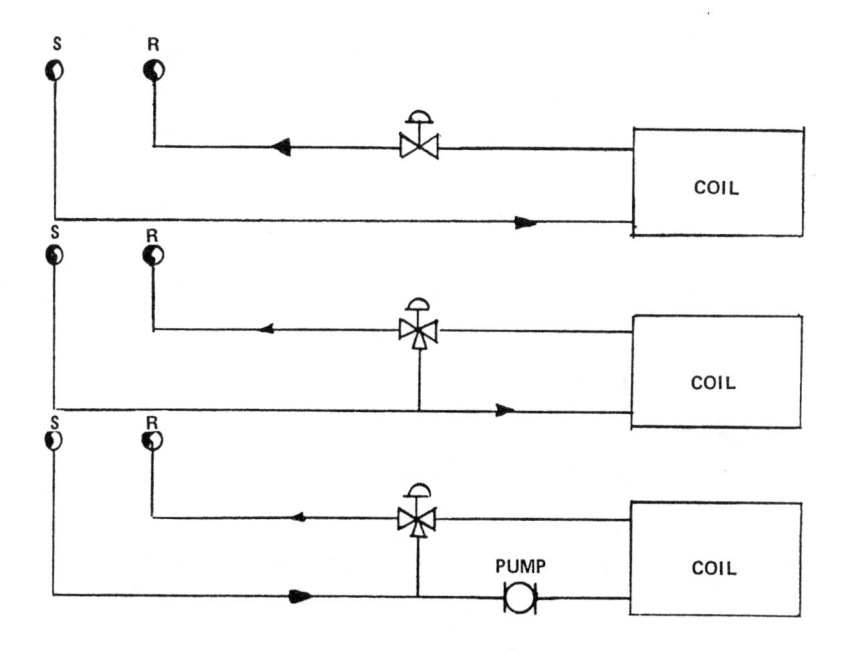

Figure 9-1 Typical Secondary Water Applications

Dehumidification must be considered when chilled water is affected. The highest tolerable chilled water temperature must not exceed the cooling coil control set point for dehumidification.

The optimization set-up involved requires stable local controls to ensure that final control elements of the distribution system (valves, dampers, mixing boxes) are measured accurately. If local control loops are not stable, this optimization technique cannot be used.

Although this optimization method is applicable to a variety of secondary loop arrangements, the system's layout must be analyzed closely. In some cases, energy saving in chiller efficiency achieved by raising chilled water temperature may be wasted on the additional pumping required to provide the same amount of total heat transfer. If the primary hydraulic loop has pressure head control at the by-pass, however, increased flow to the coils only means a decrease of flow in the by-pass, so no additional pumping energy is required. This optimization scheme then is applicable.

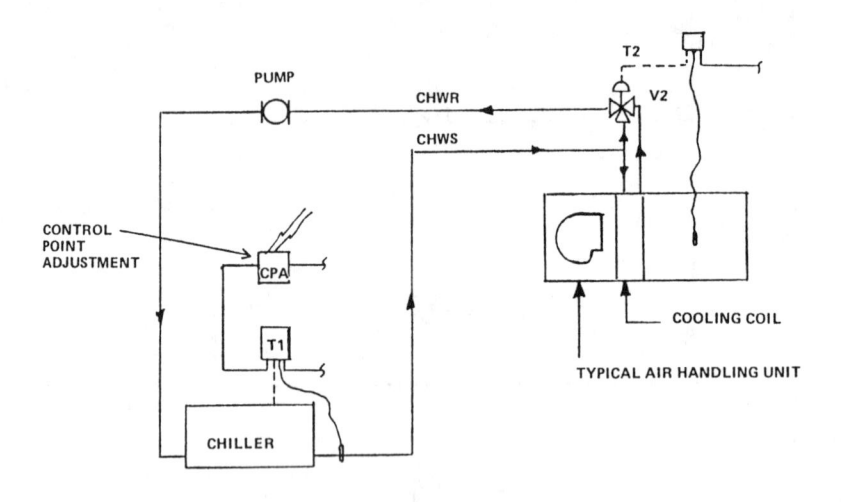

Figure 9-2 Typical Secondary Loop Chilled Water Optimization

Figure 9-2 shows conventional secondary loop optimization of a chilled water system. Optimization is achieved by monitoring all valve end switches and adjustment of the set point. The set point is adjusted by the CPA (control point adjustment) until contact status voltage is lost. When this occurs, at least one valve is near wide open, indicating the need for "decreasing chilled water temperature." The system then decreases the set point T1 by a small value every fixed time interval until the contact is made. In this way the system balances itself dynamically to maintain a minimally sufficient chilled water temperature at all times.

End switches of the valves can be wired individually or in series. Individual wiring enables the DDC to know which coil is demanding more cooling, so a proper weighing factor can be used in the decision logic. A valve stroke transmitter can be used not only to indicate the near open position but also the actual control position of the valve. More sophistication may be incorporated in the software to determine the actions to be taken. Ultimately the cost of DDC has reached a point that allows all of these sequences to be incorporated into one DDC device. The controller mounts at the air handler and communicates with other controllers over a network.

DUTY CYCLING

This is a technique that was popular in the early days of computer based control. It is still used to some degree today, however the reader should note that many equipment manufacturers are opposed to this sequence. Prior to implementing a Duty Cycle the reader is encouraged to seek the advice of a consultant, and to consult with the manufacturer of any equipment that is to be cycled.

Duty cycling consists of shutting down a system for predetermined short periods of time during normal operating hours. Normally applicable to HVAC systems, the sequence is based on the theory that HVAC systems seldom operate at peak output. Accordingly, if a system is shut down for a short period of time, it still has enough capacity to overcome the slight temperature drift which may occur during the shutdown. Although the interruption does not reduce total cooling load, it does reduce the next auxiliary loads, such as fans and pumps. It also reduces outside air heating and cooling loads since the outside air intake damper is closed while an air handling unit is off.

Systems generally are cycled for a fixed period of time, such as 15 minutes of each hour of operation. The off period time length normally is increased during moderate seasons and reduced during peak seasons. Cycle fans utilized with direct expansion (DX) cooling coils or heat pumps, and cycle pumps providing chilled water flow through chillers should not be duty-cycled.

The subject of duty cycling has caused considerable controversy with regard to the damage which it may or may not cause to motors. The four primary concerns in determining whether or not duty cycling may cause damage are:
1. The number of stops and starts in a given period of time;
2. percentage of time for which the motor is off and on;
3. external inertia, or the amount of weight which the motor must move and at what speed; and
4. level of load, meaning the capacity at which the motor is operating when it is on.

Few, if any, problems will be experienced when the motor involved is relatively small; that is, approximately 50 HP or less. This experience is verified through studies which suggest that no damage will occur when the motor is 250 HP or less. In any event, if there is a question, especially when the motor drives a critical load, expert guidance should be obtained. In all cases, when the motor still is under a manufacturer's warranty or guarantee, contact the manufacturer to determine its suggestions for appropriate duty cycling and the effect which duty cycling may have on the warranty or guarantee.

Temperature Compensated Duty Cycling

The temperature compensated duty cycling sequence is essentially similar to duty cycling. Loads are assigned a maximum and minimum off time.

Each load has a related space temperature data point assigned to it. Each load serving an interior zone (cooling only) is cycled off for its assigned maximum period as temperature in the space drops to its assigned minimum value. Inversely, the interior load is cycled off for its assigned minimum period as space temperature rises to its assigned maximum value. For loads serving perimeter zones (summer-cooling, winter-heating), an outside air changeover sensor affects the algorithm used. With the outside air sensor determining mild or summer conditions, the "off" time calculation is the same as described for loads serving interior spaces. With the outside air sensor determining winter conditions, the "off" time calculation parameters are inverted, so minimum "off" times occur at minimum space temperatures. "Off" times vary proportionally throughout the assigned acceptable space temperature range.

Data Sharing

Data sharing is a concept that has already been mentioned in this text and will be covered in more detail under Building Wide Coordination later in this chapter. With specific reference to Equipment Related Sequence, data sharing is extremely valuable for ensuring that the all of the sequences discussed thus far have necessary data. In essence the nature of Distributed DDC requires that information held in various controllers is made available to the network. This way a controller that has an outdoor air temperature sensor can "share" it with other controllers that need it for reset, Optimal Start/Stop or other sequences.

At this point it is not necessary to discuss further the details of how the data is shared. Rather it is essential to note that sharing data is required with any Distributed system.

Maintenance of Equipment through Status, Alarms and Diagnostics

As noted above monitoring for current status and safety alarms on equipment is critical to long term efficiency and operation. These alarms will notify an operator of conditions that are outside the normal operating conditions for a specific piece of equipment. Upon notification of such conditions an operator may then use the power of system communication to diagnose the problem.

One of the most valuable features that DDC systems offer is the ability to assist in a comprehensive equipment maintenance program.

A DDC system cannot actually perform maintenance, but it can be programmed (depending on the specific type system involved) to provide daily printouts of regular maintenance procedures which must be undertaken, together with as-needed notification of unscheduled maintenance procedures which may be required. There are several ways in which the maintenance scheduling functions can be handled, as follows.

Calendar Time Scheduling – The calendar time method obviously is the easiest and most direct way to schedule maintenance. For example, every 20 days maintenance men could be instructed to change all air handling unit filters, or once-a-month instructions can be given to grease and lubricate certain equipment.

Machine Running Time Scheduling – Since all machinery does not run on a regular schedule, calendar time scheduling for many items is not sufficient. Accordingly, the central processing unit can be set to accumulate running time of certain equipment and, after a predetermined number of running hours for each item, print out maintenance instructions.

Efficiency Scheduling – In certain cases the amount of time equipment runs will not be an accurate indicator of its operating efficiency or need for maintenance. Since the DDC can calculate equipment operating efficiency on the basis of raw data inputs, it can be programmed to provide maintenance instructions when efficiency deteriorates to a certain predetermined level.

Early Warning Monitor – The system can be designed to provide early warnings of impending equipment failure. As one example, bearings of certain pieces of equipment can be monitored for vibration and/or temperature. Should the vibration or temperature level increase to a certain predetermined level which indicates a problem, a panel light, dial-out, etc., would indicate an immediate, unscheduled maintenance problem. It also would be feasible for the computer to stop the particular piece of equipment involved – or the entire system of which it is a component, if necessary – until the needed maintenance or repair is performed.

Maintenance Cost Management Program – A maintenance cost management program stores maintenance costs and maintenance labor for individual units, fans, motors, pumps, etc.

Maintenance Personnel Scheduling – A program can be developed to provide daily or weekly maintenance personnel scheduling, identifying what has to be done, who should be doing it, etc.

Remote Communication

Although communication may not be specifically defined as system function, it is perhaps one of the single most essential system features. Communication is the key automation feature for remote programming, optimizing equipment and DDC performance and reducing service and maintenance costs. For more on this topic see Chapter 12.

Trouble Diagnosis

By monitoring certain parameters of the mechanical/electrical systems, diagnosis of reported problems with mechanical and electrical systems can be performed at the central console location. Some of the parameters that might be monitored for the purposes of trouble diagnosis include hot and cold deck temperatures with high and low limits, leaving chilled water temperatures and hot water temperatures with high and low limits, differential pressure switches indicating fan and pump operation or excessive pressure losses through filters or condenser tube bundles, space temperatures, and improper humidifier operation. Typical trouble diagnosis applications include those for:

1. *Refrigeration compressors* with capacity greater than 20 tons, provide separate status contact for each sensor listed; with capacity greater than 5 tons but less than or equal to 20 tons, provide single status contact for all sensors listed.
2. *Air handling filter banks* with (1) media depth greater than 3 inches, (2) media efficiency greater than 40% ASHRAE Atmospheric Dust or 90% ASHRAE Weight Arrestance, or (3) capacity greater than 6000 CFM; provide a filter differential pressure status sensor.
3. *Condenser tube bundles* operating with water from "open" circuit cooling towers; provide differential pressure status sensor contact to indicate excessive pressure differential, such as that caused by tube fouling.
4. *Humidifier installations*; provide alarm status contact to indicate simultaneous operation of the humidifier and cooling coil or economizer.

Equipment Applications and Point Lists or Matrices

With a clear understanding of specific control sequences it is now possible to consider there application. There are a variety of systems and equipment used in buildings. Some examples of HVAC system components which are suitable for DDC control are listed in Table 9-1.

Table 9-1 HVAC System and Equipment Applications

HVAC systems and equipment to be integrated into DDC systems.	
SYSTEM APPLICATION	
Single Zone Air Handler	Steam Unit Heater System
Single Zone Split System	Chiller, Air Cooled
Terminal Reheat Air Handler	DX Unit, Air Cooled
Multizone Air Handler	DX Unit, Water Cooled
Two Pipe Fan Coil System	Hot Water Boiler
Heating and Ventilating Unit	Steam Boiler
Hot Water Convector Heating System	Direct Fired Furnace
Steam Convector Heating System	Hot Water Converter

This is only a partial listing of course and tends to be focused more on equipment level control. For Zone control additional lists might include VAV terminal units, water and air source heat pumps and unit ventilators. For purposes of the examples of point matrices however, this chapter will focus on the equipment level. Though this may seem to stray from the Building Wide topic it is essential to thoroughly understand these concepts to design a complete system.

Tables 9-2 through 9-18 comprise matrices which indicate the sequences applicable to the systems inputs and some outputs. True DDC often includes more outputs for complete equipment control.

Table 9-2. Single Zone Air Handler	\multicolumn{5}{c|}{INPUT}	\multicolumn{2}{c|}{OUTPUT}
	Fan Flow Status Sensor	Space Temperature	Filter Differential Pressure Status	Humidifier Status	Start/Stop Control Interface	Local Control Interruption Interface
Time Scheduled Operation	X				X	
Space Night Setback	X	X			X	
Start/Stop Optimization	X	X			X	
Duty Cycling	X	X			X	
Demand Limiting Start/Stop	X	X			X	
Warm-Up/Night Cycle	X					X
Maintenance Run Time Report	X					
Trouble Diagnosis	X		X	X		
Table 9-3. **Heating and Ventilating Unit**						
Time Scheduled Operation	X				X	
Space Night Setback	X				X	
Start/Stop Optimization	X	X			X	
Duty Cycling	X	X			X	
Demand Limiting Start/Stop	X				X	
Warm-Up/Night Cycle	X					X
Maintenance Run Time Report	X					
Trouble Diagnosis	X		X	X		

Building Wide Sequence of Operations

Table 9-4. Single Zone Split System

	\multicolumn{4}{c	}{INPUT}	\multicolumn{2}{c	}{OUTPUT}
	Fan Flow Status Sensor	Space Temperature	Humidifier Status	Filter Differential Pressure Status	Start/Stop Control Interface	Local Control Interruption Interface
Time Scheduled Operation	X				X	
Space Night Setback	X	X			X	
Start/Stop Optimization	X	X			X	
Warm-Up/Night Cycle	X					X
Maintenance Run Time Report	X					
Trouble Diagnosis	X		X	X		

Table 9-5. Hot Water Convector Heating System

	Pump Flow Status Sensor	Space Temperature			Start/Stop Control Interface	
Outside Air Temp. Cutoff	X				X	
Space Night Setback	X	X			X	
Start/Stop Optimization	X	X			X	
Maintenance Run Time Report	X					

Table 9-6. Terminal Reheat Air Handler

		Time Scheduled Operation	Space Night Setback	Start/Stop Optimization	Duty Cycling	Demand Limiting Start/Stop	Warm-Up/Night Cycle	Cold Deck Temperature Reset	Hot Deck Temperature Reset	Trouble Diagnosis	Maintenance Run Time Report
OUTPUT	Controller Reset Interface							×	×		
	Local Control Interruption Interface						×				
	Start/Stop Control Interface	×	×	×	×	×					
INPUT	Humidifier Status									×	
	Filter Differential Pressure Status									×	
	Greatest Heating Demand Sig.								×		
	Greatest Cooling Demand Sig.						×				
	Supply Air Temperature							×	×		
	Space Temperature		×	×	×	×					
	Fan Flow Status Sensor	×	×	×	×	×	×	×	×	×	×

Building Wide Sequence of Operations

		Time Scheduled Operation	Space Night Setback	Start/Stop Optimization	Duty Cycling	Demand Limiting Start/Stop	Warm-Up/Night Cycle	Hot/Cold Deck Reset	Maintenance Run Time Report	Trouble Diagnosis
OUTPUT	Cold Deck Controller Reset Interface							x		
	Hot Deck Controller Reset Interface							x		
	Local Control Interruption Interface						x			
	Start/Stop Control Interface	x	x	x	x	x				
INPUT	Filter Differential Pressure Status									x
	Greatest Cooling Demand Sig.							x		
	Greatest Heating Demand Sig.							x		
	Cold Deck Supply Air Temp.							x		
	Hot Deck Supply Air Temp.						x			
	Space Temperature		x	x	x	x				
	Fan Flow Status Sensor	x	x	x	x	x	x	x	x	

Table 9-7. Multizone Air Handler

Table 9-8. Two-Pipe Fan Coil System

	INPUT					OUTPUT
	Pump Flow Status Sensor	Space Temperature (Several Typical Spaces)	Dual Temperature Water Flow Quantity	Dual Temperature Water Supply Temperature	Dual Temperature Water Return Temperature	Start/Stop Control Interface
Time Scheduled Operation	X					X
Outside Air Temp. Cutoff	X					X
Space Night Setback	X	X				X
Start/Stop Optimization	X	X				X
Duty Cycling	X	X				X
Demand Limiting Start/Stop	X	X				X
Maintenance Run Time Report	X					
Metering	X		X	X	X	

Table 9-9. Hot Water Convertor

	INPUT				OUTPUT
	Pump Flow Status Sensor	Heating Water Flow Quantity	Heating Water Supply Temperature	Heating Water Return Temperature	Start/Stop Control Interface
Outside Air Temp. Cutoff	X				X
Metering	X	X	X	X	

Building Wide Sequence of Operations

Table 9-10. Steam Converter Heating System

	Steam Flow Status Sensor (INPUT)	Space Temperature (INPUT)	Steam Flow Quantity (INPUT)	Steam Pressure Sensor (INPUT)	Local Control Interrupting Interface or Start/Stop Control (OUTPUT)	Start/Stop Control Interface (OUTPUT)
Time Scheduled Operation	X				X	
Outside Air Temp. Cutoff	X					X
Space Night Setback	X	X			X	
Start/Stop Optimization	X	X			X	
Metering	X		X	X		

Table 9-11. Steam Unit Heater System

	Steam Flow Status Sensor	Space Temperature	Steam Flow Quantity	Steam Pressure Sensor	Local Control Interrupting Interface or Start/Stop Control	Start/Stop Control Interface
Time Scheduled Operation	X				X	
Outside Air Temp. Cutoff	X					X
Space Night Setback	X	X				X
Start/Stop Optimization	X	X			X or	X
Metering	X		X	X		

		Time Scheduled Operation	Demand Limiting, Chilled	Water Temperature Reset	Chilled Water Reset	Start Time Optimization	Maintenance Run Time Report	Trouble Diagnosis	Metering
OUTPUT	Controller Reset Interface				X				
	Chilled Water Reset Interface		X						
	Start/Stop Control Interface	X				X			
INPUT	Compressor Electric Consumption (KWH) and Demand (KWD)								X
	Dual Temperature Water Return Temperature								X
	Dual Temperature Water Supply Temperature								X
	Dual Temperature Water Flow Quantity								X
	Cooler Low Temperature Safety Status							X	
	Compressor Low Pressure Safety Status							X	
	Compressor High Pressure Safety Status							X	
	Low Oil Pressure Safety Status							X	
	Condenser Water Pump Status Sensor						X		
	Chilled Water Pump Flow Status Sensor						X		
	Chilled Water Supply Temp.				X				
	Chilled Water Return Temp.				X				
	Compressor Electrical Demand Meter (KWD)		X						
	Pump Flow Status Sensor	X	X		X		X	X	X

Table 9-12. Chiller, Air-Cooled

Building Wide Sequence of Operations

Table 9-13. Chiller, Water Cooled

	Time Scheduled Operation	Demand Limiting, Chilled Water Temperature Reset	Chilled Water Reset	Start Time Optimization	Maintenance Run Time Report	Trouble Diagnosis	Metering	Condensor Water Reset
OUTPUT								
Controller Reset Interface			×					×
Chilled Water Temperature Reset		×						
Start/Stop Control Interface	×			×				
INPUT								
Condensor Differential–Pressure Sensor Status						×		
Condensor Water Supply Temp.								×
Condensor Water Return Temp.								×
Compressor Electric Consumption (KWH) & Demand (KWD)							×	
Dual Temperature Water Return Temp.							×	
Dual Temperature Water Supply Temp.							×	
Dual Temperature Water Flow Quantity							×	
Cooler Low Temperature Safety Status						×		
Compressor Low Pressure Safety Status						×		
Compressor High Pressure Safety Status						×		
Low Oil Pressure Safety Status						×		
Condensor Water Pump Status Sensor	×	×		×	×	×		×
Chilled Water Pump Flow Status Sensor	×		×	×	×	×	×	
Chilled Water Supply Temp.			×					
Chilled Water Return Temp.			×					
Compressor Electrical Demand Meter (KWD)		×						
Pump Flow Status Sensor							×	

Table 9-14. DX Unit, Water Cooled

	INPUT									OUTPUT	
	Pump Flow Status Sensor	Condensor Water Pump Flow Status Sensor	Condensor Water Return Temp.	Condensor Water Supply Temp.	Chiller Tube Bundle Pressure—Differential Status Sensor	Low Oil Pressure Safety Status	Compressor High Pressure Safety Status	Compressor Low Pressure Safety Status	Condensor Low Temperature Safety Status	Compressor Electrical Consumption (KWH) & Demand (KWD)	Start/Stop Control Interface	Controller Reset Interface
Time Scheduled Operation	X										X	
Start Time Optimization	X										X	
Condensor Water Reset		X	X	X								X
Maintenance Run Time Report	X											
Trouble Diagnosis	X				X	X	X	X	X			
Metering	X									X		

Table 9-15. DX Unit, Air Cooled

	INPUT						OUTPUT
	Status Contact Sensor	On/Off Status Contact Sensor	Low Oil Pressure Safety Status	Compressor High Pressure Safety Status	Compressor Low Pressure Safety Status	Compressor Electrical Consumption (KWH) & Demand (KWD)	Start/Stop Control Interface
Time Scheduled Operation	X						X
Start Time Optimization	X						X
Maintenance Run Time Report	X						
Trouble Diagnosis		X	X	X	X		
Metering		X				X	

Building Wide Sequence of Operations

Table 9-16. Hot Water Boiler

	INPUT							OUTPUT
	Pump Flow Status Sensor	Low Water Safety Status	Flame Failure Safety Status	High Temp. Safety Status	Dual Temp. Water Supply Temp.	Dual Temp. Water Flow Quantity	Dual Temp. Water Return Temp.	Start/Stop Control Interface
Outside Air Temp. Cutoff	X							X
Maintenance Run Time Report	X							
Trouble Diagnosis	X	X	X	X				
Metering	X				X	X	X	

Table 9-17. Steam Boiler

	INPUT							OUTPUT
	Steam Flow Status Sensor	Steam Supply Pressure	Low Water Safety Status	Low Low Water Safety Status	Flame Failure Safety Status	High-Pressure Safety Status	Fuel Flow Quantity	Start/Stop Control Interface
Outside Air Temp. Cutoff	X							X
Trouble Diagnosis		X	X	X	X	X		
Metering	X						X	

Table 9-18. Direct Fired Furnace	INPUT					OUTPUT	
	Fan Flow Status Sensor	Space Temperature	Flame Failure Cutoff Safety	High Temp. Cutoff Safety	Fuel Flow Quantity	Start/Stop Control Interface	Local Control Interruption Interface
Time Scheduled Operation	X					X	
Outside Air Temp. Cutoff	X					X	
Space Night Setback	X	X				X	
Start/Stop Optimization	X	X				X	
Warm-Up/Night Cycle	X						X
Maintenance Run Time Report	X						
Trouble Diagnosis	X		X	X			
Metering	X				X		

A separate matrix has been prepared for each major system. Each is easy to use. For example, Table 9-2 relates to a single zone air handler. If time scheduled operation of the system is desired, reading left to right indicates that only one input is required (a fan flow status sensor), and that the output is start/stop control interface.

BUILDING WIDE COORDINATION

The discussion of technology thus far, along with the coverage of DDC sequences, will equip the reader to better understand and evaluate systems. The next critical DDC topic is Building wide coordination.

Building Wide Coordination in distributed systems will be viewed under these key areas:
- control integration,
- building wide monitoring and history and
- remote communication with a Central Operator Interface network management.

With the introduction of Distributed DDC a number of new issues were raised around the accomplishment of control integration. An entire chapter has already been devoted to integration, yet some further coverage is warranted here with reference to this level of system architecture. Specifically the Building Wide Level is charged with accomplishing a majority of the integration tasks.

Again the author uses the term integration because of the need to accomplish control via coordinating, or more specifically integrating, the functions and sequences of multiple microprocessors. Integrating control means the accomplishment of complex sequences that are implemented in part by more than one device. There are many sequences that fit into this category and these were easy to accomplish when a Central Processing Unit (CPU) based system. Examples would include Demand Limit Control or Optimal Start/Stop for the building which were simplified because the CPU initiated all control from one central "all knowing" device.

There are many design approaches taken with distributed systems. For simplicity sake this discussion will take one of the more common approaches, Master/Slave Integration. Master/Slave approaches were covered above and this section will discuss integration with this approach further.

With this approach the Master Controller, or Island Host, serves as a CPU type device for building wide functions. This device will monitor building conditions by polling network devices for current status. Based upon that data, the Master controller will initiate control sequences. For example this Host type controller will monitor building power consumption for the network, and based upon electrical demand/consumption, and a setpoint, will initiate demand shed.

The Demand control sequence was discussed earlier. More germane to this discussion is the fact that loads will not be shed without integrated control sequences implemented via distributed controllers throughout the building. A Master controller may not have any physical points connected to it, but the shed sequence must be able to turn off such loads to avoid a peak. Network DDC controllers must physically disable loads to avoid exceeding the setpoint because this is the only way that a Building Wide Controller may have access to actual devices that can be shed.

The key to control integration is the ability of the Building Wide Controller, or Island Host level controller, to either "request" the implementation of local control sequences, or to override local control and force a specific sequence. In either case the Building Wide Controller must track the actions of distributed controllers and be capable of taking appropriate steps if the sequence is not implemented. There are many schools of thought as to the best possible approach for accomplishing these types of functions. For this discussion however, it is only important for the reader to note that building wide control sequences require integrated action by multiple controllers to be achieved.

The final topic under control integration will be System data. In essence the Building Wide Controller serves as the vehicle for ensuring that key information is available throughout the network. There are two types of system data: Global and Shared. Global data is information needed by many or all of the controllers in the same format. Good examples of global data are: outside air temperature, time and enthalpy. This data is provided by one controller which has a physical point for monitoring, though for system integrity purposes backup locations may be provided. The data from that hardware point is then "broadcast" to the network and available to all controllers.

The second type of System information is Shared Data. This is a much more custom approach to using system data for control. A simple example that will also be discussed later is temperature reset. Controller A is responsible for maintaining a discharge air setpoint, but that setpoint may be adjusted based upon an indication of the load. Controller B is monitoring space temperature for local control, but this is also a good indication of the load. The Shared Data function provides a vehicle for controller A to get access to the space temperature with the same frequency as if it were physically connected to that device. This requires that a specific point be accessed by the Island Host with regular frequency comparable to the need for control, and that the data be transmitted to controller A. These types of functions are critical to building wide control.

Another building wide function of the Building Wide Controller, or "Island Host level of Architecture", is to serve as a vehicle for monitoring and operating history data. Monitoring functions will be discussed under the remote communication portion of this section, yet the most important note is that a "window" to the network must exist. This window must allow for communication with all network devices via a single modem and telephone line in the building. Through that window a variety of monitoring and alarming features must be provided as will be discussed in a moment.

Also of critical importance is historical system operating information. Typically the Building Wide Controller will serve as a central repository for data files containing trend data on various system points. This is generally the most efficient approach to ensure that the overall cost of the system is not detrimentally impacted by the need for data from all network devices. A key topic to be discussed in later chapters will be the cost per controller, or "node", to have network capability.

To ensure that the expense of these features do not impose an undue cost burden on the overall system, the general approach is to charge one network device, the Building Wide Controller, with responsibility for history. This requires that the controller access data from network devices and provides data storage for later retrieval.

The final Building Wide Controller function is to allow for remote communication with a Central Operator Interface (OI). These topics will be covered in much greater detail under the Central OI Chapter, yet a brief overview is provided here. There are three key functions which must be provided: programming, monitoring and alarming. Programming requires that each network member including the Island Host may be commissioned via a local or remote Central OI. Commissioning includes programming of control sequences, network communication parameters and alarm limits. Monitoring includes current status on all controller points, as well as access to system history data.

Building Wide alarming is perhaps one of the most critical features of any building control system. This feature requires that critical conditions may be defined in addition to default alarms, such as sensor or controller failure. The system designer may then note whether such alarms should be logged for availability when the system is polled or should initiate a dial out. The dial out function will use the modem and communication capability to initiate communication with one or more predefined Central OI's, and send the alarm data. Again all of these functions will be discussed in greater detail under the Central OI chapter.

NETWORK MANAGEMENT AND CONTROL INTEGRATION

The topic of networking and the management of networks is not the primary topic of this book. Yet the introduction of Distributed DDC has made networking an integral part of every system. As a result a full chapter in this book is devoted to the topic, and it will also be integrated into the discussions of system hardware and software. The intent is to provide an understanding of the technology while viewing it with reference to an overall architecture.

Two types of architecture will be introduced: intra-protocol system and inter-protocol system activity. Intra-protocol system activity would be defined as control or communication interaction between controllers operating with the same protocol. The simplest example would be conventional manufacturer proprietary systems. Inter-protocol system activity on the other hand assumes that there is a larger architecture within which, there are several intra-protocol systems operating.

As the network view is examined it is critical to view the way in which control of various pieces of equipment is integrated with a system. Conventional, non-distributed systems, used CPUs as discussed in the last chapter. In distributed networked systems, key data must be shared with multiple microprocessor based devices, and provisions must be made to accomplish the same level of control. In this section the groundwork will be laid for discussing these types of issues for both Building Wide and individual controllers. Also some reference will be made to the Central OI, which is discussed in greater detail later in the book. Networking issues will be viewed under headings:

- Network Manager Interaction
- Inter-protocol system Central OI
- Troubleshooting and diagnostics

Network Manager Interaction

The network manager hardware for any particular local system may take a number of forms. In many cases it is a central control and processing panel or a series of peer controllers. The hardware employed is not critical to this discussion, however as the types of functions provided still must be provided. As noted previously, the requirement for a network manager has been imposed due to the expanded requirements of distributed controllers.

Technological advances have made it possible to spread intelligent controllers throughout the building, and dramatically increase the reliability of systems. These developments have been further expanded by Original Equipment Manufacturers of HVAC equipment through a continuing trend toward factory mounted intelligent controllers. Along with the benefits of distributed control, though new demands have been placed on the system for management of the information passing between controllers. These issues are what define the requirements for network management in Distributed DDC systems.

The Central OI in distributed systems also takes on added responsibility for interaction with the network manager. This interaction includes a full range of communication issues. Most of these communication tasks are transparent to the end user, but introduce a new level of sophistication to the system. Among the tasks to be covered are:

- Network Node Status
- Global System Data
- Remote Support

Network Node Status

In conventional systems much of the communication occurred through an interface to the Central Processing Unit (CPU). It was not uncommon for systems to support Dumb terminal emulation, and to allow direct connect interface through an RS-232 port on the device. The local control panel would support a standard data interface and allow the terminal to Building System Operations types of functions. Not all systems supported dumb terminal programming but many supported this mode of monitoring which imposed the burden of controller communication and system interaction at the local panel.

Distributed Systems do not provide such features. Dumb terminal provided an early form of standard, but data integrity was poor with these systems, particularly as high transmission speeds. Also memory requirements for dumb terminal mode put a large burden on the control processor. Further the demand from customers for more sophisticated control sequences and easy to use systems required expand software.

This expanded software, and the demand for compatibility with such programs as Microsoft Windows[1] required more software power at both the Building Wide controller and system, as well as, Central OI.

Distributed systems on the other hand provide excellent data integrity, and eliminate dumb terminal mode. Compatibility with Windows and other off the shelf hardware and software also leverages the power of DDC systems and brings new tools to system owners. These systems significantly expand the sophistication of control by focusing local control computing power on the application. Also relieving part of the communication burden from individual controllers allows the distributed controllers to incorporate added control functions such as equipment safeties.

The new requirement of distributed systems is that all Building Wide and Individual Controllers, along with a Central OI, is a resident of the network or a "Node". The network manager coordinates communication between these nodes and also supports external communication requirements. At the risk of further confusing the reader, the job of Network Management is centralized in Master/Slave systems and fully distributed with Peer to Peer approaches. Further the Central OI in both types of system is generally no more than a node on the network.

The beauty of this approach is that the essential functions of a Distributed DDC system are strategically share by network members. Loss of a Central OI has no effect at all on the control of a building. Loss of power or communication to a Peer controller does not impact the entire system, only one piece of equipment. As time progresses it is more likely that redundancy will be built into Peer type systems. This would allow for another nearby controller to take over if a failure occurs.

With Distributed systems there are a new types of information to be coordinated and provided in the course of communication activity. These must be maintained in memory at the network manager and the local or Remote Central OI. Without this information, the Central OI can not support local or remote communication. These include protocol address items needed to identify the various distributed controllers. Further, access to the network allows the network manager to be a source of information that resides in various controllers. The manager can be a central repository for key data that is updated every few seconds, and may be used for control in any controller. The term for this information is Global System Data, and it was discussed earlier in this chapter.

Global System Data

The Network Manager supports access to this data by managing flow of information between devices. It is shared by the network manager through a global system data feature. This feature allows for the system designed to identify the data that must be made available and create a means for the transfer to take place. In a Master/Slave network this occurs when the Master polls specific controllers for information and stores it in a data field. The data field has a particular name, and other controllers know how to request it when they are given access to the network.

The network manager will also monitor refresh rates to ensure that the data is updated. In the event of an alarm or interruption of data flow, this device will also have a set of instructions. In addition to sending an alarm, the manager will take one of several actions. For example, it may;
- hold the last value received,
- request data from a controller that has been designated for back up data or
- suspend control.

This is a role of the network manager, yet the parameters and sequences identified above must be programmed through the Central OI. Controllers also share other data such as: a clock, which must be synchronized for all controllers requiring scheduling, and kilowatt demand status. It is also possible for the Network Manager to introduce some global data to the systems. For example, inter-protocol system architectures may, and often do, require shared data between networks. Where a Network Manager has direct access to the Host level of Architecture, it may provide this function. Examples of this may be a smoke alarm which is monitored by a fire safety system, and used to initiate smoke evacuation. Such a system would trigger a chain of control events.

Remote Support

As stated under remote communication, all of these functions must be extended to modem interface over telephone lines as well. Typically, this means that the remote interaction necessary between Central OI and the local system is transparent to the user at a remote site. A recent advance in this area is Building Wide and Individual Controllers that support "Plug and Play". This is a feature of these controllers that supports the object oriented concept, and stores all critical system information at the controller level. With truly intelligent controllers of this type it is possible to dial into a system from a remote location, quickly upload from the controllers and have full access to the system. It is not longer necessary to have a copy of the system on disk and synchronize disks to stay current. In this was the system itself becomes the final word on what information is most current.

Building Wide Controller Trends

There are a number of trends to note at the Building Wide Controller or Island Host level. Perhaps one of the most significant trends is the drive for standard communication. The demand from end users is for remote communication with multiple manufacturers via one front end or Central Operator Interface. This was the initial driver for the entire standardization movement in the control industry. Examples of early innovators who acted without general industry consensus are IBM with its Facility Automation and Communication Network (FACN), and a variety of others like the German and Canadian Governments. This book can not focus on the complexities of standard communication, the reader is encouraged to research this topic through other resources.

Another significant trend at every layer of Architecture has been not several times in this chapter, "Peer to Peer" controllers. This technique is important for Building Wide Control because it distributes the critical network functions to multiple controllers. Ultimately the approach builds great integrity and reliability into the system. Though these controllers are most complex, the cost premium for Peer control is negligible. This is thanks to the continuing strides in electronic and microprocessor based technology that result in greater functionality for less cost.

The Peer approach holds significance for standard communication, as well. One strategy that is already being implemented is to standardize at the Host level. There is also significant industry demand to offer controller level communication standardization as well. Another related Building Wide Controller trend was mentioned above is Integration. There is a distinct movement toward integrated control between Island systems with different functions, such as security and building control systems.

There are a host of other trends with Building Wide Controllers at the Island Host Level. These will only be discussed briefly to highlight the activity currently underway in this area. Many of these trends represent significant technological and market strides

The key Building Wide trends include:
- movement from minicomputers to microprocessor based systems,
- expanded control integration within the building system between the Building Wide or host level and controllers,
- easier to use (program and monitor) systems,
- Nonvolatile memory storage to avoid program loss,
- cost effective and higher speed remote communication and
- expanded PC based Central OI's with simple user interface[2].

Microprocessors are at the heart of every Building Wide or Individual controller. This trend is unmistakable, and represents a tremendous change in every aspect of the industry from manufacturing to maintenance. Five years ago less than 10% of the Variable Air Volume boxes shipped from manufacturers included any DDC controls. Today over 50% of those units have DDC controllers. That is only one example of the growing trend. At the same time it is unheard-of to even consider a minicomputer for any type of Building Automation today.

Integration has been an important topic in this book, particularly as it applies to Functional Systems. This topic of integration however refers to the sequences that are implemented within a functional system, such as HVAC. As the various DDC Sequences were discussed, a number of references were made to integrating control between devices. The trend keys into the concept of more intelligent controllers and the idea of coordinated control.

Peer controllers that are capable of continuously interrogating the network for such sequences as morning warmup are a good example. This is a feature that would have been centralized in older systems. A single condition such as outside air temperature would have been monitored, and the Building Wide device would make a decision on start time based on the data and an internal sequence. When start up deemed necessary that controller would signal initiation of the sequence. With control integrated sequences of this type, each controller can make independent decisions based on Building Wide data as well as local controller data. This trend is critical to building highly reliable and effective control systems.

Ease of use may be the buzzword of the decade. Building Owners and System Users are exposed to software from a multitude of industries every day. In many cases they form an opinion of how software should work based upon interaction with such software as Microsoft Windows[3]. For this reason there is a tremendous demand for software that has a Windows look and feel for programming and monitoring DDC systems. This means the software must be mouse driven, point and click and graphically based.

Ease of use goes beyond software however, users also want systems that will do alarm dial outs to pagers and telephones with voice synthesis. Trends in this area are focused on simplicity and improved productivity. As a result manufacturers are beginning to develop special software specifically for the end user to incorporate many of these types of features. Simpler commissioning and programming tools are also made available to the contractor or installer. At the same time Expanded PC based Central OI's with simple user interface are also on the horizon. Features with these types of systems will be discussed under the Central OI chapter. They include anything from integrating preventative maintenance to using on-line diagnostic for troubleshooting building HVAC issues.

Nonvolatile memory storage to avoid program loss. This has been a trend for quite some time, however the technology to achieve it continues to improve. The early minicomputer approach often used Uninterruptible Power Supplies (UPS) to eliminate power failures. This was effective but costly. Earlier microprocessor based systems used battery backup with Random Access Memory (RAM). This was effective as long as the battery lasted. The next trend was Electrically Erasable Programmable Read Only Memory (EEPROM). This was a nonvolatile chip that did not require either a battery on power to hold memory. This was extremely effective and cost effective, but EEPROMs has a maximum read/write capacity before the chip is no longer functional. This actually impacted controller life and that was less effective. The last and most exciting trend thus far is Flash memory. This is nonvolatile memory and has no limitation on read/write capability.

Cost effective and higher speed remote communication is a major area of importance for DDC systems. This is also the best argument known for incorporating off the shelf technology in control systems. Early systems often contained proprietary modems. This was problematic for service and the modem industry was always one step ahead. Today most systems are compatible for industry standard modems that are commercially available. As advances in the industry occur, the user need only change out the modem. In this way the limitations a DDC systems faces are only those imposed by the telecommunication system and the modem industry.

CONCLUSION

This chapter has covered a tremendous amount of information. Many of the topics discussed here are central to the readers understanding of DDC systems. These concepts, such as DDC Sequences, will be integral to discussion of Equipment and Zone level controllers. Further, these concepts will be essential to many of topics covered through the balance of the text. Of critical note is the fact that Building Wide Control is essential to the overall effectiveness of a system. Further integration with lower level controllers forms the basis for truly reliable Distributed control. With this understanding it is now possible to discuss in more detail the Individual level controllers and their characteristics.

FOOTNOTES

1. Microsoft Windows is a registered trademark of Microsoft Corporation, Redmond, Washington.
2. "Survey on Energy Management Systems - Survey # 208", Future Technology Surveys, Inc., Lilburn, Georgia.
3. Microsoft Windows is a registered trademark of Microsoft Corporation, Redmond, Washington.

Chapter 10

Equipment Level DDC Control

There are many features and capabilities available with distributed controllers for more complex applications. Microprocessor based control becomes common quickly with large Heating, Ventilating and Air Conditioning (HVAC) equipment and with general purpose control. Equipment control and General Purpose control are distinct applications for DDC, and each will be addressed in this chapter.

The term equipment level is derived from the use of these controllers with such equipment as centrifugal chillers and air handlers. The word "level" in this context defines its role within the Architecture. As discussed in Chapter 8 there is a hierarchy of devices in any systems. The rules that define the hierarchy and each components role in the system are often referred to as the Architecture.

General Purpose control uses the same hardware, and operates at the same level of Architecture, but implement different control sequences from Equipment controllers. These sequences include such applications as start/stop of non-HVAC loads and on/off control of lighting and other electrical equipment.

Distributed control with equipment level and general purpose controllers is cost effective due to application complexity. In these applications there are also greater requirements for control integration due to the distributed nature of the control system. Also the equipment cost has an order of magnitude that can more easily support the cost of distributed controls. In fact at this level, as well as the Building Wide or Host Level, Peer-to-Peer controllers are most common.

The Building Wide Controller has been discussed as the integrator of global control sequences that affect the entire facility. Yet the benefits of distributed control are truly realized when intelligent devices are deployed at a particular control application. Equipment/General Purpose controllers form one of the two levels of distributed controllers noted in the Architecture discussions under Chapter 8. They provide full local control requirements, and integrate with both the Building Wide Controller and, as appropriate, Zone level controllers to provide building wide functions.

In reviewing these key components of a Distributed DDC system, the topics which will be covered under this heading include:
- Introduction to Equipment Level DDC
- Equipment Level Hardware,
- Ancillary Components (sensors, motors, transducers, signals)
- Firmware and Software
- Sequence Of Operation
- OEM Integration
- Networking
- Installation and Commissioning and
- Equipment Level Control Trends.

INTRODUCTION TO EQUIPMENT LEVEL DDC

In Chapter 9 the topics of Direct Digital Control and Distributed processing were introduced and briefly discussed. That section offered valuable system related detail, yet there were a number of controller related topics that could not be covered. This chapter will expand that discussion with specific focus on Equipment Level DDC Control.

The Equipment Level is defined by the applications of control, as well as the characteristics of control. The applications are the most complex and energy intensive equipment in the building. These Equipment level applications include: Air Handlers, Chillers, Boilers, Hot Water Converters, etc. The characteristics of control include extensive data point monitoring, multiple control sequences such as: temperature control, reset and warm up. They typically require faster processing for these control loops and several types of control points.

There are other key components of a Distributed DDC controller. Prior to looking more closely at Equipment Level Control, it is important to define DDC further and clarify the equipment necessary to accomplish the technology. The controller or microprocessor is the heart of a DDC controller, and leverages its power to execute Equipment Level Control.

Direct Digital Control (DDC) has become possible because of the **microprocessor.** Within this chip are the brains of all DDC systems. In the microprocessor, logic and **data processing** are executed. The processor generates programmed outputs for digital control. All sensor signals are processed within the DDC controller and output are executed to relays, devices, and information systems.

In essence, a computer is nothing more than a logic unit, a control unit, **random access memory (RAM), auxiliary memory,** and **read-only memory (ROM)**. The microprocessor controls activities both inside and outside the chip.

A **microcontroller** is a device that uses a microprocessor to introduce Direct Digital Control into environment. A basic microcontroller also contains RAM, auxiliary memory storage space (EPROM and EEPROM), away to input and output data, and a clock. The frequency of the clock determines the rate at which data moves in an out of the microprocessor. Figure 10-1 shows a basic microprocessor controller's architecture.

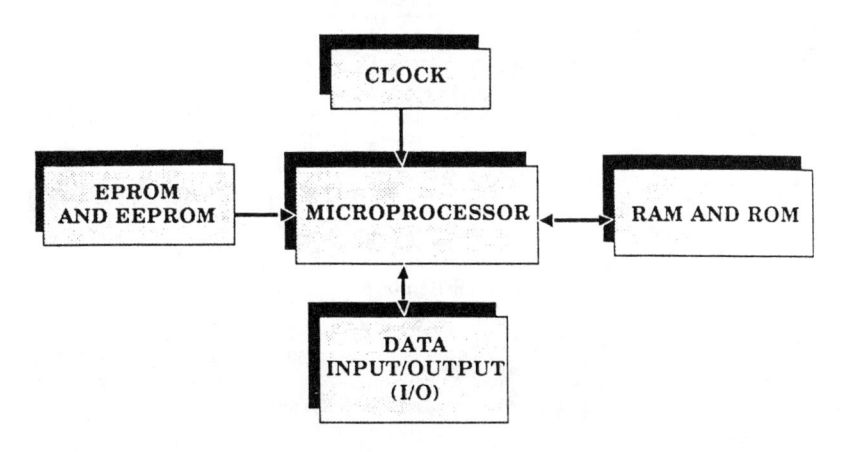

Figure 10-1 Microcontroller architecture

CONTROL LOOPS

Building Wide and Individual DDC control systems rely on control loops to regulate different parts of the control hardware. These have been categorized into two types, closed and open loops. As discussed in Chapter 9, the presence or absence of feedback, respectfully distinguishes them.

An **open loop**, for example, consists of a controller that starts and stops HVAC equipment at given times. There's no feedback to the controller about the unit's status.

An advancement to the open loop is the **closed loop**. The controller responds to continuous environmental data and compares it to programmed environmental desires (lower temperature, higher humidity, better air quality, etc.). Settings for these values are called setpoints. Figure 10-2 shows a block diagram of a typical closed loop system.

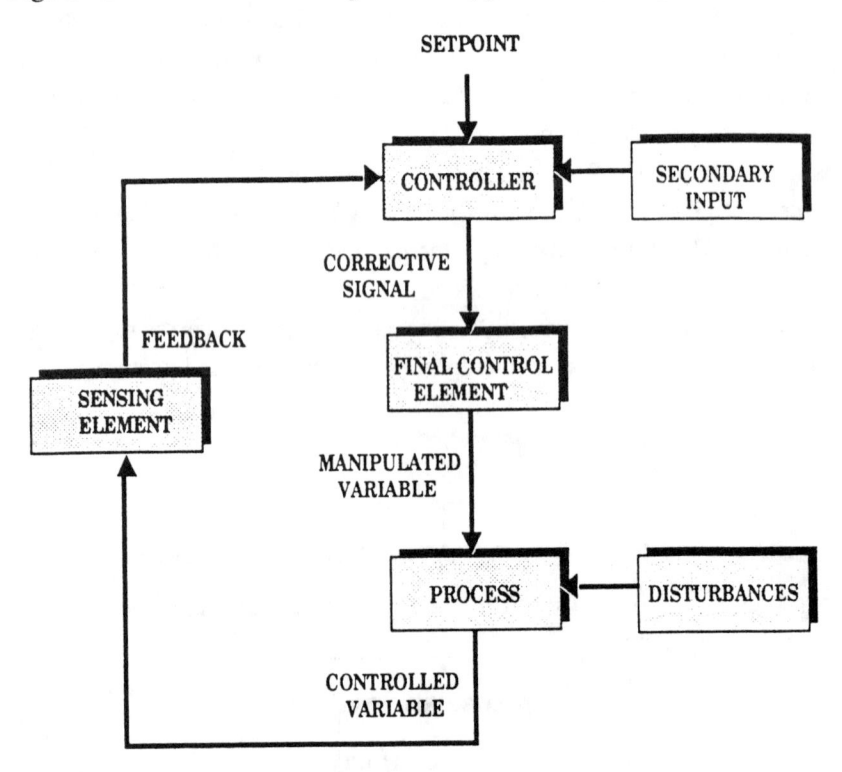

Figure 10-2 Closed-loop system

Equipment Level DDC Control

In a closed-loop system, the sensing element feeds back information to the controller. This data is compared by the controller to a setpoint that's been entered by the user. The controller sends a corrective signal to the system hardware. The sensing element continues to monitor the environment and outputs new data to the controller. The controller supervises the space's environment, by monitoring sensors continuously and activating devices to compensate for environmental requirements.

The manipulation of environmental data and setpoints is called **processing**. Based on processing, the controller sends signals to devices for HVAC corrective actions.

DIGITAL VS. ANALOG

Computers think and respond in terms of "yes or no", "one or zero" "open or closed", "voltage or nonvoltage", "on or off", "true or false". This binary logic converted into electrical impulses is defined as a **digital** signal.

The digital signal we encounter in HVAC applications are usually in the form of open or closed contacts. An input device tied to the microprocessor will provide an open/closed contact that either supplies or does not supply a DC voltage. For example, a DDC controller may interpret an input device signal as a request or non-request for heating.

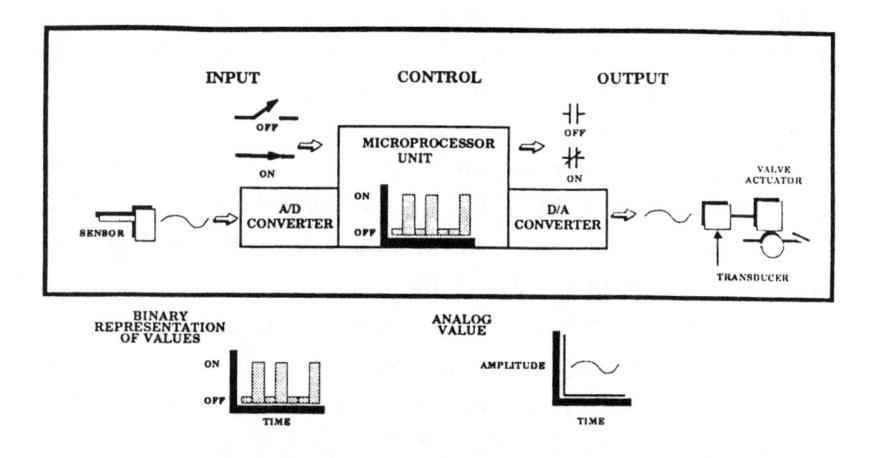

Figure 10-3 A/D and D/A system conversions.

A digital signal on the output of a microprocessor usually activates a contactor (relay) to either open or close. The system can use this open or closed contact to energize or de-energize an HVAC output device with an AC or DC voltage.

For some applications, the digital signal voltage levels are used directly instead of interfacing with open or closed relay contacts. A controller can be connected directly to another controller without using open/closed contacts.

An **analog** signal has a continuous range of values. An electronic thermostat sends out an electrical signal that corresponds to temperature values ranging from 0° to 350° F. The thermostat signal is analog because it can generate an electrical quantity that can have many values.

In DDC systems, the controller receives analog signals from sensors, and converts them to digital. The controller processes the signal and may convert it back to analog. This signal alteration occurs because some HVAC equipment (sensor, motors, etc.) function with analog signals. The controller (microprocessor) functions with digital signals.

Using sensor and setpoint data, the digital controller performs mathematical computations from a software program to manager HVAC equipment. The results are digital signals generated to turn on/off hardware.

A digital controller receives input from a sensor. When this data is in analog form, an analog to digital converter (A/D converter) is required. When the controller completes processing, a digital signal is transmitted to the digital to analog converter (D/A converter). The motor, actuator or transducer in the HVAC equipment understands this signal sent by the converter. See Figure 10-3.

Digital controllers sample data at set time intervals rather than reading it continuously. This method is called discrete control signalling. Efficient and uninterrupted environment controls result when the sampling interval is precisely calculated.

Digital controllers work with a set of measured control variables and a set of control algorithms. Control variables include temperature, humidity, and air quality settings that you maintain in an environment. Algorithms are explained below.

ENHANCED CONTROL WITH DDC

DDC uses a control loop much more efficiently than conventional control methods such as pneumatic or electro-mechanical. Because of the computer's operating speed, it can sample multiple sensor devices in milliseconds. This sampling speed enables it to control many control loops simultaneously.

Digital controllers contain **software** made up of a series of DDC sequences, discussed in Chapter 9, that issue commands/instructions. When engineers design a software program, they break down the control loop into steps that comprise its operating sequence. Programmers translate these steps into a series of symbolic commands that a computer understands. These symbolic commands constitute an algorithm or sequence. A sequence is a list of steps that perform a task. A recipe is a simple sequence. A control loop is programmed into a sequence.

A digital controller contains a sequence which processes the feedback data from the sensors. Sequences allow controllers to perform extremely sophisticated calculations. This results with effective and efficient control of hardware and environment.

Once the software is transferred to the memory (EPROM), the DDC sequence is ready to execute. The sequence instruction will control the closed control loop of the system. See Figure 10-4.

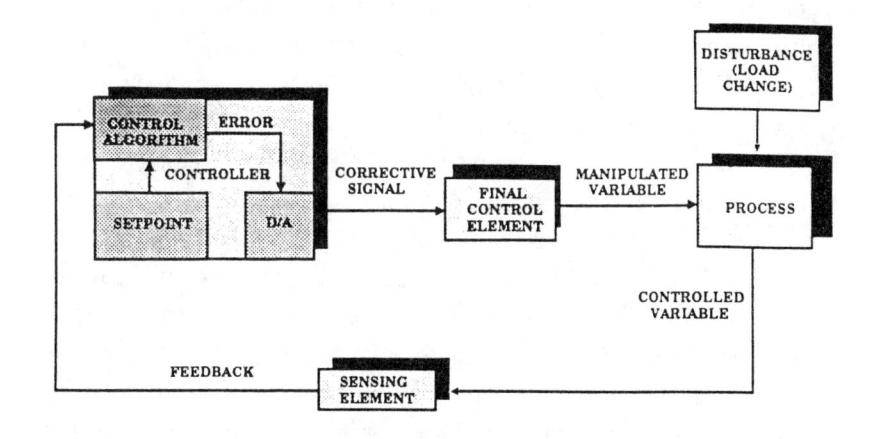

Figure 10-4 Closed loop system

As the controller receives data from sensors, it performs computation. The result of this processing, is sent out as corrective signals to the HVAC equipment.

LEVELS OF CONTROL

There are three levels of operational control to work with in DDC. They are zone-level, equipment-level and Building Wide controls. This chapter will again focus on equipment-level controllers.

EQUIPMENT LEVEL HARDWARE

The fundamental requirement for Equipment and General Purpose control is a microprocessor based device with capability for both stand-alone and networked control. One of the essential benefits of these systems is reliability, and the on board microprocessor ensures that control is not suspended for any reason short of catastrophic failure. Catastrophic failure could be power, communication or some other major interruption in normal operations.

The added controller reliability of DDC is a direct function of the microprocessor. Another important factor is the improved control that is possible. The power of the microprocessor determines two important factors: Execution time and Scan Rate. Scan rate is the process required to scan a point and update the control sequence on current status. This aspect of the controller can be seen understood by looking at the diagrams depicting closed loop control. Precise control is directly reliant on the availability of timely information from Inputs. These inputs are the source for temperature, pressure, etc.

Execution time is the second key component to ensuring effective control. The other requirement of closed loop control is to execute a command to outputs controlling equipment. In a DDC controller this relies directly on the microprocessors execution time. Execution time is the interval that elapse between each evaluation of the control sequence. For example, the control sequence for a temperature loop may execute every 10 seconds, where pressure may require execution every 2 or 3 seconds. The combination of rapid access to fresh data, and execution time that is appropriate for the application, results in reliable, efficient equipment control.

A factor that can have dramatic impacts on issues like Execution Time and Scan Rates is the power of the microprocessor. A direct indicator of power is the speed of the microprocessor. Speed is measured by the amount of information that can flow through and be processed by the chip. Typical Building DDC microprocessors are 8 bits per second, or 8 Bit, or newer products are 16 Bit. Ultimately think of this as defining the throughput of data for the processor.

The trend for DDC microprocessors is for devices from process and other high level control applications to be integrated into the Building industry. In this vein there have already been 32 Bit microprocessors introduced with the most sophisticated DDC controllers.

The other factor that is impacted by processor powers the support of multiple control loops within one controller. It is not uncommon for DDC devices to have multiple control loops for temperature control, pressure control and reset of both set points. For this reason the processor must be capable of executing multiple sequences with the speed and accuracy necessary to meet the applications needs.

In addition to control, which is the primary mission of these devices, the reader is aware by now that communication is also a requirement of the hardware. The processor must allow for integration of hardware to implement networking, typically an add on module. The processors must also be able to orchestrate control sequences using data from physical inputs and outputs, as well as through sending command request for data and control to other controllers. There will be more discussion of communication issues for Equipment Level later in this chapter.

Perhaps the second most distinguishing feature of equipment and general purpose level hardware is the on board inputs and outputs. Devices applied at this level typically provide a mix of both analog and digital inputs and outputs. Input/Output (I/O) requirements tend to be more demanding at this level due to the expanded applications. There is generally a demand for at least eight of each point type: analog and digital inputs and outputs. Due to the fact that these controllers may be applied as general purpose devices or application specific controllers, it is not uncommon for controllers to be expanded up to as many as 64 total points. This may be done through add on modules, cards, etc.

Expanded I/O capacity is also necessary because most equipment applications require more than one control point. This control is implemented through digital outputs and analog or pulsed modulating outputs, as well as monitoring of both analog and digital input values. Analog values are generally used to monitor temperature or pressure, while digital points provide contact closures typically indicating an alarm. Device outputs then respond to the commands of the controller based upon the appropriate input value and the sequence of operations. These outputs will either be enabled or disabled, or may be modulated to a commanded position.

A common trend with this equipment/general purpose hardware is for manufacturers to develop one hardware device or "platform", and use this with all applications. The platform will provide a base input/output capacity, and in many cases allow expansion through any of several techniques. Application of the controller is accomplished by changing software in the device. Software issues related to this approach will be discussed further under sequence of operations. Hardware design issues to be resolved also include the requirement for a full line of ancillary component products that will be discussed in the next section. Ancillary components include both input and output devices. The complete equipment level product includes the:

- controller,
- sensors, transducers, other input devices and
- electrical relays or contactors, motors, actuators, transducers and
- other output devices to enable control points.

There are two additional points that should also be mentioned under the category of hardware: controller size and ambient operating requirements. Controller size is typically not an issue with general purpose devices because they are often mounted in a mechanical room. The assumption is that these devices can be acquired with enclosures that meet standards agency approvals and local codes.

With equipment controllers however, there is an ongoing trend to pre integrate DDC devices within the control panel of HVAC equipment. Pre- Integration, as this is sometime called, is a trend that will be discussed further below. Quite simply it means that Original Equipment Manufacturers (OEMs) of HVAC will mount a DDC controller on their products at the factory. This approach is very effective and saves both labor and cost in the field. Yet it also imposes a number of issues. In the area of hardware the most significant issue is a size constraint for smaller units. Obviously a small packaged unit may be limited in space for the controller. This may also be a consideration if the building owner is buying equipment without controls. It will be critical to ensure that the equipment and field applied hardware integrate well. Clean power for the unit, as well as field commission termination of wiring, safeties, etc. will all be important. The idea of pre-integration will be discussed further in this and the next chapter.

A final topic under hardware is the absolute requirement for equipment level control devices to be able to operate under outdoor temperature conditions. A good example of this is a VAV roof top air handler. It is typical for this equipment to be provided in a fully self-contained manner. A single enclosure containing the HVAC equipment and controls is provided in one package that is mounted on the roof. The package will be exposed to the full extreme of outdoor temperature and humidity, and may be shipped to any location in the world, further adding to the potential extremes.

ANCILLARY COMPONENTS (SENSORS, MOTORS, TRANSDUCERS, SIGNALS)

This is an essential section of this chapter. Though these components are ancillary and therefore not at the heart of a DDC System, controllers can not carry out there sophisticated control without sensors, motors, etc. All DDC Controllers require access to information regarding temperature pressure and a wide variety of conditions. This information is provided via sensors and transducers. At the same time the DDC controller must also be able to initiate control action via, relays, motors and transducers.

There are two types of input sensors: Analog and Digital, sometimes called binary. These inputs are channels on the devices such as those described under the introduction. Each channel or set of terminals accepts wires, typically a signal and a common. The simplest form of input information is Digital. This is simply an on or off condition, typically monitored via a simple dry contact. The contact allows the controller to identify a true or false condition. This often identifies such conditions as fan or compressor failure, high or low temperature, etc.

With analog points, the controller must detect the signal and then, using Analog to Digital converters, interpret the data. The industry standard signals used with typical analog inputs include 4 - 20 Milliamps, 0 to 10 Volts and some range of resistance. For example a space sensor may have a range of 0 to 4000 Ohms. The characteristics of the sensor element determine the signal that is returned to the controller at any given temperature. The controller must read the signal coming back, convert the information to a digital value, and update the control algorithm. The control algorithm can then execute and determine the action necessary to maintain desired conditions.

An important note is that there is basic concept with typical sensing elements. These devices use a material that varies the amount of electricity that passes through the element based on changing conditions, such as temperature. There are two trends that have been under research for a number of year; sensor bus and wireless sensors. These are actually related concepts because they both rely to some degree on communication.

Both of these trends are driven by the desire to reduce cost, and yet has additional unique advantages. Sensor bus is an idea that brings networking communication technology to ancillary devices. The concept is that the cost of both labor and wire traditionally used for sensors could be reduced. This would allow for greater flexibility of installation and also for cost savings. It would also simplify relocating sensors that are not providing good information on actual zone conditions. Industry research has indicated that 20 to 40 percent of Zone sensors are installed in the wrong location[1]. Yet these are never moved due to the labor and difficulty of fishing wires through walls, etc. to move a sensor Sensor bus could simplify initial installation, relocation and also the entire process of data access for programming, etc.

Bear in mind that the nature of DDC systems requires access to information from points in sequences that exist in multiple controllers. This requires additional programming for data sharing and global points to identify the physical residence of these points and the controller that requires the point. With sensor bus all sensors would be on a communication network and any controller could simple request those points. Obviously it is still necessary to identify a pseudo point in the controller to keep the most current value for each point.

Wireless sensors are another new concept. There have been a number of attempts to bring technology from other industries to DDC systems. Wireless sensors have been introduced that use either infrared or radio communication. There are a number of benefits to wireless sensors. Cost savings are available, as with sensor bus, but even greater savings are possible because wire can be eliminated all together. Other benefits include the ability to assign sensors to any piece of equipment or to group multiple sensors (minimum, average, etc.) for an individual device. Further it is easier to move sensors, replace failed sensors and redefine any sensor to any given device.

There are a number of characteristics that must be considered with all wireless sensor technology. It is important to ensure that the distance over which these sensors operate meets the application. Further some devices require a clear line of sight for transmission of sensor data. With any type of radio based sensor or network approach interference is a critical topic. Interference that changes or interrupts communication of data would make this technology unacceptable for DDC systems. The importance of current information is one of the most essential requirements for DDC systems. No sensor that is prone to interference that impacts reliability of the data is acceptable for use with DDC. Wireless sensors are still in their infancy and it may be some time before their viability is proven.

There are also the same two types of output sensors: Analog and Digital. These outputs are again channels on the DDC devices. Each channel or set of terminals again accepts wires, typically a signal and a common. The simplest form of output control is again Digital. This is simply an on or off command, typically accomplished via a relay or contact closure. This contact allows the controller to enable or disable equipment, generally electrically via a motor starter, electrical contactor, etc. As the devices imply this is generally used to control equipment such as lighting, fans, etc.

For Analog control DDC uses a modulating analog signal from a Digital to Analog (D to A) converter. In actual reverse of the input process, the control takes data from the algorithm in digital form and converts it to a signal for output devices. The output is one of three types of signals, Pulse Width Modulation (PWM), 0 to 20 milliamp or o to 10 volt DC. PWM is actually a simulated analog that is achieved by pulsing a Digital Output. The pulses are rapid and incremental, and approximate true analog control. This is common with older generation DDC systems that tried to meet cost competitive pricing targets by eliminating true analog control.

True analog outputs are either current (O to 20 MA) or voltage. These outputs take the D to A value and send a signal directly to the output device. It may be an actuator motor, a transducer or some other type of modulating device. Again the input will update the control algorithm thus allowing for fine-tuning of the closed loop control.. The control algorithm can then execute control again and maintain desired conditions.

A key related issue with these products is whether the manufacturer will fix the ancillary components that may be used with a device. To "fix" these devices means that the user must purchase a specific manufacturer and model of component or the Equipment Controller will not interface correctly. An mismatch of components results in a non-functioning system.

A relatively common approach with more intelligent Equipment Controllers is to allow flexible application of a wide variety of industry standard equipment. This means that the controller can accept inputs that provide a current, voltage or resistance signal. Typically the added flexibility allowed with choosing your own sensors will also require more sophistication in commissioning the device for operation with a given sensor. The benefit of this feature is that the cost of system retrofit may be greatly reduced because existing field devices can be reused.

FIRMWARE AND SOFTWARE

The topics firmware and software get to the real heart of a microprocessor based control. Under previous discussion there has been a great deal of reference to sophisticated sequences, etc. These are higher level control programs, but for them to exist Firmware and software must be in place. First some definition. Firmware is a term used to define the basic level knowledge in any device that is not changed in the field. The term Firmware grew out of the use of Electrically Programmable Read Only Memory chips (EPROMs).

EPROMS can only be changed with a PROM burner, usually too expensive for most field operations. Since it could not be changed outside the factory, it's contents were considered to be "firm", thus "Firmware". Software on the other hand refers to information that may be changed in a wide variety of ways. Further software may generally be changed by any number of different people, more on that in a moment.

Firmware is an important topic because it is thoroughly essential to any DDC controller or, for that matter, any computer. Firmware is a term used somewhat loosely in reference to DDC controllers. It is often used to identify anything that the average user or technician can not change. Just like any computer, Firmware is resident in the controller. At its simplest level Firmware contains the Operating System (OS) for the DDC device. As with all else regarding these controllers however, Firmware must be versatile and much more complex than with the average personal computer.

DDC Firmware can be thought to consist of at least five basic components:
- Digital DDC Operating System (D DOS)
- A to D and D to A Conversion
- Communication
- DDC language Constructs
- Libraries and Parameters

D DOS is the basic requirement for computing. It manages all the required functions necessary for computing. It provides a hierarchical structure of functions that manage the flow of information and control activity throughout the controller.

Digital and Analog conversions have already been clearly explained. Yet the requirements to manage and initiate these sub-programs, as well as, the flow of information are the Firmware's job.

Communication has been identified throughout this text as truly essential to DDC. The Firmware is again essential in ensuring the instructions are in place to maintain communication.

With DDC systems Communication means both controller networking and remote interface. These are separate tasks, but an exciting new feature is to offer both at the Equipment / General Purpose level. This means that the designer has two architecture options. For larger jobs with multiple controllers it is possible to network multiple controllers, and interface to them through a communication module that is a Peer on the network. For application where the power of DDC is required but one panel offers enough inputs and outputs, some new controllers are capable of direct communication via modem to the remote location. In this case the controller must have Firmware to support the protocol in use and maintain a communication session.

In the past, Firmware for Remote was held in one controller, as in the network option above, and the others had to communicate through the designated device As noted, Remote communication for individual DDC controllers is a somewhat new feature but growing in desirability.

One final digression with Networking Firmware has is that there is a growing trend toward integrating D DOS and Network Communication software in the same computer Chip. In the past control manufacturers would develop proprietary operating systems and communication protocols for DDC lines. This meant that there were no standards for communication in the industry. Under the Networking chapter there will be an expanded discussion of standard communication and the reasons for standardization. That is only a part of the demand for this trend however.

Many readers are familiar with the computer industry so a simple analogy may be helpful. For those who are familiar with Personal Computers and Local Area Networks (LANs) this is like buying one chip that integrates the latest and most powerful versions of both Microsoft Disk Operating System (DOS)[2] and Novell Netware[3].

Using a single chip for both Operating System and Communication holds even greater implications for the control industry. The latest of these chips such as the LONWORKS[4] technology from Echelon provides a complete set of system development tools.

This chip technology includes most of the items defined as Firmware including D DOS, A to D converters, and Language Constructs. Perhaps what makes them even more attractive to manufacturers is that they are developed by operating system and communication experts for a variety of microprocessor based products. This means the economy of scale keeps the chips cost down. It also makes it cost effective for the chip manufacturer to develop a wide variety of software and hardware tools.

The author will state more than once in this book, that DDC systems must make use of off the shelf technology wherever possible. To ensure optimum reliability, cost effectiveness and leverage the power of products that are designed for larger markets this is essential. A good example of this concept is the use of Microsoft Windows[5] as the basis for PC interface software. This give everyone using the system access to all the compatible software and offers many new tools to improve productivity and enhance the final product. At this point the only benefit of this information to the reader is an awareness that this is a growing trend to improve your understanding of the overall technology. Also this would apply at every level of the architecture, not only the Equipment Level Controller. Yet this discussion of firmware was the most appropriate to address this trend.

DDC Language constructs are another aspect of controller firmware. This is a real departure from the computer world however. Everything discussed above could be viewed as a behind the scenes requirement or a housekeeping function. At this point the discussion strays to the support of programming languages. These were discussed earlier in Chapter 9 and are building blocks for system users that must create control sequences. A language construct is very similar to a computer language like "Basic" or "Pascal" but it is typically proprietary and specifically focused on a particular manufacturers control hardware. An example of a construct is the equation driven programming approach previously discussed. These constructs are in place already and provide a framework for delineating the tasks that the controller is to perform.

Libraries and parameters are the last form of Firmware, and are similar to language constructs, in essence these are completed segments, or sequences of operation. They exist using some specific control language and most importantly, they are developed so that the controls are simpler to apply for the designer/installer. A library is generally a complete control sequence for a given piece of equipment, etc. These are desirable because they save time and because they are proven to work. Parameters are fields that hold data for use with libraries and elsewhere in the system.

At this point the reader might ask, how does Firmware differ from Software. The key here is that software must be easily modified in the field. Remember Firmware can not be changed, or not easily changed. Software on the other hand is designed with the intention that it will be dynamic.

Software resides in the controller. It is important not to confuse controller software with Central PC Interface software. The most visible part of a DDC system that Industry members think of when they hear the word Software is that PC interface.

PC software is discussed in great detail under the Operator Interface chapter. Hear the focus is on Equipment Level and General Purpose controllers. As with Firmware there are several categories of Software including:

- Language Files
- Parameters
- Setpoints and
- Schedules

Language Files may include a wide variety of types of information. For example, a fully programmable device will have a language in firmware that allows sequences to be written. These sequences once written will be stored in the form of files. Bear in mind the sequence may be start with a blank sheet or it may be a modification of an existing Library application. Either way, the file will reside on a Central PC based Engineering tool and will also be downloaded into the controller where it again resides as one or more files. By its nature this software must remain accessible because there is an ongoing need to modify, enhance and fine-tune the control sequences. Though these software files vary based on the programming approach the concept is quite simple.

The information critical to a specific controller or project is held in software and accessible, but it must also be protected. This is where

EEPROMs and other non-volatile memory become important. Language files are not only important to fully programmable and Library program-based version of programmable controllers however. These are also critical to Application Specific controllers.

To refresh the readers memory these are controllers that are designed for a particular application, Air Handler. The Sequence of Operations is embedded in firmware, and the only way to change functionality is to modify one of a limited number of software characteristics. These controllers may offer less programmability but they still have the same requirements for software: easy access and modification. The Software files that apply in this case are Parameters and setpoints.

Parameters are Software based pieces of information that may be modified to customize a controller. Parameters are important for both Programmable and Application Specific Controllers.

With Application Specific devices Parameters are critical pieces of information such as the number of stages of cooling in an Air Handler. The internal sequence within the controller need only have values programmed into a key set of parameters to implement the DDC. With programmable devices Parameters are used in a similar manner, except they are usually created by the designer or programmer. For example, this individual may create a set of parameters for reset of setpoint of a chilled water loop based on return water temperature. The parameters allow the designer, user, etc. to then modify how much the setpoint is modified and the range of return water temperature that will cause the reset to occur. In this way the parameter is dynamic and allows the systems to be fine-tuned to the control application.

Parameters are accessible but are usually protected by Password because it should not be too easy to modify a parameter. These are generally programmed once to customize the controller to the application and then never again. However, a major change in equipment or an upgrade to the system may result in the addition of new points, etc.

Parameters often require a higher level of Interface Password by the user. Further they may only be accessible through a Central PC based Interface, or require a special technique/password from a hand held device. Parameters are very important to the long-term success of a system and the system owners ability to respond to changing environments overtime. There is another type of software information that is quite similar to Parameters, Setpoints.

Setpoints are even more dynamic than Parameters because they may be changed on a day to day basis. Setpoints define the control conditions that are to be maintained for a given application. Setpoints may be temperature, pressure, light level, etc. The key is that these Software data points are expected to change on a regular basis. Therefore changing setpoint software files may be done from a Central Interface, a handheld or even a Zone sensor. Often Zone sensors incorporate a thumbwheel potentiometer or Digital readout to allow the occupant to modify setpoint.

Actually the reset example used above under Parameters could apply equally as well to the setpoint category. However, there is a fine line of complexity that necessitates that the modification of a setpoint does not counteract the control program. For example an unreasonable change to cooling setpoints in a Zone, for example 65 degrees F, could result in inefficiency, higher cost of operations and in general waste. Therefore a feature applied with most setpoints is to have a Parameter that limits the amount of authority an occupant has to change a setpoint. Much like range stops on a Thermostat this ensures that unreasonable changes are not made to setpoints.

Nonetheless setpoints are intended to be changed, and quite simply are available to building occupants. Setpoints are again software files and are held in non-volatile memory so that a loss of power or communication will not interrupt control. Again this is a simple concept, but the setpoint is at the heart of closed loop DDC control. The sequence of operation will continually compare the setpoint to actual conditions, execute control action and reevaluate status to determine appropriate action.

Another use of setpoints for both Equipment and General Purpose Control is Alarming. Setpoints will be established for warning and/or critical alarm. These setpoints will initiate a local notification and perhaps a telephone dial out. Either way it is important for the system managers to have quick access to alarm setpoints.

Schedules are the final type of Controller Software that will be discussed. Scheduling is essential to all DDC control. This is because a fundamental requirement for cost reduction is reducing the number of hours of operations. In addition to reducing energy cost, this also extends the life of equipment because there are fewer run hours. Obviously there should be maintenance savings as well.

Schedules make savings possible and are essential to nearly every type of Equipment and General Purpose Control. Equipment Control in most applications, excluding Hospitals and some specialized facilities, requires control sequences for both Occupied and Unoccupied modes. A Schedule is defined to identify when each mode begins and ends. At the same time setpoints are keyed to those schedules that customize the desired conditions to the time of day. This is equally applicable to a chiller or a unit ventilator.

Scheduling is particularly important for General Purpose control. By definition General Purpose sequences are almost totally directed at scheduling of equipment. Very often this is electrical equipment that can simply be scheduled on and off at a particular time. General Purpose controllers may handle lighting, pumps, motors and other types of equipment that is not necessary 24 hours a day. Another aspect of Scheduling that is used with Alarm setpoints is to program different Alarm conditions by time of day. This ensures that the building manager does not receive nuisance alarms. An example would be alarms regarding low or high temperature during Unoccupied mode when equipment is shut down.

Scheduling is typically integrated with sequences such as occupied and unoccupied modes and time of day based alarms. Another example of integrating scheduling with other sequences is Demand Limiting with time of peak and time of use rates. Across the country electric utility rates vary, but a common practice is to charge more for electricity that is consumed during peak times of the day or season. Scheduling functions are integrated with many demand schedules to identify the beginning and end of such periods and change setpoints accordingly.

As with setpoints Schedules must be readily available to the user because Building use in not always static. This is particularly true of certain types of building such as schools and churches. In these buildings schedules change daily and seasonally. Quick access is therefore essential.

SEQUENCE OF OPERATIONS

The topic of control sequences really defines the equipment and general purpose controller. The author has heard this concept referred to by Doug Franz, of Honeywell, Inc., with a simple analogy. It may be related to a civic auditorium, for example Madison Square Garden. The hardware platform, ancillary equipment, etc. outlined above make it possible for that controller to be applied in any number of applications, much like an auditorium.

The sequence of operations is the specific program that tailors this controller to the application. Correspondingly the marquee and associated interior modifications in seating, etc. within the auditorium tailor it to any given event. This section will briefly touch on the techniques that are used to accomplish that programming.

For the most part DDC products on the market today use the programming approaches and control sequences that were described in Chapter 9. The purpose of this section is to describe how the individual sequences are applied to provide a cohesive control solution for a particular piece of equipment.

By definition the Equipment Level Controller (ELC) has been described as fitting the requirements for larger most complex applications. General Purpose Control (GPC) also involves complexity, but often these are varied and somewhat unique. GPCs may be used for anything from simple start stop to low level process control. As a result the Equipment Level Controller is a better example particularly because of the complex series of sequences that must be coordinated to achieve good control.

It is not feasible, or even desirable, in this chapter to provide a detailed sequence of operations for every piece of equipment that can be controlled by an ELC. Instead the idea is to break down a Sequence of Operations into the many individual sequences that must be coordinated for one sample application.

Covering a sample ELC application will give the reader a clearer understanding of what a typical controller includes. Keep in mind the mission of this book. It is to equip readers from any segment of the

Equipment Level DDC Control

DDC industry with the essential information necessary to ask intelligent question and make informed decisions.

A review of a good sample sequence should provide the reader with necessary insight into the characteristics of this Architecture level. The Variable Air Volume (VAV) Air Handling Unit (AHU) is an excellent example of the complexity that is required at the ELC level and the diversity of control requirements that must be coordinated.

As noted, ELC control sequences will vary dramatically by type of equipment, and VAV represents a broad and diverse example of typical ELC applications. These devices require complex and sophisticated control sequences tailored to the specific applications. The programming approach is not critical to this discussion, rather the focus will be on the key control sequences. In addition, consider that the variable air volume (VAV) air handler must implement control sequences with varying levels of priority.

The VAV control sequences are summarized in Figure 10-5 in that order of priority, and each will be discussed briefly.

It is important to note that all of these sequences have been described in Chapter 9. This chapter will not restate those definitions. Instead it will clarify the function and interaction of these sequences within the ELC.

This brief outline should provide the reader with an understanding of the complex interrelationship that must exist between sequences to achieve the overall intent of the a DDC controller. It should also reinforce the critical aspects of processor execution time, and the need for thorough testing of custom programmed sequences.

**VARIABLE AIR VOLUME AIR HANDLER
SEQUENCE OF OPERATIONS**

Priority Control Sequences
- PID Closed loop control of temperature and pressure (includes occupancy and optimization)
- Alarming

Secondary Control Sequences
- Morning Warm up
- Reset
- Return Fan Control
- Space Pressurization
- After hours override - tenant billing
- Trending
- Safeties and diagnostics
- Preventative Maintenance

Integrated Control Sequences
- Indoor Air Quality
- Fire/Smoke Evacuation
- Demand Limit Control
- System Data

Figure 10-5

PRIORITY CONTROL SEQUENCES

PID Closed loop control of temperature and pressure

VAV equipment ELCs must provide discharge air temperature control by sensing temperature and modulating an outside air damper for ventilation, and possibly free cooling, along with heating and cooling equipment. Often these units are cooling only, but it is not uncommon for air handlers to provide heating as well.

The primary control sequences are PID control sequences for supply air temperature and duct static pressure. Under temperature control there are specific "loops" or sequence components. The other Primary control sequence is Alarming.

The Temperature control sequence has several "control loops" within the overall sequence of operations. This means that the control requires one PID loop to make control decisions on the mode of operation, heat or cool. Two additional PID control loops, one for heat and one for cool, then determine the amount of heating or cooling to provide. The amount is based on staging or modulating of independent capacity, but may also include an outdoor air economizer sequence, as noted.

This fourth loop within the Primary control sequence is actually a set of loops. Ventilation is the first consideration and controls will ensure a minimum percent of air flow is fresh air. The Outdoor Air Ventilation and Economizer sequences are also coordinated to mix return air from the building with fresh air from the outdoors. In larger VAV Air Handlers this set of sequences modulates the outdoor and return air dampers to maintain a mixed air temperature setpoint.

In nearly all VAV Air Handler applications control of outdoor air dampers is integrated with the cooling sequence. This sequence provides a higher percentage of fresh air based on another input, outside air temperature. At outdoor temperatures below 50 degrees, or higher depending on the climate, it may be possible to satisfy the buildings full need for cooling through the economizer. Economizers are generally sized to allow 100 percent fresh air under optimum conditions. Particularly in more humid climates, the economizer may be required to consider enthalpy, rather than temperature alone, to make a decision about modulating the damper to provide free cooling.

Duct pressure control is the second major segment of the Primary control sequence. The intent of this application is to improve efficiency and comfort by varying air flow to the space based upon demand. An increase in demand at the zone causes a damper to open thus changing the duct static pressure. Increase in demand from multiple zones will drop the static pressure below a setpoint and require that VAV air handler fan volume be increased. This increase is accomplished through a PID loop that monitors duct pressure and modulates fan motor speed, vortex dampers, etc. to increase flow.

The Equipment level controller must integrate temperature and pressure control. These sequences are coordinated in ELC logic to ensure that a constant temperature of air is delivered in sufficient volume to meet space conditioning needs.

The final topic under primary control sequences is the requirement for occupied and unoccupied modes along with optimization. These sequence were described earlier and are essential to ensuring building comfort at the lower possible operating cost. Time of day schedules and Optimized Start Stop must be implemented in all buildings that are unoccupied for any portion of the day. There are some exceptions such as computer facilities, but this should be carefully evaluated.

Alarming

One of the decided benefits to equipment level intelligent control is the ability to monitor a variety of alarm conditions. For the most part these become simple control loops that monitor the condition of a contact closure or an analog value. Based upon a change in state of the digital point, or an analog point exceeding a set of range limits, an alarm is initiated.

Though the sequences are simple the function they provide is critical to ensure reliable equipment operation and reduced downtime. The financial value of the VAV air handler is one issue here, but perhaps more important is the activity taking place in the space. Equipment downtime is extremely costly in terms of productivity, opportunity cost and overall reduced effectiveness of a building occupant.

SECONDARY CONTROL SEQUENCES

Morning Warm Up

This control sequence is only applied with VAV air handlers that have heating capacity. Most VAV air handlers are cooling only, however in some buildings the design or the climate are such that heat is installed in the air handler. Since Zone level heating capacity is typically expensive, especially when electric reheat is applied at the VAV box, this strategy was devised.

The greatest building wide demand for heat during the day is in the warm up mode prior to occupancy. As the building temperature may be 10 to 15 degrees below setpoint, a great deal of energy is consumed to warm up the building.

The sequence is quite simple. VAV boxes are sent a signal to cause them to modulate dampers to 100%. The setpoint in the air handler is also modified to a heat setpoint, generally between 80 and 140 degrees Fahrenheit. In this way full heat is driven to the Zone level and the building warms up quickly. After the warm up sequence is complete the VAV air handler will usually revert back to a cooling setpoint. Any additional heat required during the occupied mode will normally be provided by reheat at each individual zone.

Reset

The first general DDC control loop in this section is reset. Reset is common with DDC systems and functions to change the control setpoint, temperature, pressure, etc. based upon a change in related conditions. For example discharge air temperature reset for a VAV air handler is design to consider one of several temperature conditions, i.e.: space temperature, and reset the temperature setpoint.

The intent is to identify a change in the demand due to reduced or increase load and react accordingly. The energy and cost saving impact on VAV air handlers comes from modifying setpoint under mild conditions and avoiding wasted energy. This can further save costs associated with fan horsepower if the volume can be reduced due to a decrease in load. Reset is an extremely effective strategy, and is implemented on both supply air and mixed air sequences in a Variable Air Volume Air Handler.

Return Fan Control

If a return fan is installed in the VAV air handler the DDC controller will provide some form of fan control. The simplest return fan control is to interlock the fan starter with the supply fan and both always run at the same time. There are also more elaborate control sequences that use control logic to track the return fan to the supply fan. With a Variable Frequency Drive to control motor speed, this allows for energy savings in return fan horsepower. It also may be integrated with space pressurization to provide Indoor Air Quality Sequences and enhance comfort. Each of these are discussed further below, and will not be elaborated more in this section.

Space Pressurization

Another control loop that is becoming very common with VAV air handlers is monitoring and controlling space pressure. This sequence has long been common in lab environments where a negative pressure is maintained to avoid flow of contaminated air outside the room.

In VAV space pressurization is primarily focus on cost/energy savings, comfort and Indoor Air Quality (IAQ). IAQ will be discussed further below, for now remember that it is important to maintain tight control on pressure to enhance Indoor Air Quality. Maintaining a slightly negative pressure induces the flow of fresh air into the building. Many problems contribute to IAQ problems, but DDC systems can help with sequences that enhance ventilation. It is important to keep building spaces from being starved for fresh air if sick buildings are to be avoided.

Perhaps the key reason for this sequence is energy and cost saving. By maintaining a continuously equalized pressure it is possible to reduce equipment consumption. Fans are used more efficiently because they do not condition more air than is necessary. Also they do not deliver a volume of air that is unnecessary and ends up being exhausted.

Comfort is the final issue and it relates to space temperature but also to noise. Maintaining neutral space pressure avoids problems like have the doors of a building stay ajar and forcing expensive conditioned air outside due to positive pressure. The basic sequence enhances the building environment by adjusting supply air based upon the desire to maintain a positive or negative pressure. Positive pressure also can result in a great deal of noise around diffusers that may create occupant complaints.

After hours override - tenant billing

Override is an essential feature with most commercial buildings. Spontaneous access to the building by any tenant is necessary, and override ensures that conditioning is provided during these times.

Override or "bypass" must be available through several means to allow spontaneous access to building facilities. It is generally initiated by a push button for the override. This may be initiated by a button, switch or even via dialup with the Central Operator Interface or by touch tone telephone interface to the controller. There is normally a fixed duration of time that the override will last. Also it is possible to monitor and summarize the amount of time that a controller, etc. remains in override. The use of time to log the status of various points and the duration is a valuable tools to expand the capability of this sequence.

Many building owners have started to change the way they write their leases. They include access to the space on a 24 hour basis, but they only provide space conditioning during normal working hours. Occupants who use the space after hours are likely to pay a use charge for the lighting and space conditioning. This process is called after hours tenant billing, and has become quite common with office building owners.

Trending

Trending is not really a sequence, rather it is a method of collecting information that was described in Chapter 9. The importance of trending for applications is to allow for a user to have access to data for analyzing equipment operations. This may be used to fine-tune control sequences or to identify mechanical problems. Trending provides data to do a variety of diagnostics that can indicate actions necessary to increase the overall effectiveness of the VAV Air Handler and the DDC controller.

Safeties and diagnostics

At first glance the reader might think that safeties should be primary sequences. However, safeties are not alarms. Safeties may result in alarms, but in general these are control sequences that are implemented to protect a piece of equipment. With the continuing trend toward Original Equipment Manufacturer (OEM) pre-integration of controls it has become common to include more safety sequences with DDC systems. With a VAV air handler these might include Duct high static pressure alarms, compressor alarms and freeze, or low discharge air temperature, limits.

Again safeties will often result in alarms. Specifically, this subject is covered here by highlighting the existence of these additional sequences within the DDC controller.

Diagnostics are a related set of sequences. One definition of diagnostics discussed in this book is using access to data to identify problems. A user can interrogate a system and diagnose problems by reviewing temperature conditions, alarms, etc. Diagnostic also may be defined regarding the controller. This topic is included here because it refers to problems that may arise with the control hardware. Diagnostics are self test routines that are included with DDC controllers. With these sequences the controller tests itself and can identify a variety of hardware, software and communication issues.

The overall benefit of both safeties and diagnostics is to optimize controller operations and reduce downtime.

Preventative Maintenance

This is another case of non-control related sequences that are implemented by DDC controllers. Preventative Maintenance (PM) is universally proven to increase equipment efficiency, extend equipment life and ensure operation with a minimum of downtime.

The involvement a VAV air handler controller has with PM is in the area of information. DDC controllers can be a primary source of critical data that is necessary for input to a PM program. This data includes; run hours on all key equipment and components, operating data to evaluate the effectiveness of filters, coils, etc. and alarm logs to indicate any particular problems that require immediate attention. These and other data inputs allow for an effective PM program, which in turn contributes to the overall effectiveness of a DDC system.

As the reader is now aware, DDC control for equipment such as the VAV air handler is quite sophisticated. The DDC controller must implement a large number of both simple and complex control sequences. All of these routines must be combined with the primary temperature and pressure loops described above. Beyond implementing these primary and secondary sequences with the equipment directly under control, DDC must also integrate into its sequence of operations external data and control. External data has been discussed as both global and shared data may be used in VAV sequences. Further, Building Wide control may require execution of control sequences requested at the Air Handler level. It is also important for the reader to consider these additional sequences under the topic of Local Level Integrated Sequences.

INTEGRATED CONTROL SEQUENCES

The concept of a Integrated Control Sequence requires a little discussion. Integrated sequences have already been discussed at some length in this book and will be covered further as well. These discussions, however have focused on integrated control sequence between unrelated systems. This type of integration is essential to successful implementation of DDC systems.

There is also a second form of integration that will be called Local Level integration is coordinating control using either information from other controller or outputs on other controllers. Thus achieving a desired action at the local controller level with assistance from other controller on the same functional network. The critical sequences that are important in discussing Local Level Integration are:
- Indoor Air Quality
- Fire/Smoke Evacuation
- Demand Limit Control
- System Data

Indoor Air Quality

Indoor Air Quality sequences have been discussed under both primary and secondary sequences. The key here is note how these sequences are implemented using system wide data and control points.

IAQ tends to focus in two key areas.
1. Control sequences introduce additional fresh air to maintain parts per million of Carbon dioxide (CO^2) .
2. Control sequences that reduce the amount of fresh introduced to the building if undesirable contaminants are detected outside the fresh air intake.

Access to current information on both CO and outdoor contaminants is often provided through the network. For example a controller located near the outdoor air intakes may monitor for carbon monoxide, or other contaminants, and reduce or eliminate fresh air intake if these are detected. At the same time, Zone controllers are typically identified to monitor CO^2 and these will send a signal to the air handler if additional fresh air is required to purge the space and reduce parts per million of CO^2. This is particularly important as the most recent American

Society of Heating Refrigeration and Air Conditioning Engineer (ASHRAE) standard identifies this strategy. Building designers are offered the option to provide a specific volume of CFM per person of outside air, or to maintain space conditions below a specific limit of CO^2 in parts per million.

Again these are integrated sequences because of the coordinated action that is necessary between zone and equipment controllers. Further integration may involve exhaust or return fans and dampers as well.

Fire/Smoke Evacuation

Another integrated set of sequences are fire and smoke evacuation. In the integration chapter these sequences were discussed. With regard to a VAV air handler the smoke pressurization sequence would be executed under a fire condition.

The Smoke Pressurization sequence requires integration with all VAV air handler controllers as well as with the zone controllers. To provide a positive pressure on floors above and below a fire will require overriding the sequences discussed thus far and controlling those floors to a positive pressure. The fire floor on the other hand is controlled to a negative pressure with all zones shut down, and return/exhaust fans at full capacity.

Smoke evacuation is a much simpler sequence. Yet again integrated sequences may be necessary to enable exhaust and return fans on all units. Another approach is to have separate smoke evacuation fans that are enabled if a fire occurs. In this case all VAV air handlers will be shut off to avoid introducing more air or interfering with the smoke evacuation fans.

Both of these types of sequences are becoming common in buildings today. DDC controllers much be able to integrate with these sequences or to enable them with internal logic. This again adds further complexity to the DDC controller.

Demand Limit Control

Again Demand Limit Control (DLC) has been described. It is a sequence that must be integrated with DDC air handlers if shed sequence will involve any of the equipment that they control. For example, the shed sequence may include direct expansion (DX) compressors in the air handler or electric heat coils, which are much less common at the air handler level.

DLC sequences may also raise a cooling setpoint to cause the unit to turn off stages of cooling. In any case, DLC is a Building Wide sequence that requires integrated controllers to turn equipment off and avoid a demand peak.

System Data

This topic of shared and global data has been discussed a number of times already. It is mentioned here again to emphasize that DDC involves sequences that are implemented directly at the piece of equipment, but also involves network interaction. This interaction is valuable, if not essential, to ensure that controllers are able to integrate Building Wide data with local control sequences.

The use of Global data is also a cost effective way to avoid duplicating sensors. Some obvious duplications with many systems are outdoor air temperature and enthalpy.

Control integration at the equipment level is not likely to be limited to interaction with lower level devices. Integration, as noted, also includes a wide variety of other control sequences for Building Wide coordination. These are coordinated at the Building Wide Level and are implemented with Equipment Level Controllers to carry building wide functions such as optimal start/stop.

ORIGINAL EQUIPMENT MANUFACTURER (OEM) INTEGRATION

This is a topic that will not be covered in great detail here because it is a marketing and distribution topic. The focus of this book is to provide readers with an understanding of the technology rather that the marketing of DDC.

OEM integration is an important topic to at least briefly discuss though, because it dramatically impacts flow of these products to the market. In years past, a Building Automation or DDC system was viewed as a completely discrete technology. These systems were always field applied and were never introduced to HVAC until it arrived at the job site.

A new trend has developed over the last decade, and this involves OEMs mounting DDC controls at the factory. The reasons for this are many, but quite simply OEMs viewed this as an effective way to differentiate their products and offer new features and benefits to customers. In the past OEMs typically purchased controls direct from a manufacturer that specialized in that technology. As time progress, customers asked for more and more features that were not related to the equipment itself but to the way that they control and interface to the equipment. This trend led OEMs to believe that it was necessary for them to develop proprietary controls and thus enhance the marketing of their products.

Pre-integration of controls has taken an evolutionary trend. OEMs started by developing controls for larger equipment. At the Zone level however pre-integration of other manufactures products has been common for some time. In particular with VAV terminal units these tend to be devices selected from a field of as many as 15 to 20 different manufactures.

Typically the equipment is ordered from OEM's along with a written request for a particular controller from some representative of the owner. With larger equipment however OEMs have taken a variety of approaches. As noted, some OEMs have developed their own controls and mounted them on all equipment whether the customer orders DDC or not. Other manufacturers have selected a control vendor and standardized on a specific set of control products.

The significance of this discussion to the reader comes in the need to ensure that control products meet their needs. Ultimately the owner needs to secure assistance from a consultant, etc. to ensure that all controllers arriving on the job meet the following conditions at a minimum.

1. They must provide an acceptable sequence of operations for the owners equipment and
2. Devices shipped by the OEM must be compatible with any other purchased or existing DDC products.

Compatibility means at a minimum that integrated control sequences are possible, that they may be monitored and programmed by the same software tools. Unfortunately these conditions can not be met by most systems. Combining systems often eliminated one or both of these capabilities.

Further discussion of this trend is not necessary. The reader must simply be aware that the trend is underway, and take whatever action is necessary to ensure effective DDC systems are installed.

NETWORKING

This is the final topic under current Equipment Level Controls, and has already been positioned as critical to building wide control. A complete chapter is dedicated to networking. It is essential to implement control with distributed controllers because of their inherent need for data from other controllers in the building.

The requirement for networking with Equipment Level Control, as with other DDC devices, is to provide data flow between controller and allow for remote communication. All of these types of network issues were discussed earlier in this chapter, and will be covered further later in the book. This section however will briefly note some key networking issues that relate to ELCs specifically.

The functions discussed under the Building Wide Controller and Operator Interface are essential for programming, monitoring and remote communication. These devices must have access to the ELC. Therefore ELCs must have complete communication capability. In spite of that statement however, it bears noting that the devices at this level may not require the same level of communication available at the Building Wide Controller. However, do not loose site of the important trend toward Peer to Peer controllers which have communication capabilities similar to a Building Wide Controller.

The essential requirement for an Equipment Level Controller is to be able to support network communication. The nature of Distributed DDC requires that these devices are networked for two key reasons. ELCs must be able to communicate for monitoring, programming and commissioning. Networking ELCs is also critical for access to shared data and for implementation of integrated control sequences such as those described above. Again the concept of networking will be covered in more detail later in this book.

INSTALLATION AND COMMISSIONING

These are two complex topics and yet it is not possible to cover them in great detail here. Limited coverage is due to the focus of this book on understanding DDC technology. Installation is typically done by an Electrical Contractor. Actually connection of the hardware is essential to DDC operation, but it not extremely complex. The controllers require low voltage power and networks are generally simple twisted pair installations.

Installation of these systems should be contracted to an experience contractor, and all manufacturer guidelines must be followed. Though these installation were said to be simple, there are a number of hazards to the equipment that can be caused by installation mistakes. For example, the microprocessors in these devices are susceptible to electrical noise when wires are run together with line voltage, near lighting ballast's or in proximity to microwave transmitters.

An experienced contractor will look out for all the typical DDC installation problem areas, and ensure quality completion of projects. Some other important side notes for the first time systems buyer are to verify whether shielded wire is required, and to ensure that an effective commissioning phase is completed. Beyond that installation is an extremely straight forward process for the experienced contractor.

Commissioning may be one of the most important initial phases of system start up. The chapter on Ensuring DDC Success covers many important techniques applied during commissioning, and to manage the system after installation, in some detail. The focus here will be on providing a simple overview of the process.

The focus here is to identify only the key steps that are required to successfully "Commission" a system. The simple fact here is that DDC systems are extremely intelligent, but like all computers they only execute commands as programmed. A process, commissioning, is very simply a double check to make sure that ever ancillary device, controller and program sequence is implemented as designed. Simple mistakes are the major enemy of any complex system, DDC is no exception. Therefore this section will briefly identify the key tasks necessary to commission a DDC system:
- verify hardware installation,
- simulate local control conditions to verify sequences and
- simulate network control conditions to verify integrated sequences.

Hardware installation is of key importance because of the large number of devices, controllers, wiring terminations and potential sources of interference. In this phase both installation and operation of every input and output device, control module, firmware and software revision is verified. Each device is checked according to its functions. For example temperature sensors are checked and calibrated using external sensing devices, and outputs are forced to a specific condition, observed and verified to operate as intended. In the VAV example this would ensure that the correct temperature sensors, valve and damper operators, etc. were installed and operational.

At its most basic level Direct Digital Control is implemented to control a piece of equipment. This requires local control sequences such as those describe throughout this chapter. The second phase of commissioning is intended to verify that these sequences operate as intended. Much like the hardware phase this done through simulation and observation.

The process is to simulate conditions that determine if a primary or secondary sequence responds appropriately. For example a unoccupied to occupied start up would be initiated to observe the morning warm up sequence. This can be complex because of the overlap between many control loops and the need to ensure that the right response occurs to various control sequences. This process can be time consuming and also can be impact by the time of year. It is difficult to test cooling loops in January. However, it is essential to the long effective operation of DDC. One final note is that the process is somewhat simplified for application specific controllers and those that use proven library control loops. However, this remains a critical component of systems commissioning.

The final stage of commissioning is integrated sequences. The process of using simulation and observation is exactly the same, but in this case the power of the network must be applied. Simulation requires access to data points from elsewhere in the system. Further it may be necessary to observe control points connected to other controllers for verification of control. Again part of the power of DDC is in networking and integration, yet it does add complexity to the checkout and commissioning process.

EQUIPMENT LEVEL CONTROL TRENDS

There are three important categories of trends with Distributed Equipment Level Controllers (ELCs):
- Communication Trends,
- Control Integration and
- Control Trends

COMMUNICATION TRENDS

With Communication, as mentioned under the Island Host, standard networking. is the critical trend. Again this book contains an entire chapter on Networking and this chapter will not reiterate that data. However important trends will be noted with a brief explanation of the significance for ELCs.

With ELCs the focus must be on controller to controller level communication standards. Standard communication for DDC Local Area Control Networks can be implemented at several levels. A common implementation is at the Island Host or Building Wide Controller Level. When that is the case, the ELC remain essentially unchanged from earlier generation DDC systems and communicate up to the Building Wide Controller and down to the Zone level with a common protocol. Another option is to standardize on communication at the ELC level and offer the option of communicating with Building Wide Controller level systems that are providing various functions such as Fire, Security, etc.

Though the standards level varies, the ELC may still provide communication through either Master/Slave or Peer-to-Peer controller implementations. Though the controller implementation is not entirely critical to discussion of the ELC standard communication, it is most likely that Peer type controllers will be necessary to implement a controller to controller standard. This is not a pre-requisite, but it would simplify the process by ensuring that the ELC has enough power to support a higher level protocol.

With or without communication standards, it is important is that ELCs must have fast, reliable communication access to all information on the network to implement necessary control strategies. Ultimately this is the essential requirement for long term effectiveness with DDC. This is a requirement for a number of other ELC network issues. Hand held and Laptop PC based interface at the zone level is essential today.

These tools are used for field commissioning, and also for diagnostics and trouble shooting after the installation is complete.

The current trend is to be able to panel mount or hand carry an operator interface for these tasks. With both hand held and Laptop based interfaces the future clearly is to be able to plug into any ELC and access that device or any other controller on the bus. Further, this will include software tools that allow diagnostics and fine-tuning control loops to actual equipment through analysis of live system data.

Another important communication trend with ELCs is the ability to dial in and out directly from an individual controller architecture. This is particularly important in ELC applications where a controller is installed stand alone, but there is a need for dial in and dial out on alarm. An example of this might be a chiller or boiler plant.

The trend with ELC communication trend is also toward high speed communication Local Area Controller Networks. These networks will be capable of 1 Megabit per second and higher. Also important from an Architectural perspective, it is just beginning to be common for a variety of media to be used in wiring these networks. The most common network medium for the past several years has been standard twisted pair wiring. A new media implementation that is again just beginning is the use of Fiber Optics with DDC systems.

ELC Control Integration

The reasons for control integration have already be cited several times in this text. At the ELC level this is an important topic because this is the level where much of the complex control occurs. This means it is also the focal point for a great deal of control integration. Numerous examples of this have been provided thus far, but this integration requirement involves both higher and lower level controllers. Zone reset has been noted several times and is a good example of ELC integration with lower level controllers. Space Pressurization sequences for fire control and coordination with other functional system would be a higher level controller integration example.

With communication standards there will be a continual evolution toward more complex ELC control sequences. In spite of this trend, it bears noting that there are a number of reasons that ELC level standards are slow in coming. There are some key concerns with control integration, especially if more than one manufacturer's device is involved.

It becomes extremely difficult to ensure overall system integration of such functions as demand limiting because every manufacturer takes a different approach. Some manufacturers override the distributed control and force a load off, while others embed smart strategies at the controller that have the option of implementation or responding with a "no shed" message. In the short term, a communication standard for the control industry is likely to slowly implement networking and control integration starting with the Building Wide Controller. ELC integration will be implemented first with equipment specific sequences like reset and eventually with all practical sequences.

Control integration is something of a hybrid topic between communication and control. In fact many of the trends that contribute to more sophisticated control will require higher level communication and preferably a standard industry protocol. Some of these key trends represent significant technological and market strides.

ECL Control Trends

Application specific ELC's have been discussed already and will continue to common with a wide variety equipment. Expanded I/O capacity will also be possible with application specific controller to increase the flexibility of the controller. Another related trend is highly interactive approaches to apply standard library programs with free programmable ELCs. Ultimately the benefit here is in reduced engineering time to make the technology viable in more buildings.

Implementation of multiple control sequences and control integration are hardly future trends as they are happening today. However these approaches will continue and lead to much more sophisticated interaction between controller for such routines as smoke pressurization during a fire.

Continued development of more sophisticated control sequences can also be expected. Fuzzy logic type approaches sometimes called Adaptive PID are an important trend. The concept with these sequences is for them to be self learning and tuning to continually adapt to changing building conditions. This will result in great energy and cost savings, as well as improved comfort and effectiveness of the process.

Ease of use is a wide ranging concept. Future Technology's[6] new survey says that factors contributing to a competitive edge include:
- Ease of Use in general,
- Ease of Installation and
- Easy to understand, maintain and upgrade.

The author firmly believes that users in our industry is in a difficult position with this issues. Users have an impression of what systems, microprocessor based devices and software can do that is based on the PC industry. What is possible with a piece of PC software that may have 2 million users is not always possible with a Control software package that may have 50 to 100 thousand users over it's life. Yet the control industry must meet this expectation, and not price itself out of the market in the process. The focus must be on key areas like programming. It is essential to make systems easier to program using graphic, object oriented based approach rather than the more cumbersome techniques of the past.

There are a host of important trends with controller delivery such as more OEM mounting at the ELC level. In fact many manufacturers of equipment will charge more to ship it without DDC controllers. Along with this is a trend toward lower cost controllers, but the user must be careful to evaluate more than the controller cost. Particularly with OEM shipped controllers it is critical to ensure that integrating those controllers into a complete automation system does not involve cost prohibitive programming and network hardware.

From a pure hardware point of view, ELCs are becoming more generic. The power of these devices is in the software and the control theory knowledge base of the manufacturer and installer.

There are some important hardware trends to note before closing the discussion of ELCs. Non-volatile memory storage including EEPROM, and more recently Flash memory is important to avoid program memory loss on a power failure. Faster processors with more power for control sequences and communication has been noted several times in this discussion, and will continue. Modularity in components is also essential to avoid having unnecessary hardware cost in points, etc., that are not needed for the job.

Expanded compatibility for worldwide use is also important. This means controllers that operate at 50 or 60 HZ, and are able to use a variety of engineering units including metrics. The wireless communication trend discussed earlier will also impose some hardware modifications on the ELC.

This chapter has been a detailed review of Equipment and General Purpose Level Controller technology. There are a number of topics that have been covered here to give the reader a better understanding of control technology as well. Particularly given the trend toward Peer to Peer controllers networks, this level of architecture is critical to understand. Coordinated activity here is likely to be required to accomplish many of the Building Wide sequences discussed in Chapter 9 along with the Equipment Level sequences outlined here. In any case this chapter should have provided the reader with a much more comprehensive understanding of DDC control. The final key component in a Distributed DDC architecture is the Zone control. With an understanding of ELC functionality it is now possible to explore Zone control further and round out our discussion of controllers.

FOOTNOTES

1. Air Conditioning, Heating and Refrigeration News, February 15, 1993, page 24.
2. Microsoft MS DOS is a registered Trademark of Microsoft Corporation, Redmond, Wa.
3. Novell Netware is a registered Trademark of Novell, Inc., Salt Lake City, Utah.
4. LONWORKS is a registered Trademark of Echelon, Palo Alto, Ca.
5. Microsoft Windows is a registered Trademark of Microsoft Corporation, Redmond, Wa.
6. Future Technology Surveys, Survey # 208, Lilburn, Ga.

Chapter 11

DDC Zone Control

This section covers one of the fastest growing component technologies in the DDC industry. Historically Zone control was handled by either pneumatic or electro-mechanical controls. This could include anything from a mercury bulb thermostat to a simple pneumatic thermostat providing a change in pressure to a zone pneumatic actuator or receiver controller. These controls accomplished the task, but had a number of drawbacks. Some of those included the use proportional control resulting in marginal comfort, the need for regular calibration and lack of access to information about zone conditions.

With CPU based systems Zone control remained essentially the same. The cost of processing power, control point and labor to run wire long distances to central panels made zone control cost prohibitive. A number of approaches were implemented but we will only note two, those used in large CPU based building and those for Load Controller type applications.

The most common Zone control approach in CPU based systems was to leave the local pneumatic stat in place and control the equipment only. The interface is still common, electronic to pneumatic transducer are applied at the equipment level. When the air handler etc. were operational, the Zone controller could get conditioning. Otherwise the pneumatic stat, etc. would generally modulate to whatever position would normally meet the setpoint. For example a VAV box control would modulate full open, but since no air was in the duct it would simply remain in that position until the next occupied mode. At the beginning of occupied mode when the air handler was started, these devices would then reach there desired setpoint.

Zone control with load controllers was much simpler. These buildings usually had a limited number of zones and the predominant piece of HVAC equipment was a packaged roof top unit. The technique use was to interface a relay on the "R" leg of the thermostat between it and the unit. This was often called "breaking R". The relay was controlled by a Digital Output from the load controller and was enabled and disabled based on time of day. This technique was very effective because it achieved the optimum savings possible from equipment shut down. Of course an additional loop had to be implemented to allow units to run during unoccupied heating mode. This would happen if the temperature in the space dropped below the unoccupied heating setpoint.

With DDC Systems there has already been discussion of the technique. An intelligent controller is interfaced at each zone to provide full control with optimum comfort, scheduling and information. Bear in mind that there are still many installations today where the approaches described above are used. These are still cost effective techniques in many buildings. However, the focus of this chapter will be Zone Control in a complete Distributed DDC System.

The topics to be covered for this Distributed Zone DDC discussion include:
- Introduction to Zone Level DDC
- Zone Level Hardware, Ancillary Components and OEM Integration
- Firmware and Software
- Sequence of Operation
- Networking
- Installation and Commissioning and
- Zone Level Control Trends

INTRODUCTION TO ZONE LEVEL DDC

In an attempt to avoid confusion it is important to start this chapter defining the two central topics that will be covered: Zone Applications and Zone Controller.

Zone Applications are best typified by a Variable Air Volume (VAV) terminal unit or "box". This is a self contained piece of equipment with the ability to control air flow into the space and also to possibly add heat as well. We will discuss the device sequences more later in this chapter. The reason for starting with this concept here is to highlight some key characteristics of Zone Control applications.

1. Zone applications involve small, low cost, and relatively simple equipment such as VAV boxes and unit ventilators.
2. These applications are highly interdependent with other building controllers, consider the relationship between a VAV box and air handler.
3. As there is a somewhat loose definition of a "Zone" application a wide variety of equipment is included in this category such as packaged roof top unit control. This further adds complexity to Zone control sequences.

Zone Controllers are obviously the topic of this chapter, but it is important to clarify the nature and architecture of these devices. The Intelligent Buildings Institute in Washington, D.C., USA did a good deal of research to clarify the role of control equipment within an overall system. The results of their work that have been covered in chapter 8, and elsewhere in this book, separate controllers by function.

The reader will remember the Building Wide or Host level controller that has been discussed at length. Below that level in the architecture was the level of distributed controllers. The author of this text has presented in technical papers and books, an enhanced view of that architecture. This view accepts that there may be one level of intelligent controllers for architecture purposes, but it is not possible to have one level of hardware. The industry obviously shares this view because nearly every manufacturer of control systems has one hardware platform for equipment and a second hardware platform for zones.

The characteristics of Zone Level Direct Digital Controllers (ZLCs) are the focus of this chapter. In this introduction it is simply essential to note that these devices utilize all of the same technology described under the introduction to Equipment Level DDC. These are microprocessor based devices with extensive data processing capabilities. Zone controller incorporate all the same RAM and ROM characteristics for A to D conversion, operating system and sequence memory as the ELC. The result is full Distributed intelligent control at the Zone level. However, there are a number of key distinctions in the way that this technology is applied at the zone level. These distinctions and a thorough discussions of Zone Level Controllers (ZLCs) are the focus of this chapter.

ZLC HARDWARE, ANCILLARY COMPONENTS AND OEM INTEGRATION

Zone Level Hardware introduces new features and capabilities to the zone through the availability of distributed processing. The cost of microprocessors, and the associated electronic technology, has allowed for significant inroads to be made in applying smaller more powerful microprocessors with distributed Zone Level Control (ZLC). ZLCs are critical to the evolution of distributed processing. As noted, the distributed concept is to provide control that is tailored to a specific application, and is located as close as possible to the controlled equipment. ZLCs fit that definition in every detail.

The nature of zone products imposes a number of requirements on any controller which is applied to this equipment. As with other control devices processing power for Execution Time and Scan Rates is still critical for optimum control. In addition, Input/output (I/O) capability is also critical particularly due to the impact this can have on cost. I/O must be considered carefully with these devices as they may have fewer inputs and outputs than ELC's. This being said it is essential not to short change I/O capacity and to also ensure that there is some expansion capability.

Zone Level hardware must be well suited to a variety of applications, such as Variable Air Volume Air Handlers and Water Source Heat Pumps. At the same time these devices also must be cost effective for those applications. As with ELCs it is important to consider controllers, ancillary devices and relays, etc. for end equipment control. The discussion of these issues will not be repeated from ELCs, but there are some specific ZLC issues that are important. These will be addressed through this chapter as appropriate to this level of control. In addition there are several hardware related areas that are equally as important. Among these are: component packaging, physical dimensions and ambient considerations.

Input/Output (I/O) requirements are generally less stringent than with the ELC due to the limited number of points required with zone applications. Overall there will typically be fewer points, however analog and digital inputs and outputs will be necessary. This is because most zone applications will require both monitoring and equipment control. Control is accomplished through digital and analog or pulsed modulating outputs for such equipment as dampers and valves. A common trend with the ZLC, also noted at the Equipment Level, is that manufacturers will typically develop one hardware platform. This piece of hardware will then be used with all zone applications. This is accomplished by changing software in the controller. Software issues related to this approach will be discussed further under sequence of operations.

With specific regard to inputs and outputs (I/O), an important zone hardware design issue is the available number and type of I/O channels or points. Designers must first determine the number of points that is appropriate for monitoring and control. Consideration must also be given to whether a fixed number of I/O will be available, or if expansion is to be provided. Expansion I/O may be critical if multiple applications are to be accommodated by one platform. Expansion is usually provided through boards that may be added to the base device or via a module that is applied external to the base device. There will be trade-offs associated with any I/O expansion scheme.

Ancillary Components are particularly important with ZLC hardware design. At the ELC there was a discussion of flexibility with these devices, but at the Zone level this is generally not possible. A high degree of flexibility adds overhead in the controller in both memory and hardware required. Therefore at the ZLC it is most common to require a full line of compatible products including the: controller, sensor or thermostat, static pressure measurement device and damper or valve motors.

These devices must be optional and fully compatible with each other and the ZLC regardless of application software.

Again, as with the ELC there is a concern regarding manufacturer flexibility in selection of these ancillary components. Typically as noted however, this flexibility is limited or non-existent with the exception of some key components. For example, VAV terminal units may require standard interface to pneumatic output devices, while a fixed set of electronic sensors may be perfectly acceptable. These devices still use industry standard signals such as 0 to 20 Milliamps, however usually it is not possible to select any off the shelf device. This does limit flexibility, but it offer some advantages by making these systems easier to specify, order and program because of the combined technology approach to the offering. The combined technology approach will also be referred to here as "self contained", meaning that all necessary components for Zone Level Control are available in a single offering. As a result of this approach to ancillary components further discussion of these components themselves is not necessary. The focus through the balance of this chapter will be to discuss Ancillary Components in the context of the "Self Contained ZLC" offering or package.

The costs of the complete controller offering including ancillary components must be carefully evaluated. This is because Zone Applications are extremely cost sensitive, in fact there are cases when the zone control products cost more than the HVAC equipment. The outcome of these decisions will also impact the next hardware issue, component packaging.

In the discussion above the concept of combined technology was discussed. Component packaging expands on this concept, and is important because Zone control products are self contained in nature. Unlike the ELC, many zone control applications like VAV are applied by Original Equipment Manufacturers (OEMs) in the factory. Each manufacturer will offer a variety of control products with their HVAC equipment. There are actually two distinct issues that are involved here:

- packaging as it relates to easy application of zone products and
- packaging as it relates to physical design.

Easy application is an important facet of packaging because the zone HVAC equipment is also self contained in nature. In addition, zone

OEM's generally install or "mount" a large percentage of the controls for this equipment at the factory. To date about 50% of the 400,000 VAV boxes shipped each year are fitted with ZLCs. As a result, the VAV box remains the most viable zone control device applied followed closely by the Packaged Roof Top Unit.

In general application of these controls is also growing rapidly with all ZLC control applications including: water source heat pumps, small self contained single zone packaged units, unit ventilators and fan coil units. Using VAV as a component packaging example, there is a growing demand to provide one easily installed package containing: the controller, sensor or thermostat, duct static pressure measurement device and output devices for dampers and valves. This is the combined technology approach noted above, and OEMs like it because they prefer to simplify their interaction with the control.

Unlike the ELCs applied with larger equipment, the ZLCs have tended to be highly discretionary sales that are requested by about half the users. This requires that the OEM be able to install and verify operation of a variety of manufacturer's ZLCs on any device. The concept of component packaging is targeted at satisfying the customer needs, of both the OEM and the End User, by offering a fully packaged control product. This package as touched on above incorporates all the components in a tested hardware product. That means the controller and all ancillary components are shipped together, often in a self contained unit.

Self Contained ZLC packages simplify processes in the field for commissioning, support and for component replacement now or at the end of its useful life. Given the wide range of control products OEMs apply this is also an important option to reduce the installation time at the factory. These issues indicate the need for careful evaluation of all ZLCs to understand component packaging issues with specific applications in the installation, design and long term management of DDC systems.

The Physical Design aspect of packaging relates to the concept of ensuring that a control meets the space constraints of the application. This is actually a hardware design and selection issue. In fact space constraints are always a concern with any control product, but added mention with the ZLC is important due to the very limited space available.

VAV boxes, for example will vary greatly in size based upon the zone CFM requirement and the type of box. A pressure dependent box with no fan or zone reheat and low CFM requirements affords very little space for the ZLC. Yet both OEMs and building owners will require that the same ZLC be applied for all size boxes, though each for different reasons. The OEM is concerned with standardizing all facets of production including: installation, manufacturing procedures and test/checkout for the ZLC. While the building owner is concerned with interoperability and system control integration among all VAV boxes. Further, as discussed above, VAV boxes are not the only application that will be controlled by a ZLC. As a result, use of any ZLC will be dictated to some degree by its ability to meet space requirements for any intended application.

Ambient consideration is another important hardware consideration. Again this is a critical hardware design and selection issue, as is physical dimension. Key to the discussion of ZLC products is that the variety of applications for these devices requires that they meet a range of operating conditions. In nearly all cases, ZLCs which are mounted with VAV boxes need only operate under standard interior ambient temperature conditions.

Single zone packaged units, which may be fitted with the same ZLC hardware platform, are likely to be exposed to exterior ambient temperature and humidity conditions and must be able to operate in either application. As may be obvious this dramatically increases the requirements, and perhaps the cost, for these devices due to the potential for application in Alaska or Florida. Also the control business continues to be more global, so ambient conditions must be considered in the same manner for Malaysia, Northern Canada or anywhere in the world. Therefore a ZLC could conceivably be applied at the Arctic or the equator. These are valid issues for consideration by any designer or end user selecting a DDC system.

As the reader is now aware there are some very specific requirements and characteristics for both the ZLC and it's compatible Ancillary Components. Given those characteristics it is highly unlikely that the larger more complex ELC panels could not be applied effectively in this market. Further cost has been noted as a critical issue for the ZLC and an ELC would be cost prohibitive on a VAV box. As a result the ZLC hardware platform is typically a completely different hardware platform. This allows the ZLC to be tailored to both the application and the controller requirements for zone level equipment.

FIRMWARE AND SOFTWARE

The topics of Firmware and Software were covered in extensive detail under the ELC. Though we have clarified that the requirements for ZLCs eliminates the use of higher level controller hardware, the software details are nearly identical.

The ZLC must support Digital Operating System, A to D and D to A Conversion and Communication. All of these are critical requirements and vary little in their implementation from the ELC. There is no change at all to the DOS and in fact that is seen as a standardization benefit.

Analog and Digital conversion varies somewhat because the Ancillary Components are generally fixed as discussed above. This means that a limited number of fixed conversion parameters are embedded in software and the programmer simply selects the appropriate component. In some cases certain channels will be completely fixed and the software will only accept a specific component, such as a space temperature sensor, on a particular channel.

Communication and Networking requirements are essentially the same except the ZLCs are much more reliant on the rest of the system. As a result this process is directed to higher levels of the architecture for information, commands and requests for data. As a result direct modem communication to the ZLC is not appropriate or useful. As with the ELC however, the integration of DOS and Network Communication holds great potential benefit.

At the Zone Level nearly all controller sequences are in Libraries. There are free programmable ZLCs on the market, but these have a major drawback. Programmable ZLCs are cost prohibitive due to the engineering, programming, testing and additional commissioning time required. As has been discussed, programming with a "blank slate" approach is time consuming and requires extensive verification in the field.

The high number of Zone Controllers on any job makes custom programming nearly unacceptable. There are still those who engage in this approach, and there are also hybrid devices that provide both proven libraries and a limited amount of memory for custom programs. The key here is to carefully evaluate the cost associated with these approaches and verify that it is appropriate on a specific job.

The Library and Parameter approach is used extensively with ZLCs to address the issues noted above. It is not necessary here to restate the components of Control Software. The same types of files exist, though in some cases they are simplified due to the nature of a ZLC. The typical ZLC in fact is basically like the Application Specific approach to the ELC. This approach is focused on simplifying ZLC installation and commissioning, as well as long term management. A uniformity at the ZLC level makes diagnostics and troubleshooting much simpler when the system is complete and a problem develops. With this clarification of ZLC Firmware and Software it is now possible to move to the heart of the controller, its Sequence of Operations.

SEQUENCE OF OPERATIONS

As with the ELC, the topic of Sequences truly defines these devices and allows the reader to more fully understand the equipment. These devices are small, and self contained, yet much more powerful than any Zone controller applied in the past.

In fact those characteristics can be both major benefits and a source of apprehensive as well. The key thing to remember with a ZLC is that it is performing all the same sequences as a familiar thermostat. However at the same time it is integrating a number of functions that used to be carried out by other devices, or were not available at all. These new sequences and features can be extremely powerful once the user understands the product. As a result this discussion of sequences will incorporate a broad range of operations that are found in the most sophisticated Zone Level Controllers.

To set the stage for discussing high powered ZLCs it is important to note that many who interface with these devices are caught off guard initially. One of the most frustrating aspects of DDC for new users, both ELC and ZLC, is looking at a box and trying to understand what is happening inside. Many HVAC mechanics and facility operators are accustomed to being able to use a meter for measuring current, etc., some

DDC Zone Control

jumpers and a few tools to troubleshoot controls. With all the sequences embedded in a microprocessor, it becomes confusing to perform the same task. With ZLCs there are new tools however, the hand held interface and the Laptop PC. These tools can be much more useful that those above, but the meter can still be a valuable tool. These are important tools and their use will be discussed somewhat here as appropriate with ZLC Sequences, and also throughout this book.

Moving on more specifically to control sequences, this topic varies dramatically by type of equipment. This common fact is true of any HVAC or load control application, yet there are added distinctions between the ELC and ZLC. At first impression, a reasonable case could be made that ELCs must maintain more complex and sophisticated control sequences than would be encountered in zone applications. In many cases this may be true, yet the zone device must be able to meet the requirements of all applicable equipment. This chapter will again use Variable Air Volume (VAV) as the example application. In this case sequences for the VAV box will be the example.

VARIABLE AIR VOLUME TERMINAL UNIT (BOX) SEQUENCE OF OPERATIONS

Priority Control Sequences
- PID Closed loop control of temperature and pressure (includes occupancy and optimization)
- Alarming

Secondary Control Sequences
- Morning Warm up
- Space Pressurization
- Trending
- Safeties and diagnostics

Integrated Control Sequences
- After hours override - tenant billing
- Indoor Air Quality
- Demand Limit Control

Figure 11-1

There are a number of control implementations with VAV at the Zone Level. The simplest VAV box is pressure dependent, and additional features are added to move toward more sophisticated approaches such as pressure independent dual duct applications. As an extreme case the pressure independent system will be considered, but for simplicity sake the sequences here will focus on single duct. Note that the sequences outlined are only an approximation and used for example rather than as a direct control application.

As with the ELC these sequences not be restated, as in Chapter 9, rather they will be clarified for the Zone Level. It is important to gain an understanding of the complex interrelationship that must exist between sequences to achieve the overall intent of a DDC controller. It is further important here to note the coordination that is required between the ZLC on the VAV box and the ELC on the Air Handler.

PRIORITY CONTROL SEQUENCES

PID Closed loop control of temperature

To set the stage for all of these control sequences it is important to first define the example application more clearly. Single duct pressure independent systems also vary in implementation. For example they may include fan assisted systems, and may also provide reheat. Consider one of the more common systems: a single duct application with fan assist and 1 stage of electric reheat. Assume that control is provided by a dedicated ZLC, also typical.

As with the Air Handler, this system would require primary control loops for temperature control and static pressure control. Temperature control in this case however, is space temperature rather than Discharge Air Temperature. The Space Temperature control sequence has several "control loops" within the overall sequence of operations.

The first loop discussed under the Air Handler was mode of operation. Unlike the Air Handler the sequence does not switch between heat and cool modes, rather it controls to a specific Zone Setpoint. a cooling only box has no ability to provide heat so is limited in actions to maintain the space setpoint. The controller can modulate a damper to increase or decrease the flow of conditioned air into the space. As you

DDC Zone Control

know the air handler usually provides that air at 55 degrees F. The only other control action available is to enable or disable electric or hot water reheat. In this example there is one stage of electric reheat, and so the box will have a simple heat sequence based on space temperature and setpoint.

Typically the ZLC temperature setpoint is adjustable from the space, but with DDC systems there are generally setpoint limits in software. These limits work like range stops on conventional thermostats to limit the range of settings. Maintaining setpoint means that the ZLC requires one PID loop to make control decisions on whether the air coming down the duct can meet the space needs or the reheat stage must be enabled. This is done based upon the PID control loop for space temperature control. The amount of action taken based on this loop is determined by the independent capacity of the box.

There is no independent outdoor air economizer sequence because air reaching the box is a blend of return and fresh air form the air handler. Ventilation requirements are satisfied in this way, but there is an opportunity to increase fresh air based on the Indoor Air Quality will be discussed later in this section.

PID Closed loop control of pressure

Operating concurrently with the temperature control is a static pressure control. The intent of this application is to improve both efficiency and comfort in the building. This is done by varying air flow to the space based upon demand. This is a "Pressure Independent" box which means that a constant pressure of air is delivered regardless of the air handler.

ZLC static pressure sequences allow for variations in cubic feet per minute (CFM) delivered to the box by measuring pressure at the box and modulating the damper according. The intent of this pressure control is to ensure that the space does not suffer from noise or discomfort due to variations in delivered CFM from the air handler. An increase in demand at the zone causes a damper to open thus changing the duct static pressure. Increase in demand from multiple zones will drop the static pressure below a setpoint and require that VAV air handler fan volume be increased as discussed in Chapter 10. This sequence controls based upon data from a duct static pressure measuring device, and will be integrated with other ZLC sequences to respond to CFM variations.

The Zone level controller must integrate temperature and pressure control. These sequences are coordinated in ZLC logic to ensure that the setpoint is maintained. This is done by delivering air in sufficient volume and temperature to meet space conditioning needs.

The final topic under primary control sequences is the requirement for occupied and unoccupied modes along with optimization. With a fully distributed DDC system, the space temperature and pressure control sequences provide occupied and unoccupied operating modes. The unoccupied mode sequence for our example would most likely compare space temperature to setpoint and take action accordingly.

VAV box ZLCs generally allow space temperature to float during the unoccupied cooling mode. They will also provide a sequence to bring on reheat at an unoccupied heating setpoint. With the fan powered box in our example the sequence would enable the fan and the stage of reheat. The fan would draw air from the plenum above the ceiling and the air handler would not be started. The only other unoccupied sequence required would be "override" which is discussed later in this section.

Time of day (TOD) schedules and Optimized Start Stop (OSS) for VAV boxes are implemented in all buildings that are unoccupied for any portion of the day. There are some exceptions such as computer facilities, but this should be carefully evaluated. These sequences are very much integrated with the Air Handler, but they are part of primary ZLC control so they will be discussed here.

The current trend with VAV boxes is to identify all boxes associated with a particular Air Handler and to coordinate both TOD and OSS. Individual TOD for each Zone is possible, both the Air Handler must be started to condition any Zone so the benefit is limited. There has been a move for some time toward floor by floor VAV air handlers to make allow for more coordinated control. Due to a smaller area served and fewer boxes it is easier to coordinate, and this helps with occupied start. In any case the ZLC will share zone temperature data with the air handler and based upon a schedule for the Zone will initiate occupied mode. The Air handler will usually calculate OSS based on a representative number of zones.

Alarming

One of the benefits discussed for DDC is general is the ability to monitor a variety of alarm conditions. As with the ELC these are simple control loops that monitor the condition of a contact closure or an analog value. However at the Zone Level there are many fewer conditions to monitor.

The primary area of concern is space temperature alarm limits. Another alarm that is becoming common is Indoor Air Quality at the Zone discussed below. As the control application is extremely simple there is generally no need for any type of equipment alarming. The key point is that comfort is critical and therefore space temperature alarms are extremely important to the building owner.

SECONDARY CONTROL SEQUENCES

Morning Warm Up

This control sequence is only applied with VAV air handlers that have heating capacity as described in Chapter 10. The ZLC Warm Up control sequence is quite simple. The VAV air handlers will typically send a message to all Zone Controllers commanding them to modulate to 100% open. Generally the sequences commands boxes to stay full open until an occupied heating space temperature setpoint is achieved. As noted in Chapter 10, Zone level heating capacity is typically expensive, especially when electric reheat is applied at the VAV box. This strategy is used to improve the efficiency of Zone warm up and avoid prohibitive energy costs for heating.

After the warm up sequence is complete the VAV air handler will usually revert back to a cooling setpoint. The ZLC will also revert to a cooling control loop, and satisfy any call for heat in the Zone through reheat.

Space Pressurization

The space pressure control sequence that was described with VAV air handlers has some interface at the zone level. The primary integration is by using space static pressure sensors in the Zone. The typical approach is to install one representative space pressure sensor for each air handler. These devices measure static pressure in a narrow range, typically plus or minus .25 inches of water column. The ZLC does not implement any control sequences for space pressurization however, as this function is typically provided directly at the Air Handler.

Trending

Trending is accomplished at the ELC or Building Wide Level as described in Chapter 9 and 10. Trends are not maintained at the Zone level, but it remains critical to establish a history of temperature conditions. The overhead required for trending however, makes it necessary to accomplish this function in a higher level controller. As always the information is available through an on site or remote Central Operator Interface.

Safeties and diagnostics

Again controller overhead in memory and hardware is limited for safeties and diagnostics, however there are some sequences that are appropriate. As with the ELC, safeties may result in alarms, but in general these are control sequences that are implemented to protect a piece of equipment.

Diagnostics function very much as the sequences described under the ELC. Information is available through the ZLC so that a user can interrogate a system and diagnose problems by reviewing temperature conditions, alarms, etc. Again ZLC hardware diagnostic routines are implemented as well to allow for self checking of the controller itself. With these sequences the controller tests itself and can identify a variety of hardware, software and communication issues.

The overall benefit of both safeties and diagnostics is to optimize controller operations and reduce downtime.

This discussion of DDC control sequences for the ZLC should reinforce in the readers mind the complexity of Zone Level Control. These include a large number of both simple and complex control sequences. Beyond implementing these primary and secondary sequences with the equipment directly under control, Zone DDC must also integrate into its sequence of operations external data and control. External data has been discussed as both global and shared data may be used in VAV sequences. Further, Building Wide control may require execution of control sequences requested by both the Zone Controller and the Air Handler. It is also important for the reader to consider these additional sequences under the topic of Local Level Integrated Sequences.

INTEGRATED CONTROL SEQUENCES

The concept of a Integrated Control Sequence was covered under the ELC and this discussion will simply highlight the Zone role with Integrated sequences. Those sequences that are important with Local Level Integration are:
- After hours override - tenant billing
- Indoor Air Quality
- Demand Limit Control
- System Data

After hours override - tenant billing

Override has been discussed as an essential feature with most commercial buildings. Spontaneous access to the building by any tenant is necessary, and override ensures that conditioning is provided during these times. The ZLC plays an important role in this sequence because the override button is nearly always located at a Zone sensor. The override is usually incorporated as a momentary contact button on the sensor.

Specifically with ZLCs the sequence required is to identify the input initiating an override and send that message over the network to the ELC. This control sequence will then enable occupied setpoints and maintain normal temperature conditions for a pre-programmed duration. The balance of the override sequence resides in the ELC, and no further action is required at the Zone level.

Indoor Air Quality

Indoor Air Quality sequences have been discussed under the ELC and here with Zone Level Controllers. The key sequences of note here is monitoring conditions and then implementing action using system wide data and control points. The primary IAQ Control sequence of concern with ZLCs introduces additional fresh air to maintain parts per million of Carbon dioxide (CO^2) .

Access to current information on CO_2 at the Zone controllers is typically provided through a sensor on the ZLC. The ZLC sensor monitors CO_2 and these will send a signal to the air handler if additional fresh air is required to purge the space and reduce parts per million of CO_2. Again these are integrated sequences because of the coordinated action that is necessary between zone and equipment controllers.

Demand Limit Control

Again Demand Limit Control (DLC) has been described. There is a limited number of electrical loads available to shed at the ZLC. This is because most Zone Level Application have very few electricity intensive components. There are some components such as refrigeration compressors and electric reheat stages. The key issue with DLC at the Zone Level is that the capacity of the units is sized to the load in the space. This means that maintaining temperature in the Zone requires the ability to enable that equipment as necessary.

The DLC sequence therefore must allow for the Zone Controller to override a request to shed loads. This is particularly true given the critical nature of zone temperature control and the fact that conditions can become unsatisfactory quickly. If shedding loads under a DLC condition is likely to sacrifice space temperature, then the ZLC should be able to override a DLC command.

The reason for this somewhat detailed discussion of one application is to raise the readers awareness of zone control sequence complexity. Given that zone applications typically involve smaller less sophisticated equipment, and smaller numbers of I/O, it would be easy to assume that sequences are simplistic as well. After this reading it should be obvious on the contrary that ZLC sequences tend to be complex, and that there are also interactive sequences to be considered.

The ZLC for VAV box control also requires network communication to allow integration with the air handling unit (AHU) for simple functions discussed above like unoccupied override to start the AHU. This means that using the full power of a microprocessor in the ZLC requires communication capabilities that will be discussed further below.

NETWORKING

This topic has been positioned throughout this book as an integral requirement of any distributed system. As noted under the ELC, the requirement for networking with Zone Level Controllers, as with all other DDC devices, is to provide data flow between controllers and to allow remote communication.

The need for networking with the ZLC was reinforced above through discussion of several control sequences that are carried out between several modules. ZLCs are distributed processors, and networking is essential to distribution. As with Building Wide and Equipment Level Controllers distributed control require networking to achieve control integration. This is equally true, and perhaps a little more so with Zone Level Controllers. As the processing power, memory, etc. are on a smaller scale and applications like VAV tend to involve interdependent equipment operation, networking is absolutely essential. The ZLCs are highly reliant on network interaction for two reasons: to provide remote communication and to allow control integration. Remote communication is essential for monitoring, programming and commissioning these controllers.

Control integration is key to the ZLC because the sequences outlined above require a great deal of information, and it is not possible to terminate every necessary point to this panel. Also it may not be feasible to build every control sequence into the ZLC, for example time of day scheduling. In some cases a clock does not reside in the controller, and an occupied command is sent from an ELC or other device. These and other data points fall into the system data category. Given the wide variety of sequences that must be implemented at the ZLC networking may also be necessary to implement building wide control such as optimal start or demand limiting. All of these functions make the ZLC extremely reliant on a distributed network.

INSTALLATION AND COMMISSIONING

The installation and commissioning process was described in some detail under Chapter 10. Installation should always follow the manufacturers documentation to ensure optimum controller performance. As the reader is aware many ZLCs are applied at the factory by OEMs, and installation requirements are limited. These devices have generally been tested and the only installation remaining is to connect the controller to a DDC network in the field.

Commissioning is so critical that it could consume an entire chapter. However it is simple none the less, as a means of systematically checking and verifying every aspect of the controller. This is not the most glamorous component of DDC, but its importance to successful installations can not be stressed enough.

ZONE LEVEL CONTROLLER TRENDS

It is somewhat difficult to target specific Zone Level trends for discussion. This is because the very existence of ZLCs has been considered as a trend for several years. The author conducted an informal survey on ZLC installation by OEMs with new equipment in the late 1980s. Results showed that ZLCs were being installed at the factory on around 10% of the VAV boxes sold each year. The recent market research conducted by Future Technology Surveys[1] states that 29% of the DDC systems sold in 1992 included ZLCs. Further it projects that by 1997 45% of systems installed will include ZLCs. The author believes based on other data that this may even be understated.

Given the dramatic growth in ZLC installation it is critical to understand the nature of these controllers. It is also essential to consider specific trends for the ZLC and how these will effect systems in the future. As a result this chapter will discuss trends in the same categories as the ELC:
- ZLC Communication trends
- Control Integration and
- ZLC Control trends.

ZLC Communication trends

The trend major trend under communication, has been discussed under Building Wide and Equipment Level Controllers, standard communication. This trend has relevance here but at a very fundamental level. This means that a standard communication protocol may be used

---

[1] Survey on Energy Management Systems, Report #208, Future Technology Surveys, Inc., Lilburn, Georgia, 1993

to develop ZLCs in the future, but initially there will be extensive use of communication gateways. The evolution of standard communication is expected to be the Building Wide Controller and ELC first, then at the ZLC.

The approach noted above will allow DDC systems to meet the requirements of standard communication, but phase in new products. Several options for phasing in standardized products were outlined in Chapter 10, and this chapter will not repeat that discussion. It is important though to reemphasize that chapter has presented the ZLC as network reliant. Networking is essential for ZLCs to both receive and transmit data and control requests from specific ELCs. The interaction between the ELC and ZLC must be coordinated for effective control. As a result the concept of interoperable Zone controllers that could be applied with several different manufacturers DDC equipment is possible yet not a reality today. This means that ZLCs for the near future will always be applied with a corresponding ELC from the same manufacturer to ensure integrated control operation.

Connectivity is another issue however. In the near term it is likely that an ELC/ZLC controller combination from one manufacturer could easily talk to Central Operator Interfaces from other manufacturers.

Another important communication trend that was noted in Chapter 10 is the use of hand held and Lap top PC interfaces to the ZLC. These interfaces are essential today for field commissioning, and also for diagnostics and trouble shooting after the installation is complete.

At the ZLC level interfaces are not panel mounted, the user will typically hand carry an operator interface for these tasks. The trend with ZLCs is to allow the interface to be plugged into any zone sensor through a jack like a simple RJ11 phone jack. Once this connection is made with either the hand held and Laptop based interfaces it is then possible to access the ZLC for that sensor or any other controller on the bus. Further, this will include software tools that allow diagnostics and fine-tuning control loops to actual equipment through analysis of live system data. In addition, it is becoming common to equip these interfaces to do complete air balancing at the zone using the interface and the sensors connected to the ZLC.

The trend toward higher speed communication for the bus itself will likely impact ZLCs, but it will be evolutionary and follow implementation at the ELC. A final trend of interest for the ZLC is the intelligent sensor and actuator, or the sensor bus. These wireless and communicating sensors were covered in Chapter 10 but they will effect the ZLC as well as other DDC controllers.

Control Integration

The reasons for ZLC control integration with ELCS were clearly delineated in the chapter. Numerous examples of this have been provided thus far such as the use of zone temperature data for reset of supply air setpoint at the ELC. The Morning Warm up sequence is another good example of integration between the ZLC and a higher level controller. There are also a number of examples of the ZLC being commanded to take action as part of functional level integration between such systems as HVAC and Fire Life Safety.

With communication standards there will be a continual evolution toward more complex integrated control sequences. Yet in spite of this trend, there are some key concerns with control integration when more than one manufacturer's device is involved. It becomes extremely difficult to ensure overall system integration of such functions as demand limiting because every manufacturer takes a different approach. Further at the ZLC level there is a limited amount of capacity to shut off loads. If the zone temperature becomes unacceptable it is extremely difficult to regain control.

A key issue with the ZLC is that unsatisfactory space comfort is a major problem for many end users, and therefore the ultimate power to implement a sequence. For example even if zone temperatures are outside a given comfort range, some integrated sequences would override the ZLC and force a load off. While other ZLCs will embed smart strategies at the controller level that have the option of implementation or responding with a "no shed" message. As already noted, ZLC integration will focus on coordinated sequences with the ELC for specific sequences like reset and eventually with all practical sequences.

As noted in Chapter 10, control integration is something of a hybrid topic between communication and control. Many trends contribute to more sophisticated control, but will require higher level, and perhaps faster communication, between the ZLC and ELC. Some of these key trends represent significant technological and market strides.

ZLC Control trends

Since the centerpiece of ZLC technology is control this chapter can not be closed without a discussion of zone level control trends. The ZLC is by nature an application specific device. Expanded I/O capacity will also be possible with application specific controller to increase the flexibility of the controller. Another ZLC trend is implementation of highly interactive approaches that apply standard library programs and also allow some limited programmability. Ultimately the benefit here is in reduced engineering time to make the technology viable in more zone applications.

Implementation of multiple control sequences and control integration are hardly future trends as they are happening today. However these approaches will continue and lead to much more sophisticated interaction between ZLCs and other controllers for such routines as smoke pressurization during a fire.

Continued development of more sophisticated control sequences can also be expected. Fuzzy logic type approaches sometimes called Adaptive PID are a possible trend, but will not occur at the ZLC until they are well established at the ELC.

In all cases the memory that holds the ZLC sequences can be expected to be non-volatile. Non-volatile memory storage was discussed in Chapter 10, and includes EEPROM and Flash memory. Faster processors with more power for control sequences and communication has been noted several times in this discussion, and will continue. Modularity in components is also essential to avoid having unnecessary hardware cost in points, etc., that are not needed for the job. This is particularly critical with ZLCs because cost is extremely important to viability of zone controllers.

Ease of use is as important a topic with ZLCs as with all other DDC controllers. As noted with ELCs, Future Technology's[1] new survey says that factors contributing to a competitive edge include:

- Ease of Use in general,
- Ease of Installation and
- Easy to understand, maintain and upgrade.

That discussion will not be repeated but ease of use is particularly critical for ZLCs. This is because a typical job may include anywhere from 10 to 1000 zone controllers. Simple and easy programming, installation and commissioning is an absolute requirement if ZLCs are to be widely installed.

The continued trend of OEMs mounting ZLCs is important to cost effective and extensive implementation. Along with this is a trend toward lower cost controllers, but the user must be careful to evaluate more than the controller cost. Particularly with OEM shipped controllers it is critical to ensure that integrating those controllers into a complete automation system does not involve cost prohibitive programming and network hardware.

As with other Direct Digital Controllers, expanded compatibility for worldwide use is also important. This means controllers that operate at 50 or 60 HZ, and are able to use a variety of engineering units including metrics. The wireless communication trend discussed earlier will also impose some hardware modifications on the ZLC.

This chapter has been a detailed review of Zone Level Controller technology. This chapter has also expanded on the control technology covered in Chapters 9 and 10, as well The intent of this chapter has been to merge these topics and give the reader with a much more comprehensive understanding of DDC control.

Chapter 12

Communication and Networking with Distributed DDC Systems

Control networking is one of the most complex topics in the industry today. The intent of this chapter is to provide a brief history of both the motivation for DDC networking and the communication technology applied with DDC systems. This is an intentionally detailed chapter because of the importance of this area in systems today. This summary is intended to serve as a resource for anyone who is manufacturing, specifying, installing or managing automation and control systems that must communicate.

The author has published a complete text on this topic. Readers who seek more information on control networking are encouraged to read "Networking for Building Automation and Control Systems", The Fairmont Press and Prentice Hall, 1991.

Rather than delve into the more complex aspects this subject however, the chapter at hand will focus on the critical issues. This chapter is intended to provide the reader with a general understanding of the key issues. Armed with this information it will be easier to make intelligent decisions when faced with the many options and buzzwords used by vendors in the industry today.

For nearly two decades the control industry has made increasing use of microprocessor based systems. The culmination of this trend is

Distributed DDC, and requirements for communication with these systems is the focus of this chapter. With these systems come a host of benefits and features for users. Many of these including: expanded and more sophisticated control sequences, and system communication have been discussed in previous chapters. Achieving these benefits however means that a full set of system requirements be met.

Application of Distributed DDC systems has become the norm for Building Automation, and these systems are inherently dependent on both network and remote communication. These facts along with the customer demands for easier ways to communicate with systems has directed a tremendous amount of interest on the topic of Open Protocol or industry standard communication. This interest has resulted in both attention to the issue, and a great deal of confusion in the industry. There have been countless articles, symposia and discussions on "open" or "standard protocols" for control networks.

For clarification, an open protocol is simply an openly published set of rules (in computer language) for exchanging information between computers. The key issue is that openly publishing the protocol does not mean that everyone will use it. Therefore the real demand in the industry is for a standard means of communication that will be used by all participants in the business. This would greatly simplify remote communication with systems, and application of control industry networks. Full and detailed coverage of DDC networking is beyond the scope of this book. Again, the reader should seek information from other resources to get full coverage of issues and trends underway in the industry.

CONTROLLER COMMUNICATION AND NETWORKING

It is worthwhile to review the evolution networking as an issue. This includes the concepts of open or standard DDC protocols or "standardization". To do this we must briefly consider how this technology has evolved, and what are some of the key industry concerns that highlight this issue. For nearly ten years, end users have voiced concerns regarding communication with automation and control systems.

This industry is technology driven and as a result there are several key trends that have become inherent features of DDC systems. Many of these features and trends that have fueled concerns about communi-

cation. Many of these key trends were introduced in Chapter 8, and will be readdressed here to show how important communication and networking are to state of the art DDC. These critical trends include:
- system integration,
- Distributed Direct Digital Control,
- user friendly interfaces,
- Personal Computer Central Operator Interfaces,
- flexible systems that are easy to use, and
- expanded functionality at every level of systems including implementation of intelligent buildings.

The discussion of standard networking and interface in this chapter will be addressed as it involves each of these trends.

Each control trend adds complexity to the issue as new systems are implemented. Also, end users note, when more than one manufacturers control equipment is installed in the same building the problem is further highlighted. This problem has been referred to as a communication barrier and is illustrated below.

To the user, this issue has traditionally meant that: there will be more than one Central OI system, data can not be shared between systems and control can not be integrated among systems. In addition, controller networking throughout a building system, with multiple manufacturers, for information and control integration is impossible. Because of these complications, many end users do not find their systems easy to use, and the issue of a standard, sometimes referred to as "open", protocol developed.

Many users see the concept of a standard as a solution to the networking and interface issues discussed. Such a standard or open protocol, employed in the design of all new control systems, could allow end users to mix and match various manufacturers components in the same system. This standard could also allow remote system communication from a single operator interface. The control trends noted, along with existing day to day systems issues that create problems will be discussed in brief under the issues section below. Yet there may be some more basic obstacles to overcome before the technical issues are considered.

BASIC OBSTACLES TO A STANDARD

Resistance to change may be one of the most basic human traits. This concept addresses a major change that is necessary to resolve a set of specific problems significantly effect the industry. The solution requires standardizing a day to day function that has become essential to the success of these systems. That is a standard for use by computers in "talking to" or transmitting data to one other, specifically computers for Building automation and control. These computers include individual controllers, host-based systems and any PC or mini-computer operator interface. It appears that the automation industry is not unique in its need to develop a standard means for computer systems to share information. This problem is faced by every segment of the computer industry, from simple microprocessor based devices through personal computers (PC's) and mainframes. As a result some discussion will refer to other industries and may discuss examples of how this standards problem may be approached.

Standard is a word that is defined by the context, it may mean uniformity, high quality or a basis for comparing the characteristics of our life style. In this context standard refers to a uniformity or commonalty, in the language or communication protocol that is used to transmit information from one computer system to another. Such a standard does not exist in the Building automation and control industry, a fact that is painfully evident to those individuals who work with systems on a day to day basis. In fact there has been a significant effort underway in the industry to address this issue since 1987. The effort has been led by the American Society of Heating Refrigeration and Air Conditioning Engineers (ASHRAE). A document called the Building Automation Control Network (BACNET) will be discussed later that is intended to begin the process of addressing the communication needs of Automation industry.

Each manager and technician should take the responsibility to understand this process, the proposed standard for communication and where potential vendors stand with regard to communication standards and interface. At an even more basic level this chapter will continue the discussion began earlier in this book to expand the readers basic understand of the issues and technology. This knowledge will be useful in making interim buying decisions as products offering solutions and quasi-solutions hit the market. This was particularly true when users had to make decisions regarding several vendors who came to the marketplace proclaiming compatibility with ASHRAE BACNET even before the standard was published. These same responsibilities for knowledge and understanding are shared by every member of this industry. These members include the engineer who specifies equipment or lays out a system, the contractor or installer that implements the system, and the end user who ultimately lives with that system into the future.

Setting aside for a moment the technical issues regarding a communication standard there are also business and human oriented issues. One notable business issue is the base of controls equipment which in currently installed throughout the world, and may never be compatible with a newly developed standard. There is also the substantial investment that controls manufacturers have in existing equipment which is not likely to be compatible with a standard. The perceived loss of revenue or competitive edge which these manufacturers expect, if forced to conform to an interface standard, also must be considered. Lastly on a more human note, is the inertia of an industry which is accustomed to technology as it currently exists. This industry must invest a substantial effort in climbing the learning curve.

The inertia discussed above gave way over the decade between the authors first article on this topic and publishing of this book. In fact a recent Energy User News[1] article documented the inter-compatibility between manufacturers. Not only have manufacturers recognized the requirement for standard communication, but they have begun to share communication languages or protocols. This is obviously an interim phase in the move toward standards. At some point in the future when a standard communication is fully implemented this approach will only be necessary to implement older generation systems.

To return to the basics though, historically control system manufacturers each adopted communication protocols for reasons that are deemed sound and justifiable. Industry competition led to the desire for

[1] Energy User News, Radnor, Pa., August 1993.

proprietary protocols for networking and communication. This was an obvious area to embed personality into control devices. The idea of standardizing these protocols presented some major obstacles. People involved were committed to their method or style of interface, therefore likely to resist change. In considering the issue of an investment in technology, along with the commitment given styles, it was obvious that the transition would not be easy.

The issues discussed above along with technical obstacles presented a serious, though achievable, task. Of possibly greater concern upon initial evaluation, is cost. There are two elements to consider: Research and development investment to implement a standard and incremental controller cost for products which support the standard.

THE COMMUNICATION PROBLEMS

Issues of inertia or resistance aside, the topic of system communication and protocol was recognized as critical for the building automation and control industry. The issues and problems involved in evaluating systems and communication products, and in choosing an industry standard are complex. This chapter can only begin to introduce the key issues or problems. As noted above there are a number of issues encountered with both existing and emerging technologies in the industry. Actually the problem has existed for as many years as the systems have, but in the mid to late 1980s it came to the forefront of the industry. Throughout the time since then the issue has been a source of considerable confusion. This chapter will touch on some answers to the questions that are usually posed regarding the issue.

The topic of standard communication and "standardization" in this context refers specifically to controls systems. There are a number of other related topics to be covered, including a great deal of discussion that revolves around personal computers (PC's), other computer systems and networking systems used in other industries. As a result this chapter assumes that the reader has a general understanding of the control systems concepts discussed thus far, and at least a superficial knowledge of computer systems.

Because the standard communication issue has reared its head in nearly every industry which relies heavily on computers, the automation industry has been able to learn a great deal from past experience. These industries are not the topic of this chapter, though some references may be made through the course of the text.

Some significant problems have spurred all this talk about standard communication protocols. These problems have been highlighted over time due to several basic desires of automation system end users. Upon examination, these problems may be summarized in three categories: Control, Central Operator Interface software and controller networking. To simplify the discussion, issues regarding both existing and emerging technologies will be covered under each of these sections below.

CONTROL

Control is the essential discussion because that is the purpose of our systems. Ultimately, the continued viability of the systems business is based upon the benefits of control. Energy savings spurred dramatic growth in the industry, however a number of factors, including fluctuating energy cost, utility Demand Side Management programs, etc. have varied over time as factors in continued system sales. Certainly cost is still a key concern, and programs which began as energy management efforts remain viable due to cost reductions. The ability of companies to develop programs that contribute operations dollars back to the bottom line was welcomed by top management. Energy savings is a desirable side benefit, but cost reduction is one of the key reasons that Energy Managers did not become extinct. The resurgence of utility rebates and added environmental, indoor air quality and other issues such as those discussed in Chapter 8 have resulted in ongoing growth in the industry.

Beyond cost however, control is a key issue. It has become possible to meet cost reduction targets through efficient control of more complex HVAC systems. It is also possible to improve comfort while saving energy through sophisticated control. Some of the more important developments in this area include: Direct Digital Control (DDC), Distributed control, System Integration and global/shared data control points. Emerging technologies such as adaptive or self-tuning DDC also play a part in this discussion. This discussion will be confined to the above issues however, because implementation of these features has the most bearing on our topic.

Direct Digital Control is not new to our industry. The definition published by ASHRAE states that: DDC is microprocessor or computer based open or closed loop control of an output device based upon input data and a sophisticated control algorithm, typically Proportional, Integral and Derivative. DDC provides efficient, sophisticated strategies for the optimum in control and efficiency. The impact of this trend on Central Operator Interfaces is limited when it is applied through Host-based controllers, meaning that one device is in control of the entire system. It becomes a major issue however as intelligence and DDC are distributed throughout the system.

Distributed DDC via peer controllers is state of the art in our industry, and the choice of users to an increasing degree. This technique, as discussed in earlier chapters, relies on individual microprocessors applied at the controlled equipment. Often called "stand-alone", these controllers maintain a complete control program, have inputs and outputs connecting them to all critical data and equipment, and they operate independently of the rest of the system. Critical however to this technique is the ability to provide a communication network to tie these systems back to a central front end. This is the key driver for Central Operator Interface and networking features.

It is most important here to note that distributed systems must share information with a central point of interface for local or remote communication. Without this feature, these systems can not match the overall functionality of the traditional CPU based system. The benefits of a distributed system are that the control remains in tact, even if a power or communication failure occurs elsewhere in the system. Yet this benefit is not sufficient to change the industry to distributed control if each controller is an island, and the information in the system is not available to the end user. There is however, another concern in this area: global points.

Global points are an essential feature of any distributed control system. Consider for a moment the CPU based control system. Their are obvious system integrity concerns because that the system success or fails relies on CPU operation, yet there is a major benefit: all system information is available at one place. To provide and improve upon this feature, distributed system must share data among controllers. More importantly, systems must provide a means to control output points in one controller from input points terminated to other controllers. This technique involves the use of "Global Points", in fact there are a number of other uses for global points. They may be used for data that is shared by all controllers, time synchronization, outside air temperature and many other pieces of information. Another use for global points is to create a pseudo point that combines information from one or more controller. Data is extracted and a controller performs some type of math, averaging, minimum/maximum operation, etc. to acquire a piece of data that may be used throughout the system.

System integration is the final topic under control, and it assumes that more that one type of system is installed in a building: automation, security, fire, etc. In such cases it may be desirable to integrate control from multiple systems. For example a control sequence could use information from the security system, and allow the automation system to take action. When the security system sends a signal to a controller that the security has been breached during the unoccupied mode, the DDC system turns on all the lights as a crime deterrent. This simple example provides impetus for thought in devising enhanced building automation functions that would make use of data from other systems. For example, smoke evacuation, the DDC controller does not need to meet code as a fire safety system to provide this function. Again a signal from the Fire panel could allow the controller to initiate a sequence to exhaust air from the building and aide in starving a fire for oxygen. The examples are many, but both of those above share a common technique, they are initiated by a controller input. A standard communication protocol solution offers many more options. It will then be possible to actually share data between systems for control and to monitor all the systems from one Central Operator Interface.

Central Operator Interface

As with the control issue, there are several considerations regarding Central Operator Interfaces (OI). It has already been stated that the concept of a single point of interface for one or more systems remains a central theme in the discussion of communication standards.

This is because the benefits of automation systems have been demonstrated beyond control and comfort, information is now a critical commodity. Information is available through a Central OI, typically a IBM compatible personal computer (PC). There are two key areas of discussion which illustrate the importance of front ends: remote communication and program management.

Remote communication in this case means that a system installed in any building may be accessed via a PC and a modem that is located around the block or around the world. This feature has become one of the most valuable to building managers for staying informed about system status. The benefits of remote communication include information management, remote troubleshooting and diagnostics, as well as remote programming and monitoring. These are all significant to many building owners, in particular those who are managing buildings located miles away. Again direct cost reduction is possible. Maintenance costs can be reduced by troubleshooting a problem and dispatching a service person with a clear idea of the repair that is necessary. Capital equipment dollars may be saved through alarms. Owners can avoid permanent equipment damage due to a problem that would be unannunciated because the building was unoccupied.

Program Management of existing Automation systems also provides real benefits to a user. The example above was of a critical problem, but there are also a number of tasks centering around system performance. The bottom line is that the cost savings we have discussed do not miraculously appear, they are the result of program management. Much as a poorly maintained automobile will ultimately fail a motorist, a neglected automation system can not hope to achieve the projected benefits. The discipline of program management includes regular system verification to ensure operation and system fine-tuning for enhanced efficiency.

The basic problem with Central OIs when used for remote communication and/or program management is that they are typically proprietary, meaning that they only interface with one manufacturers system. As a result, the need to maintain a number of front end systems, or at least software packages, to interface with multiple automation systems has created a major industry problem. A standard communication protocol between all OIs and systems, particularly via modem, is the intended solution.

NETWORKING

The final area of concern that contributed to the standard dilemma is networking. This concern was held for last because an understanding of the other issues is required to understand the role of networking. In discussing the controls issues, networking was mentioned as the primary vehicle for sharing data between distributed controllers. In fact the only real problem with networking and data sharing is that they are essential to meet our future control needs and yet, as with OIs, existing controller networks are typically proprietary. What does this mean? Communication between controllers is only possible if they are manufactured by the same supplier, and they share the same network protocol. Some proponents in the industry have always viewed the standard issue as a need to allow multiple manufacturers controllers to reside on the same network and act as a system. There are two key drivers for this concept which deal with problems in expanding existing systems and in specifying new systems.

Expansion of existing systems, prior to a standard, was a problem because system compatibility typically required that the same manufacturers equipment be purchased. Otherwise, systems and controllers could not communicate and multiple Central OIs were required. This situation has never been acceptable to users because upgrade bids tend to be expensive when they are limited to a single vendor. In addition to the Central OI issue, this problem points up the issue of controller to controller communication. Control panels from different manufacturers utilize different protocols to exchange or share data, and therefore cannot function as a system.

Specifying new systems is a problem for many of the same reasons as existing upgrades: number of options, cost and compatibility. This primarily effects users with multiple systems or sites which require remote communication to a central front end in one geographic location. A standard allows for common controller to controller communication to ensure that building owners make the optimum use of existing technology, and that it is possible to integrate different equipment with new systems.

It becomes more obvious as we examine these issues that networking is central to the entire discussion, and the heart of any network is a message protocol. The need for a standard protocol is unquestioned, the issue that has remained is how do users manage their programs effectively during the development and implementation of this standard.

Given an outline of the problem, it is now possible to discuss the role that communication and protocol play in this issue. Actually there are several related topics which may be considered in outlining the issues. These topics along with proposed solutions to the problem are outlined later in the balance of this chapter.

NETWORK ISSUES AND SOLUTIONS FOR THE FUTURE

The basic issues and problems leading to the call for a standard protocol have been covered thus far. For the most part the Control Industry position is that the concept of a standard protocol no longer requires examination or justification, rather it requires a well conceived solution. Equally as important, the various players in this industry must acquire an understanding of the issues and options available to ensure the long term success of their programs.

This section of the chapter will cover the most common solutions that have been proposed for a standard communication protocol. These must cover basic issues, and address all applications: multi-site users, large complexes, large central station systems, local networks of host to host systems, distributed controllers, etc. The most viable options for the future address the problems already discussed.

The topics of control, Central Operator Interfaces and networking have been covered as they impact a protocol standard. The need for communication with automation systems, which was also highlighted as a key issue, will also be covered. System communication, however is only a part of the issue. Though there is a definite need for standardizing on this process, the industry and its technology are changing.

Distributed processing systems have become state of the art, and therefore communication is required with numerous controllers. In fact, it will not be uncommon in the near future to have thousands of distributed controllers in a facility which previously used only one. Therefore the issue networking can not be overlooked.

Networking is at the heart of the standardization issue. It would be cost prohibitive to support multiple phone lines and modems in one facility, each accessing one of many stand-alone controllers. Through the network, one modem and phone line provides access to all the distributed controllers in the system. A central source for information is now possible through one Central OI for the entire network. It will not be long before end users note that building wide networks offer the same savings with other systems: fire, security, access control, etc.

Personal Computers as noted are a key issue to be discussed because the communications and networking features rely heavily on PC's. This is where the two key issues of remote communication and networking directly relate, because in most cases interface sessions will be carried out on a PC. It has become even more common for this interface to be initiated from a remote site, and a PC on the network at the local site is becoming more common as well.

Again this section discusses options and alternatives available to the industry for solving the standards problem. Each of the key technologies discussed above play a part in the options listed. As these discussions unfold it will also become evident that a number of major and minor issues must be resolved to facilitate the development of a meaningful standard.

Prior to discussing options specifically, a review of several terms used in the media everyday regarding this subject is important. These terms are: Software links, Open Protocol, Connectivity, interoperability and manufacturer alliances. These words are also used in the context of a number of other industries, but that is beyond the scope of this book. The focus here though is only there meanings in the control systems context.

Software Links: Several years ago this author created one of the terms mentioned above, "Link". This term referred to the need for a standard interface among personal computer (PC) front ends for control systems. These were being used extensively for communication with control systems of multiple manufacturers, and each front end was proprietary. Use of a common protocol or communication "Link" between a

PC and multiple systems was posed as one solution to this issue. But the process became more complex with the advent of distributed control systems.

Technology developments in the controls, electronics and data processing fields have contributed to making Distributed direct digital control (DDC) systems cost effective on individual pieces of equipment. This requires a network for communication between these controllers, other control products and a front end, typically a PC. In addition, system integration requires that we consider that there may be more than one level of sophistication or architecture required. More complex systems may require higher level architectures to accommodate there needs where such complexity would overburden simple systems. These types of issues will be discussed further under protocol definition and requirements.

PROTOCOL DEFINITION

The term protocol has been used quite extensively in this chapter, but what is the definition of a protocol, and how does it become open? In the simplest of terms, a protocol is a set of rules which allows one computer to understand what another one is saying. The key elements of a protocol define:
- format of the data,
- information necessary for data conversion between machines and
- timing to define the data transmission speed and sequence.

The written computer instructions which make up these elements of a protocol are generally called source code.

A protocol exists wherever two systems must communicate, and it is common practice for the protocol source code to be proprietary. Interestingly, this is not unique to the controls industry, the issue may be found in every aspect of the computer industry. The search for a solution in other industries has led to development of independent research organizations and to significant corporate investment. The Corporation For Open Systems is one such research group focused on standard communication between mainframe computers. Manufacturing Automation Protocol (MAP) is the result of a major expenditure made by General Motors Corporation to ensure that all GM production control systems use standard communication. In each of the above cases communication guidelines were provided by the International Standards Organization (ISO).

ISO has developed a model or guideline for development of communication protocols and standards for various industries. This model is called the Open Systems Interconnect or OSI. Detailed coverage of this model is beyond the scope of this book. It is important to note however that OSI outlines a seven layer reference model that is being used by our industry to develop a protocol standard. However, an analogy may be drawn between OSI and a map of a large metropolitan area.

In the OSI map analogy, a driver wants to get from point A to point B, and knows what direction (north, south, etc.) he or she wants to go. They consult the map for a route. Any number of options may present themselves; take one or more interstate freeways, use surface streets or possibly a combination. In much the same way OSI presents the model as a guideline or map, but does not dictate one specific protocol or route. Evaluation of the protocol development task should not be simplified because OSI has been referred to as a solution. OSI significantly limits the task of protocol design, but does not eliminate the need to draw the map.

The definition of a protocol aside, what is an open protocol? Open protocols differ from other protocols only because the source code is not proprietary. The source code is published, or the protocol is made available in a format that is not copy protected. Such a protocol could be designed into more than one manufacturers system, and allow for a standard in controller communication. End users in the industry have indicated the need for such a standard to be available for control system communication. Therefore, the issue at hand appears to be establishing a standard protocol for the controls industry, rather than an open protocol.

For the best possible results, the Control industry has initiated an effort through a cooperative and objective group of industry professionals. This effort is focused on meeting the critical needs of building automation, without becoming entangled in an open versus proprietary protocol battle. In an effort to re-focus the topic, the term open protocol will be replaced with standard or standard protocol through the balance of this chapter.

A final issue regarding a standard protocol, is that some method of compliance to the standard must be devised. Standards organizations exist within the related disciplines that make up our industry for many reasons. Underwriters Laboratories, for example has developed a series of tests to document that devices meet the minimum requirements for safety, operation, etc. In this case, a compliance test for approval would assure end users that a piece of equipment meets the industry standard for communication. Selection of the appropriate agency or organization to provide this service and development of the compliance procedure is an important step in implementing the standard. Based upon recent media coverage it appears that the National Institute of Statistics and Technology (NIST) is willing to take on this role.

INTEROPERABILITY AND CONNECTIVITY (MULTI-SYSTEM INTEGRATION)

It important to discuss both protocol requirements and options. Prior to that discussion there are two terms to cover that are often discussed with protocols and systems: "Connectivity" and "Interoperability" Connectivity and Interoperability involve the same issue which is an extension of the protocol discussion that I call Multi-system Integration (MSI). The basic issue is *connecting* controllers from different manufacturers and having them *operate* together as a system. MSI actually involves two concepts, one at the controller level and one at the systems level.

Controller MSI mixes more than one manufacturers controller within a system, and the controllers are intended to operate as though they were designed to be a system. At the system level, multiple complete systems are integrated. In most cases the system level MSI does not integrate controller level functions. Rather it uses a single front end for programming, monitoring and other PC functions with all the systems. This type of MSI is very similar to a gateway which is discussed under options for the future.

MSI introduces a number of confusing variables into the discussion of standard protocols. Some of the most critical concerns include: warranty, service, maintenance liability and control integration. Warranty is a question that arises with these systems because each manufacturer would be hard pressed to identify legitimate warranty claims. Legitimate claims would be those involving traditional problems that could not be blamed on other controller interference, design error, field installation problems, etc. Service and maintenance liability are similar to the warranty issue, however the key here is who does the owner call for service. Further the challenge is to ensure that unnecessary site visits and finger pointing do not result in extended downtime.

Control integration is the last and perhaps the most critical problem. In order for these controllers to work as a system, the designer and installer must plan for control interaction. This means that the control loop in one device could overrule the internal design algorithm in a second device. This is extremely dangerous, particularly where the second device does not have sufficient data to provide effective control. These problems present significant obstacles to multi-system integration. Options for developing such system can not be addressed until the basic standard issue discussed here is resolved. These issues also present an additional level of testing that would be necessary during standard compliance testing.

Manufacturer Alliances

Perhaps one of the most aggressive recent strategies undertaken by control and equipment manufacturers to meet users needs has been the alliance. Under this approach, two companies agree to share their proprietary protocol with the intention of meeting their customers needs. This process is perceived to provide mutual benefits for manufacturers. Initially it was done by control and equipment manufacturers so that the OEMs pre-integrated controllers could be seamlessly integrated into the owners system.

In the last few years alliances have also been done among control manufacturers as well. There are several motivations. Some entered into alliances because they were missing control components in their line, for example a low cost VAV box offering, etc. This has also been viewed as an important political step in showing that a company is open to, and supports, the standards process. The actual mechanics of this process are twofold; legal and technological. The legal aspects involve non-disclosure and defining agreements to ensure that companies respect each others intellectual property. The technological aspects are basically a communication interface, and that technique is discussed further below under options for the future.

OPTIONS FOR THE FUTURE

With the discussion of issues provided earlier and the previous definition of protocols, a framework has been provided for discussing where the industry is headed. There have been any number of short and long term solutions posed to address the industries concern over the number of proprietary packages in use for networking and communication. For the most part these proposed solutions fit into the 4 categories below, and each will be discussed.

1. universal front end software that can talk to any system,
2. a hardware or software conversion package to translate between protocols,
3. industry standardization on an existing protocol, or
4. development of a state of the protocol for controller networking with a direct component to enable standard remote communication

Universal software

This first option offers a short term solution, particularly for users with existing systems, through a universal front end. This is the use of one product, normally designed to operate on a personal computer (PC). The product typically provides local and remote communication, but is limited to monitoring and programming. Packages of this type are often characterized by a host of advanced PC features for: graphics, windowing, point and click mouse operation and expanded Central OI features like database management and maintenance tracking.

Universal software packages have generally been developed by companies outside the controls industry who specialize in software. This limits the scope of their products, and further complications may also arise due to their lack of control application experience. This may be a minor concern, but the industry should bear in mind the lessons learned from the black box revolution. During the early and mid-1970s a large number of electronics companies were attracted to the controls industry due to the crisis mentality that existed regarding energy. These products rarely performed to owners expectations in part because their designers did not understand control applications.

The universal software package provides communication with each manufacturers system through a special software routine written to interface with the controller. This software is called a conversion package or gateway, and will be discussed in a moment. The second method of effecting a universal front end is via, what I call, the "Hot Boot" technique. Hot Boot is done by embedding the control manufacturers communication software in the universal package, and writing a routine to call that software when controller interface is necessary.

The use of universal Central OIs has been implemented sporadically, but is generally popular with end users. This is because they offer an option to those with multiple systems. Many users have had to support multiple Central OI software and hardware packages, and this is not necessary with a universal package and systems are easier to use. One of the demonstrable benefits to the end user is that a total building automation package may involve several subsystems which are transparent to the user. One key note is that the trend toward distributed control is a key driver for standard protocols. To the detriment of this option the networking requirement for distributed control is not addressed through universal Central OI software.

Protocol conversion

A hardware or software conversion package would be a decided shortcut to solving the protocol issues. This method has been mentioned in conjunction with MSI and universal Central OIs. Protocol conversion would require the development of a device to act as a translator between PC OIs, host or peer level controllers and control systems. These conversion packages are often called gateways. With this option there is no change to the existing protocol. A package is developed which can interpret that protocol, convert it to a protocol which the front end system understands and pass it through to the front end. Development of a gateway is essentially the technological step taken when Manufacturers develop an alliance and share protocol.

The desirability of this option is that existing systems could be easily modified to allow communication, and that any Central OI could conceivably function as a standard. As noted, a gateway would be used for universal OI or manufacturer alliance technology, but gateways are not limited. Gateways can be integrated into host or peer distributed controllers to allow controller to controller communication. This option does not assume that all existing protocols are acceptable for the long term, but it does offer a technique for merging existing and new protocols within the same system.

Existing Protocol

The final two options for a standard offer the choice of a controller to controller protocol. The primary distinction is whether the protocol currently exists or must be newly developed. Standardization on an existing protocol offered a short term solution several years ago, but was rejected by the ASHRAE standards committee. Such a protocol which had been designed specifically to meet control industry requirements could have saved time, but a number of drawbacks were also identified. Ensuring that these requirements reflect the needs of end users is critical. Existing control industry protocols existed which could serve this function, and a few manufacturers offered these as options. There were also protocols which had been developed for other industries that were candidates for use. Some believed that the controls industry was not large enough to support a unique communication standard. However, it was evident that this industry did have a set of current and future requirements for this protocol. Once these requirements had been defined, protocols from other industries and generic local area networks were carefully evaluated.

Due to the unique requirements of our industry, it is likely that standardizing on a controls industry protocol would be desirable. Among the options were several protocols that were been designed and developed by major manufacturers in our industry. Honeywell, for example had developed the control network automation protocol (CNAP). This protocol was designed to serve as a worldwide standard for networking and communication with Honeywell control systems. In addition to control manufacturer developed protocols, there has been a good deal of activity in some unexpected areas. For example, West Germany, Canada and Japan developed national protocols for control systems.

In those countries, the government is the largest end user and invested in a protocol for the future. The German option called FND was also used in other parts of Europe, as well. FND was, in fact, a PC to PC solution offering a gateway and a standard PC OI. Evaluating an emerging protocol product and standardizing on it would also be a very timely solution to the industry need. To explore the full range of possibilities though, discussing the final option is necessary.

A new standard

Develop a state of the art protocol for controller networking with direct component to enable standard remote communication is the fourth option. This package could be designed to meet current, as well as future system needs. This option, when combined with either a universal OI and/or a gateway, would incorporate new and existing controllers to offer a comprehensive solution. Development of a new protocol therefore was a viable option. This protocol had to qualify as state of the art while meeting a full compliment of end user needs.

In this area the American Society of Heating, Refrigeration, and Air Conditioning Engineers (ASHRAE) took the lead in defining the approach and the requirements. The approach selected by ASHRAE was to develop a completely new standard protocol. Their leadership and effort in this area has defined a solution that the industry has been willing to support. The process has taken significantly longer that any would have projected. This is to be a US and potentially worldwide standard. As noted above, though there is also activity in other international areas such as Canada and Germany. However, the ASHRAE standards committee, which was established in 1987 to develop a new standard, is visible industry wide as an essential vehicle for evaluating the task at hand. In fact, the eyes of the controls world are following ASHRAE's progress because this work should identify and define key industry requirements. Identifying these requirements was the first step in selecting or developing a standard for the future. This is essential to ensuring that a communication product meets industry expectations.

ASHRAEs proposed standard product, called the Building Automation and Control Network (BACNET) is expected to be published by 1995.

Though the treatment of this particular industry option may seem oversimplified here, in fact it is the most simplistic option in theory. However there are some major issues which will be discussed in summing up the chapter. Most important to note is that the issues and options surrounding the development of a new standard as faced by ASHRAE have been significant. Each of the other options for the future identified in this section were considered as either short or long term steps toward a standard.

One of the most significant options which could be undertaken was the development of a completely new standard. Many believed that the needs of the controls industry do not warrant development of a completely new standard based upon requirements and of research and development cost. This may have been a valid point, especially when one considers the number of existing protocols in the controls and related industries that might form the basis for a standard.

To thoroughly explore this issue however is beyond the scope of this book. The author has published a separate book on that topic called "Networking for Building Automation and Control Systems", Fairmont Press/Prentice Hall, 1991. This chapter has rather been focused on introducing the reader to the basic topics necessary to provide the reader with information to evaluate the technology.

This chapter could not cover the complete set of protocol requirements due to the extensive nature of that topic. Though this topic must be critical to evaluating any standard. A short list of the requirements considered would include:
- the system architecture including the structure of the system and its components
- local and remote communication requirements of controllers and PC's
- PC compatibility with standards in other industries
- Network capacity for "X" number of controllers, front ends, etc.
- Data transmission speed
- Meet worldwide requirements for system communication and networks

- Cost is extremely important both for the protocol and the product itself.

This general coverage does not address many of the more basic issues that we considered by controls manufacturers and the ASHRAE standards committee. Yet of perhaps greater importance for the reader is positioning an effort toward an understanding of how ASHRAE's BACNET will affect your future.

WHERE THE INDUSTRY IS HEADED

Based upon the drivers and issues that have been outlined in this chapter, establishment of a new protocol standard for networking and communication is approaching reality. This standard must meet a full set of control industry requirements for each level of control system architecture.

Yet the primary issue that is unclear is when the protocol may be available. Publishing a protocol by 1995 simply means that manufacturers can then commence a development cycle. These cycles tend to run from 2 to 5 years, meaning that standard based products may not be available in this century. Again many manufacturers have already begun to tout products that are BACNET compatible and ASHRAE has stated that this is not possible. The reader should beware of such products and carefully scrutinize any offering prior to ASHRAE published the standard and defines a process and seal of compliance.

Including the proposed "new" standard approach, each of the future options has benefits and detriments. Timing and cost must also be considered because development of a new standard and compatible products has required a significant investment in time and dollars. Yet these obstacles did not deter users from making the best possible decision about the future of our industry. Support of all the efforts underway to identify industry requirements, in ASHRAE and elsewhere, has been essential to meet day to day needs. These efforts have been extremely important for developing the yardstick that can be used to judge protocol options.

In the interim, universal OIs, gateways and manufacturer alliance continue to offer users the opportunity to utilize existing technology.

These options make it possible to integrate control systems with one Central OI, thus simplifying use and interface with these systems.

The establishment of a standard protocol is exciting, challenging and necessary. As the standard approaches market availability it is now more important than ever for managers to become conversant with the language of protocols. This is because options for new and existing systems will be affected by every action that is taken with regard to this issue. It is also important because information is critical to the effective management of systems, and access to data is dramatically affected by this issue. Also, system communication remains the best means of maintaining controls and ensuring their overall performance. And the topic of protocols cannot be separated from any discussion of system communication.

The best first step that each individual, and the industry, can take is to understand our present and future option for communication via any protocol. With this information it will be possible to make intelligent and effective decisions about the communication products that make sense for our industry.

Chapter 13

DDC Operator Interface: Handheld and Central

Operator Interface may be one of the most critical features available to the DDC System users. It has become essential for them to be able to monitor and program systems from local or remote interface devices. They find that having a wide variety of information at their fingertips is essential to optimizing system performance.

Interface is primarily a communication issue and in this chapter a variety of communication devices and capabilities will be explored. Ultimately, communication is one of the primary reasons that many users choose to install DDC systems. It offers the user one of the most valuable commodities in building management, accurately, timely information. With this information, the manager can evaluate the building efficiency, monitor status from remote locations, conduct troubleshooting on equipment prior to dispatching a service person and have a direct window into day to day facility operations. A host of benefits come along with these abilities that translate to productive and satisfied building occupants, and therefore more profitable building operations

This chapter will cover a variety of interface topics, but for simplicity purposes they will be discussed in reference to the hardware that provided the interface. This hardware falls into two basic categories; handheld devices and personal computers.

HANDHELD OPERATOR INTERFACE

Handheld devices have become more and more common. These are extremely powerful devices that dramatically improve the on-site information available when compared to conventional systems.

Many early DDC systems relied on direct equipment measurements or reading for local information. In many cases pneumatic interface panels were the only local device, and this required building operators to record equipment status by going out to the panels. The only other option was a central console, which was extremely useful, but they really fall into the central interface category and will be discussed later. So, other than the Central console building operators had to rely on making rounds to various mechanical rooms, and often took readings on a daily basis.

The handheld interface offers a new level of functionality and information to the building operator. These types of devices were first seen as panel mount keypads with Liquid Crystal Displays (LCDs). They were common with early load control or "Energy Management Systems". Generally these operator interfaces were an integral part of the Central Processing Unit (CPU) or master controller. This was useful for monitoring status of system point and the user could usually change setpoints and schedules from the keypad as well.

The first true handheld devices were later developed when distributed controllers became more common. The devices were interchangeable with various pieces of controller hardware, and again allowed monitoring and varying degrees of parameter programming. This leads to the primary differentiator between handheld devices in the marketplace: Local and Network based devices.

All handheld interface devices must have a connection to a controller for monitoring and other functions. The primary distinction between these devices is that connection and the level of functionality available.

The Local Handheld interface is by definition, local. This means it is only able to communicate with one controller at a time. As you might guess, the connection is therefore local and typically involves a direct plug connection on the controller. This type of interface is usually acceptable because these devices are often used with larger equipment

DDC Operator Interface: Handheld and Central

or building wide controllers. As a result, it is possible to walk up to the controller in a mechanical room or equipment closet and plug in the hand held device. Since these interfaces only communicate with that one controller, the operator is then able to acquire a wide variety of information.

Local Handheld devices are extremely useful and often incorporate a variety of functions that enhance the operators capabilities. Obviously full monitoring is available, along with setpoint and schedule change capabilities. This allows the operator to conduct diagnostics, and in many cases correct problems from the handheld. With typical distributed systems these changes are then uploaded automatically to the host or system manager. This is done by the controller however, and not by the handheld. All in all these are valuable tools, but there are some limitations as well. What about interface with smaller equipment that is mounted in the plenum, or with large numbers of controllers. For example, a Variable Air Volume (VAV) job with 500 terminal or zone controllers and floor by floor air handlers would present some real problems for an operator with local handheld devices.

Network based Handheld interface technology grew out of applications like VAV and Water Source Heat pump (WSHP). As suggested above there are two complicating factors in such applications. First the equipment is often not readily accessible, and second multiple pieces of equipment would require extensive time for physical connection. The concept of a network based handheld device is therefore to allow the operator to gain access to the communication bus through the interface.

Network access through the handheld can occurs through a physical controller connection as with Local devices. Due to limited access however, it has become common to make this connection elsewhere. One of the popular approaches today is to allow connection through the space sensor located in the zone. This connection often is made through and RJ11 phone type jack or special purpose plug.

Physical connection allows the user to have access to the communication bus directly or to use the controller for this capability. In essence the choice is determined based upon the power of the handheld device. Initially the reader must note that this form of interface typically requires more intelligence in the handheld. If the handheld has direct access to the bus, it must function must like the communication master or peer described in the networking chapter. To support that level of functionality means that the handheld be a microprocessor based device with full protocol implementation. From a logistical point of view, the handheld must also be capable of identifying network controllers by address to request data or change program parameters.

Network based handheld interfaces that address the communication bus through the controller require less sophistication. On the other hand though, these devices impose communication overhead at the controller level and this increases controller cost. A particularly important issue that was touched on in the earlier chapters is that many zone level controllers have limited network privilege. As a result, it may not be possible to allow the full measure of required interface features required and this approach is less common.

The Handheld interface, local or network based, has contributed a great deal to building operator capabilities. These devices improve the operators ability to ensure system performance and offer a host of other features. For example, the handheld interfaces are integral to system commissioning and demonstration phases. These are also becoming quite common for doing such tasks as complete air balancing. In some cases the equipment requirements for this task are even minimized because balancing can be done using the input devices installed with the system. Functions of this type are important to producing energy and cost savings. Equally important, these functions produce a host of new benefits beyond cost saving, including improved comfort and building information that further contribute to the value of a DDC system.

One final note on handheld interface, of the network based variety in particular, is that they do not always fit the typical perception. These are generally considered to be small as the name implies with a limited keypad, and an display screen of some type. More and more today, installers and operators are using notebook PCs with PC software through the network access ports on sensors, etc. for expanded interface. These however fit more into the category of the Central Operator Interface, and such functions will be discussed below.

CENTRAL OPERATOR INTERFACE

The topic of interface can not be covered today without discussing the Personal Computer (PC). Under this heading the PC will be viewed in two key capacities, the on-site Central and the Remote Monitoring Operator Interface. In either case, the Personal Computer (PC) has become an integral part of DDC systems.

These two basic variations on the implementation of this tool suffer from a large number of names and as a result can be confusing. The on site Central Operator Interface is sometimes referred to as simple a Central or a Host, front end or console. The Remote Monitoring package version of the PC has also been called a Front End or simply remote software. The key to all of these incarnations is that the functionality remains rather constant. For simplicity purposes this chapter will refer to the onsite PC as a "Central Operator Interface (OI);' and the remote PC as a "Remote Monitoring Operator Interface (OI)."

This chapter will cover a wide range of features that are available to managers through these Operator Interfaces. The primary focus will be the PC, more specifically the software that runs on the PC, and the system features and functions this software makes available. The primary emphasis here will be the Central Operator Interface as an ancillary system distinct from the controllers. Whether the PC is on site or remote, its principal function is to provide a window to view system operation. This window is not passive however. In either case it offers the operator a powerful tool for diagnosis and response to building irregularities on a real time basis.

In conventional Automation or Energy Management Systems, the Central OI was actually a "terminal" to the minicomputer based, or CPU type, control system. Terminal in this case means basically a dumb input and output device for interfacing with the CPU. With the rapid acceptance of distributed processing systems, this role has changed, and in many respect that is where the term host originated. Essentially a Host level controller has a large level of capacity much like the old CPU based system. The CPU was actually an all knowing Host. Today PCs with high powered microprocessors offer expanded Central OI technology which is often applied initially with a number of PC software packages such as spreadsheets and word processors. This has resulted in a wide variety of Central OI implementations typically running specialized PC software for DDC and other packages to aide the operator in managing the system.

The challenge in writing this chapter is determining where to draw the line in discussing Central OIs, and particularly PCs. The personal computer has expanded dramatically over the past decade, and has becoming a standard component with many, if not most, automation systems. This has also been impacted by the changing DDC market and drivers for users to apply this technology. Market research indicates that systems are still installed to save energy, but a host of other factors may also be considered in justifying a system. In fact, these other factors are rapidly became essential to any building system.

Along with the trend toward PC based Automation Management, new features are starting to be integrated with systems: Job Design Tools, Equipment Diagnostics and Preventative maintenance along with a host of reporting options for system data. These equipment related features are ancillary to control, but contractors, building owners and end users did not stop there with the demand for integrated Central OI features. Contractors want a wide range of job design and cost estimation tools built into PC interfaces, while building owners would like to see multi-tasking systems to allow office automation on the same PC. Building owners also want to incorporate other capabilities like monitoring after hours use of HVAC to allow billing of tenants. Leases in many areas are written to allow use of the space 24 hours a day, but space conditioning only during normal business hours. Tenants wishing to use the space after hours are billed an hourly use fee for HVAC. End users have other Central OI needs, they want: access to the system data in standard formats and communication standards such as those discussed in the networking chapter. These interfaces in general, must offer end users the information necessary to generate reports summarizing system performance and results.

This chapter will provide a detailed discussion of Central OIs and their role with Direct Digital Control (DDC) systems. The key topics covered are listed below.

Central OI Personal Computer Requirements
Central OI User Interface

Central OI Core Features:
- Building System Operations,
- Remote Communication and
- Network Management

Central OI Enhanced Features and
Central OI and Standard Protocols.

It bears noting that OI features will be viewed in general. Distinctions between Central OIs and Remote Monitoring OIs will be covered under discussions of each topic. As noted earlier, these devices are extremely similar in functionality, and therefore will be covered together under the general headings noted above.

CENTRAL OI PERSONAL COMPUTER REQUIREMENTS

As rapidly as DDC technology changes, it does not keep quite the same pace seen in the computer industry. Given that the Central OI is basically a Personal Computer (PC) with specialized software, this presents some real problems in specifying operator interfaces.

In light of the dramatic changes taking place in the computer industry, it is not feasible to outline a minimum set of PC requirement here. However, it is possible to provide the reader with a general set of issues to evaluate prior to PC purchase. It is also recommended that the reader conduct independent research into computer technology. Becoming fluent with the technology and following, if only at a distance, the trends in that industry is a worthwhile exercise for anyone in the DDC industry.

The first phase of any PC evaluation by a DDC user should begin with the Central OI. It is important to know whether the OI in question is a software package only which runs on industry standard hardware, or a custom package. Custom packages, which include both hardware and software, are much less common today than in the past. Industry standard or "off the shelf" technology is much more desirable because service and support are readily available from a number of sources. This is clearly not the case with a custom package which can be very problematic for a building owner.

Central Operator Interfaces that operate with industry standard Personal Computers should be evaluated first based on the hardware required. It is equally important however to ask the Central OI manufacturer the same questions regarding software, training, service and support.

Note that Central OIs that run on industry standard hardware may also have special Hardware or Microprocessor requirements. This is particularly important if expanded features or multi-tasking functions are necessary. To aide in addressing some of these issues, this chapter provides some general PC guidelines along with a cursory discussion of computer technology. The software requirements should be evaluated with these considerations in mind, however this should not be used as the sole source of information, as noted above. Users should also consider pursuit of training in computer technology or hiring an expert to assist in decision making.

After completing the research in computer technology the user is strongly encouraged to develop a set of evaluation criteria for Center Operator Interfaces. This criteria should include at a minimum the hardware and software items listed in Table 13-1: PC Evaluation Matrix. It is beyond the scope of this text to provide explanations for the items in the matrix. Yet a general discussion of the critical items in each category will be provided to give the reader an indication of its impact on the Central OI.

Hardware

Under hardware configuration a number of issues are listed. Each of these issues should be considered with reference to the software requirements of a specific system. The microprocessor in your PC has a great deal to do with the software the system can run, the speed at which programs execute and other related issues. Initially it is important to review computer requirements for the software that you expect to be using. This should be an evaluation for both the short term and the intermediate term. Long term planning is not stressed because the computer industry is extremely volatile, and changes cannot be foreseen beyond a few years.

In spite of dramatic changes in computer industry, the user should still give serious thought to their use of both hardware and particularly software. Even if you don't know what specific software programs you will be using in two years, the type of program may be enough of an

PC EVALUATION MATRIX

Project: _____   Evaluation Date: _____

Package: _____   Evaluator: _____

HARDWARE CONSIDERATIONS   **REQUIREMENT NOTES**

Microprocessor

RAM memory required
Disk memory
Floppy disk compatible
   – 3.5 inch – 720 K or 1.44 M
   – 5.25 inch – 360 K or 1.2 M
Hard Disk Required
   – Megabyte capacity
Monochrome operation
Color or Graphics
   – CGA
   – EGA
   – VGA
   – Custom or special quality graphics
Printer operation
   – Dot Matrix
   – Laser Jet
   – Plotter
Modem Compatibility
   – Baud Rate (1200, 2400, etc.)
   – Internal or External
   – International Requirements
Network Options
   – File Service
   – Access
User Interface
   – Special keyboard
   – Mouse Compatible
   – Track ball Compatible
   – Touch screen or other special interface
Special Cards Required
Other hardware considerations
   – _____

Table 13-1 PC Evaluation Matrix

PC EVALUATION MATRIX

SOFTWARE CONSIDERATIONS	REQUIREMENT NOTES
MS DOS Compatible - Version: .	
OS2 or Multi-tasking DOS Requirement	
Special licensing required	
Other required to support system	
Compatible: other software packages	
Compatible: memory resident programs	
Other hardware considerations	
— _____	
GENERAL OPERATION CONSIDERATIONS	**REQUIREMENT NOTES**
In house training requirements	
Manufacturer warranty	
Manufacturer support	
— training offered	
— manual and users guide	
— hot line	
Other operation considerations	
— _____	

Table 13-1   PC Evaluation Matrix (Cont.)

input to your PC buying decision. For example, Database managers and spreadsheets are memory intensive and require relatively fast processor speeds due to the size of the files in use, and the number of tasks undertaken.

Future plans to use database or spreadsheet software may dictate the need for a higher speed processor and more hard disk storage. Also note that users should consider trends toward new technology. This may not be a hard and fast criteria, but purchase of a product which becomes obsolete in the near term can have a critical impact on service and support.

Once you have completed your analysis and determined the processor requirements it is important to identify the memory required. Central OI Memory will be covered in three areas: 1) Random Access Memory (RAM) needed to run software, 2) disk access required to load the program and exchange data on Floppy media and 3) hard disk memory required to store system and data files. In addition presentation is important, both from a CRT and printer point of view. Will data be displayed in monochrome or color, and will there be a special requirement to print, plot or output the data to hard copy.

Other hardware issues involve interface to other systems. These may include modems for remote dial in and dial out communication, and network communication for interchange with a Local Area Network. For on site Central OIs this interface is important for communication with other System Hosts or distributed controllers. The onsite Central OI concept has been discussed as a central that is a system master or a Control Network peer with distributed controllers.

With a Remote Monitoring OI modem communication allows an on site interface through a dedicated module, or in some cases an On site OI, to transfer information from an entire system. In each case there may be requirements for special add-on hardware, such as internal cards or external modems, etc. Each of these must be carefully considered as they affect current system operation and future expansion. For example, using your On site Central OI as a node or terminal on a local area network may require an internal card. This card plugs into an expansion slot on the main board of your PC. What that means is that you must consider how many slots are available, because future expansion may not be possible if this card fills the last slot on the board.

Expanded technology that has become standard on many systems is important to consider in evaluating Central OIs, and may pose special hardware requirements. For example features such as "point and click" mouse operation and dynamic graphic monitoring have become standard on many systems. These features often require special cards or interfaces which must be considered in the current and future plans for the Central OI.

Finally in buying a PC system the user should define whether a requirement exists for additional hardware now or in the future? Is the hardware proprietary and manufacturer specific or off the shelf gear? How do you get warranty and service work done on the PC? What about upgrades to the packages for software or hardware? How about trends in the industry? Does the manufacturer intend to obsolete this system in the short term, or introduce a new more powerful model? Is there ample software available from a variety of industry sources to meet your needs? All of these points are included in the hardware analysis and should be recorded in your evaluation.

Software

The evaluation of software is key due to the integral part it plays in the use of any PC. As with any other business decision it is important to consider the current and future uses of the equipment. In this case these uses will likely involve software that is purchased independent of the hardware. However, the hardware and software interface is only one software issue to consider.

The most basic software issue may be the disk operating system that will be used with the PC, though this decision is generally made by the manufacturer. The Microsoft Disk Operating System (MS-DOS)[1] has been standard in International Business Machines Corporation (IBM)[2] Personal Computers (PCs), for nearly two decades. New software requirements resulted in a number of enhancements to the product until it reached the multi-tasking level. Multi-tasking DOS allows the PC to operate more than one software program at the same time. This is valuable for Energy Managers because it can allow them to run front end system software while also running another package, for example a spreadsheet to access data and perform calculations on live data. Another term is "multi-user" which refers to system software that allows more than one operator to use the system simultaneously.

It is important to know if the software you intend to use requires a special version of MS-DOS. Evolution to multi-tasking led to a new version of DOS called Operating System 2 or OS2[3]. These products are only solutions from one company however, others exist and should be identified in your analysis. This is extremely important because software that runs with one operating system may not be compatible with another. This also affects data files which users may want to share between systems.

MS-DOS is not the only operating system found in DDC Central OIs, however and this further complicates compatibility. With expanded control integration on increase, there are many DDC applications with multi-functional systems. With control of HVAC, Fire, Security and perhaps other systems the use of other operating systems has become commonplace with PCs. So again it's important for the user to understand the software requirements for any application. These issues should be viewed with business judgement, and if necessary the opinion of a consultant.

Another issue of note is the license agreement. What are the users obligations and responsibilities? What are the vendors obligations and responsibilities? Also evaluate PC software support and general operating concerns. Most packages have specific RAM and disk memory requirements, and users must ensure these will be addressed. Some packages also restrict the use of memory resident or other packages during use due to memory requirements or memory locations which are reserved for the program. Ensure that current software packages that you want to run are compatible with the proposed hardware and software.

Training and Support

This key area of concern is easy to overlook when comparing the physical characteristics of a system is manufacturer support, both hardware and software. It is recommended that the user require access to factory training. This resource will allow the user to optimize use of the systems, and will provide a support mechanism when questions arise. Questions will inevitably arise, and users must get answers to avoid costly downtime and loss of productivity.

Once the user has evaluated in house training requirements, it will be easy to eliminate suppliers that do not meet the standard. The manufacturer or software developer should offer training for new users at a central location or at the customers location. Training is a somewhat broad term that may cover anything from general marketing and sales data to technical troubleshooting and repair. Determine the level of training you believe is required and compare that to potential suppliers offerings.

Manufacturer support is the final area of concern. First the supplier should provide a complete manual and users guide. Users find troubleshooting and diagnostic or reference manuals to be invaluable. Users also find a customer support hotline to be extremely helpful. The hot line is a resource when the user has exhausted the information available through training and reference manuals.

This may appear to be a somewhat lengthy discussion of computer technology and evaluation. However, it is a significant capital investment and users should ensure that optimum value is derived from that investment.

Central OI Interaction Media

Another important consideration in evaluating PC systems is operator interface. The evolution of Central OIs has resulted in expanded interface features, and careful evaluation is warranted. The evaluation here should focus on tailoring the interface for the intended application. There are a variety of tasks to be carried out at the Central OI. These tasks will also vary based upon whether this is an onsite or remote OI. There are also many levels of user capability.

It is important to target key activities and conduct an evaluation of the user interface with respect to both the tasks and operator sophistication. Three key forms of user interface will be discussed: Keyboard, Graphic and Mouse Interaction and Voice Synthesis.

Keyboard input has been the most common form of interaction with Central OIs. Trends in the industry toward graphic and voice interaction have led some users to believe that the keyboard may be an outdated. Again consider the intended task and operator sophistication. Keyboard interaction lends itself well to users that are performing everyday operations such as trending or diagnostics. Graphics can be very useful, though some users who perform regular system interrogation find these to be cumbersome and, at times, slow.

Sophisticated users will typically dial up, sign on and access all necessary system data quickly. The user may conduct several diagnostics while on line, but will try to complete the session and sign off in a short time. This allows users to improve productivity and minimize telephone charges for Remote Monitoring of systems. Keyboard interface lets the user move through system selections to accomplish this goal. Of course this also requires that data be available in tabular summary displays, as well as graphic representations. This is generally true of both keyboard and graphic based systems.

DDC Operator Interface: Handheld and Central

Graphics first became available with keyboard based Central OIs. These were accessed through a menu option or function key, and allowed the user to see a pictorial representation of equipment. Very quickly a new feature was developed called "Dynamic Graphic Monitoring". This feature allowed users to generate their own graphic representation of a system and to define fields to the graphic which displayed active point data. The screen could be set to a "refresh" or update mode and point data would be update to the display field at regular intervals. These systems still provide that feature, however they are evolving to full graphic based media.

Full graphic based Central OIs vary based on the tasks performed. First the OI described above is obviously used for monitoring of a fully commissioned system. Systems which allow the user to have animated display enhance the users ability to quickly identify changes in status. These again are monitoring systems, and provide color changes or moving objects to denote status. Again, consider operator proficiency, this may be an excellent tools for certain users.

Graphic OI functions do not stop at monitoring, however. At a minimum, provisions are made to allow quick changes to setpoints and schedules from the graphic display screen. At the same time it is possible to move between graphic screens by "clicking" with the mouse, more on this below, on a particular part of the screen. The next dramatic use of graphics will be in controller/system programming.

Graphic programming is a relatively new concept. It is discussed elsewhere in this book and will only be mentioned briefly here. The basic concept is to allow system programming with pictures. Several approaches have been taken, but generally control sequences are depicted using a "control diagram style" flow chart.

Complicated routines within the sequence, such as a PID loop or specific subroutines, such as Time of Day programs are depicted by "Icons". These are symbols that are shown with connections via lines to other parts of the routine. The user can move the mouse arrow to the Icon and click on it. This brings up a window or box that allows specific program parameters to be input to the sequence.

The "mouse" is a Personal Computer Industry Development that has been adopted widely with control systems. Under hardware in the product matrix, the mouse with "point and click" operation was noted. In fact mouse technology is not limited to graphic systems, many packages now provide "menu bars" with pull down windows for more tabular display and simple software interaction.

DDC systems that apply a mouse typically use symbols or "Icons" that were discussed above, to make selections. It is particularly interesting to consider these icon based systems. The trend to more user friendliness had some impact on this approach. Also impacting the process is the trend in computing toward "Object Based" code. Object based systems are designed to employ the types of graphic Icons noted above to represent a particular sequence or action. Quite simply the concept is to allow the user to use the power of the PC to program and monitor using "Objects". An object is a point name, etc. that refers to the actual item to be controlled or monitored.

Using Icons to speed interaction or simplify complex sequences of activity is extremely desirable. A simple example used with Apple

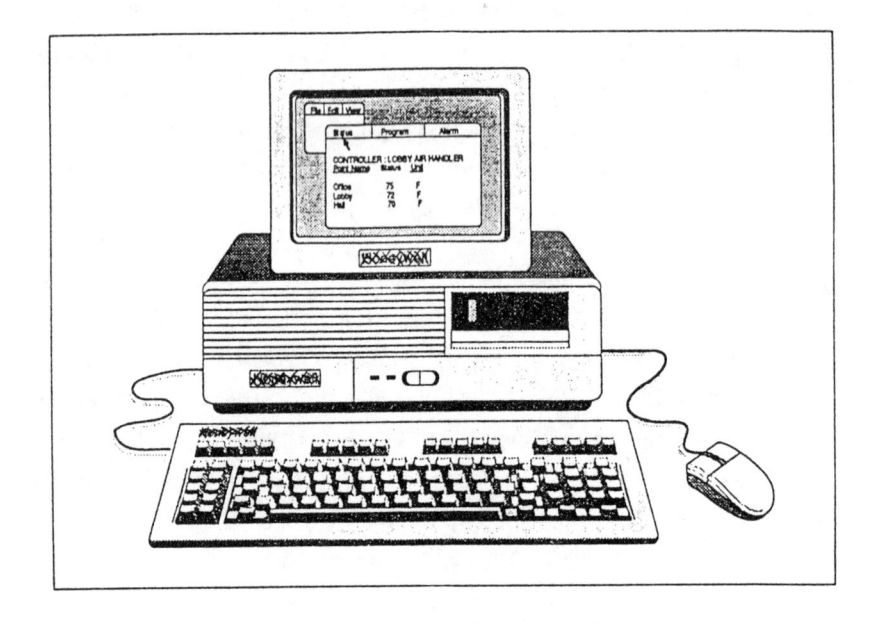

Figure 13-1 Mouse Window Selection

MacIntosh[4] is an Icon picture of a trash can. Instead of a special command to delete files, the user simply clicks the mouse on a particular data file and drags it to the trash can.

Icon based interaction goes beyond simple graphic representation of system data, it provides a simplified approach to every aspect of the user interface. This means that such a package can be used for commissioning, monitoring or any other system function. In the area of commissioning, icons will be discussed as a means of reducing engineering time for the commissioner.

The mouse whether used with icons or pull down tabular menu bars frees the user from keyboard operation in most cases. It can enhance the speed of interaction for the individual who wants to move around the system quickly. Regardless of the software characteristics, mouse interaction serves to simplify user interface and can be an excellent tool for the intended user.

In determining the appropriate interface there are a number of considerations. What types of individuals will be using the system, and how much computer expertise do they possess? These are important questions because they will help to eliminate the systems that do not meet the users needs. Experienced computer users may not want simplified menus and graphics, but they may also ask for "user definable" screens. These could be graphic or tabular, but they would allow an operator to customize their interface. Such a process can enhance productivity, but it is obviously a feature for the more sophisticated user or the system commissioner.

To enhance productivity, even the simplest user may ask for graphic monitoring with a Mouse and pull down windows for menu selection. A simpler form of this is tabular monitoring with "Keypath" to allow shortcuts from one section of the software to another. A keypath is like a macro in LOTUS 123[5]. This allows the user to commit frequently used activities to a keystroke or mouse sequence memory. Such features would in fact be like expanded function keys that are available to the user. Another option user interface option is voice synthesis.

Voice synthesis technology has been used to provide Automation systems with a limited vocabulary. The vocabulary has been used primarily for telephone interface to date. One of the more common approaches to date has been to allow systems to dial out on alarm to a standard touch tone telephone, and transmit a voice alarm message. The individual receiving the alarm has interaction capability through the telephone keypad. Such interaction might include system or output disable and some type of status request. At this time voice synthesis does not provide the full range of Central Operator Interface discussed with keyboard or graphic systems.

CENTRAL OPERATOR INTERFACE (OI) FEATURES

The Central OI, discussed here as a Personal Computer (PC), is an ancillary system distinct from the controllers. Yet Central OIs are becoming an integral part of most DDC systems due to their tremendous utility for users. As a result they are found routinely in DDC specifications. This is partially due to the high demand for carrying out the features discussed in Chapter 21 on managing DDC systems. It also results from the demand for local and remote monitoring and alarming features.

As we are discussing the Central OI here, local features are performed by an on site PC with the appropriate software. The remote monitoring features are performed via a PC located in a separate location, and equipped with both the software and modem technology to accomplish the full breadth of features. Also of note, there is a growing trend in the industry toward using both on site and remote Central OIs for non-control related functions, such as those discussed Central OI enhanced features.

This chapter outlines the key features of an Automation System Central Operator Interface (OI). Specific capabilities provided in Central OIs will be discussed in two categories:
- Core Features and
- Enhanced Features.

Core features focus on functions required for day to day operations.
- Building System Operations: - Programming
  - Monitoring
  - Alarms

- Remote System Communication
- Network Management

Enhanced Features are non-conventional capabilities that may be introduced to a system through the Central OI. From the large number of enhanced features that could be covered, a key group will be discussed including:

- Job Design and Cost Estimation Tools,
- Equipment Diagnostics and
- Preventative maintenance

One of the market drivers for Enhanced Features is greater utilization of Central OIs in conjunction with Automation Systems. This means that Central OIs are applied not only as on site system components in a majority of jobs, but also that there is a stronger demand for PC based commissioning and interface tools. Also contributing are advances in PC technology that make it possible to share data between software packages, and to operate both Central OI software and specialized application packages concurrently.

CENTRAL OI CORE FEATURES

There are a large number of topics to be covered under day to day features that users expect from a Central Operator Interface. Yet these topics may be summarized under the three categories below.

- Building System Operations: programming, monitoring and alarming,
- Remote communication: system management and data access and
- Network Management: expanded requirements of distributed controllers

Building System Operations: Programming

Programming was briefly discussed under operator interface. This section will expand that discussion as programming is a key function that includes both system and controller commissioning and setup issues. Distinguishing between system and controller issues is important because distributed processing entails expanded functionality that is resident in distributed intelligent controllers. The core discussion of programming will be limited to programming as it is required to carry out

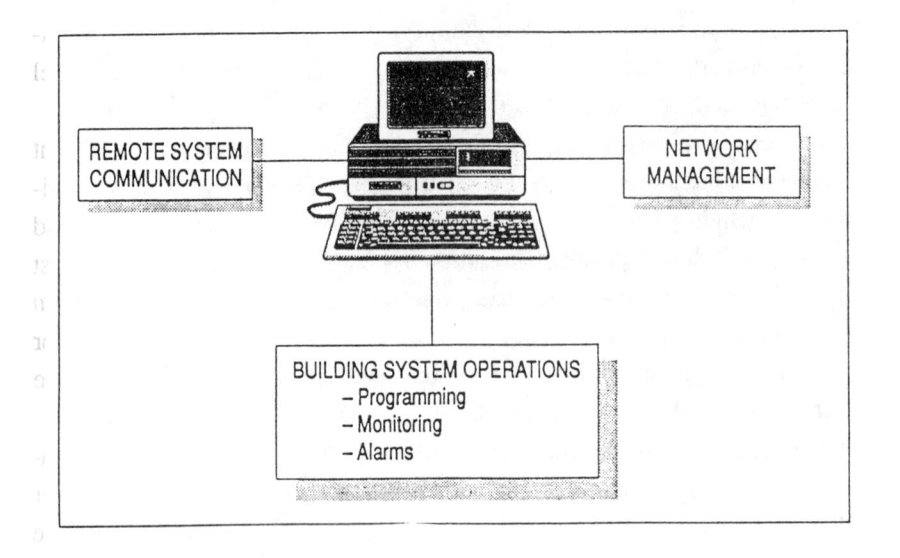

Figure 13-2 Core Central OI Features

control. All topics discussing system architecture, addressing and commissioning will be covered under network management.

Programming can be a very complex task and is dictated, to a great degree, by system control software. In Chapter 9 DDC programming styles were discussed under fundamentals and strategies. System Programming will consist of key tasks that are defined by the level of access. Four types of control software have been discussed in this book: Graphic Sequences, Equation Driven, Expanded Parameter and Setpoint Entry.

Graphic and Equation Driven Sequences and Expanded Parameter based systems provide the programmer with an ability to modify the control algorithm or sequence. Setpoint Entry Systems allow users to have data access and to change setpoints only. For programming purposes, Expanded Parameter and Setpoint Entry will appear quite similar because they both require basic data entry into specific fields. This will be referred to as "Data Field Entry". Equation driven control software, though, requires more flexibility in data entry and some added features. Programming for these systems will be called "Code Entry". Because traditionally code entry systems require the user to program in a language or "code". Graphic Sequences have very different look and feel and these will be called "Pallet Programming".

Perhaps the most critical thing for the reader to understand is that code entry and pallet programming systems offer a great deal of flexibility. This flexibility allows a programmer to address both standard and custom or non-standard applications. Flexibility is not without a cost however. These types of systems require extensive testing and system acceptance because every control loop is new. A series of features for control simulation and tracing various blocks of the control code may be essential to such packages.

Central OI's are designed by the manufacturer based upon the control software interface required for a given programming approach. Bear in mind that there are a variety of custom PC operator interfaces have been developed to interface with multiple vendors. Yet, for the most part these systems are monitor only, and the control manufacturers' Central OI still provides the full range of programming features.

There are many benefits to using this manufacturers package. For example, user access levels restrict access to control parameters through access levels such as access for:

- "Read Status Only",
- "Program Setpoint/Schedule Parameters",
- "Controller/System Commissioning" and
- "All Access levels above Plus Change Passwords".

Each level has a programmable password associated, and they are arranged in a hierarchy such that each level incorporates all previous levels, as well as a new features. Most monitoring software package provide the same types of access levels, but this section focuses on programming. For simplicity sake, the discussion of programming will be segmented into two areas: Commissioning and Setpoint/Parameter Changes.

Commissioning

Commissioning is a critical topic with any DDC system. Due to the complexity of the task and the innumerable approaches taken by manufacturers, the burden rests with the programmer to understand application of each system. As noted manufacturers maintain special software packages for use in commissioning, and these will also likely include some remote interface features as well. The trend is toward separate software packages for programming and commissioning because this allows for more features that directly fit the programmers needs. As with any other software, an attempt to include every feature a programmer/installer would want along with those that a user would want makes the package harder to use. As a result this special purpose software and the commissioning tasks are generally carried out by experienced contractors or system commissioners.

The intent here is to focus on the commissioning tool which includes all features necessary to setup controllers and conduct system start-up and shakedown. These tools also allow remote communication for dial up during commissioning. Interface features provided by this package though are targeted at a system commissioner rather than end user audience. A related issue downstream from commissioning is a requirement of many system specifications to allow full commissioning features at the Central OI. This issue is somewhat simplified however, because the commissioning tool can reside on the same PC and have the same "look and feel" as the monitoring tool. Another feature that is becoming more common is "plug and play" OIs. This means that any OI that can run the manufacturers software is able to plug into the controller. Once interfaces the use can fully interrogate the program with all the "ease of use" features found in a desk top Central OI.

Plug and play should not be confused with "dumb terminal", a feature of older generation micro-processor based control systems. Dumb terminal allow fairly universal interface, but to use this a user had to be fluent with fairly unfriendly computer languages. The key phrase for plug and play is "ease of use features". This means that any programmer who is familiar with the OI software, including the end user, can instantly get on line and program or commission. In this way the ultimate owner has all be features necessary while avoiding the complexity of having an unwieldy software OI package.

In spite of the various interface trends, this chapter focuses on commissioning with the Central OI. Under commissioning the user's primary limitations will be oriented to the control software type. Data Field

Entry programming, for "Setpoint Entry" and "Expanded Parameter" systems, will generally be tabular in fashion. The software will prompt the user for entries, and the process is fairly straight forward.

Code Entry for "Equation Driven" systems is often called the blank slate approach, and has a very different appearance. Icon based systems may be used with either of the above systems, though they lend themselves more directly to Data Field Entry. In these packages, the commissioner selects from graphic characters or "icons" to do commissioning. Finally, pallet programming is similar to using icons, but these icons are use to build graphic control diagrams. There are a variety of approaches with the most sophisticated being application based, for example programming with HVAC diagrams.

Under system commissioning the initial discussion involves features of the PC front end which will vary based upon system type. Yet there are several generic topics which are germane to all three types of systems, because they are typically carried out after the initial system commissioning is complete. These generic topics include: File Save, Retrieve, Copy and Exchange, Controller Upload and Controller Download.

In taking a controller from carton to start up it must be commissioned or initialized. This discussion will focus on initialization and the commissioning tools that are required to modify a controller. For example, the Central OI must allow for when a user decides to add or delete a control point. Basic system commissioning for initialization will obviously differ by system type. Data Field Entry is generally menu driven or mouse point and click operated, and directs a user through a series of screens for parameter selection.

Many of the "Data Field Entry" systems are actually hybrids for they allow a significant number of complex parameter choices, which modify the controller operation. These hybrids in many respects offer the best of both worlds, because a controller can be customized to an application without a concern for the viability of the control loop. The data entry process in this case is more complex yet it does not depart from the basic entry field style which requires a response to a prompt. Such systems also employ error trapping which limits responses to a particular prompt to answers within a specific range.

Icon packages are similar to data field entry except that commissioners select an Icon that represents a sequence or control strategy. With Data Field Entry the icons are fixed as they are in the tabular formats, but it is possible to speed and simplify the process by selecting the appropriate icons during commissioning.

"Code Entry" or "blank slate" system techniques vary dramatically from the menu approach. In this scenario the user has a blank screen and a language available. Equation languages have typically offered an "IF, THEN, ELSE" type logic along with a mix of arithmetic, boolean and relational operations. Arithmetic features include algebraic expression for: addition, subtraction, multiplication and division. Many systems also offer complex mathematical operations such as: sine, cosine, square root, and exponents. Relational features compare one element to another and determine if they are: "greater than", "less than" or "equal to" one another, along with several variations. More complex variations may include such sophisticated operators as calculations for proportional, integral or derivative control.

Code entry approaches have also been implemented with "off the shelf languages", such as C, which are used by computer programmers. The obvious benefit of this is that there are extensive training options and software tools available to the user. Unlike a custom language that is developed by a manufacturer and is totally unique, there are many sources for assistance.

Central OIs must also interface through the languages and allow the user to define the physical devices which are connected to input and output channels. These features combine to require that the OI user have a high level understanding of the application to be controlled, and its interaction with all other controlled equipment.

Code Entry and Palette Programming systems vary in the amount of control programming that the user must generate. For example, some systems will provide a library of strategies for the programmer to draw upon or link together, and incorporate into an individual program. These libraries may reside in the controller, but this is uncommon due to device memory constraints. It is more common to have the Central OI these library control loops for the programmer to select and download. This is preferable to writing programs or equations to accomplish each specific task because it saves engineering design time. These libraries are also more reliable because they have been proven.

Palette Programming, the current state of the art, allows Central OI based control loops with graphics. There are two types of Central OI pallets, Application based and Control Diagram based. The application based approach uses a control diagram of the actual equipment to be controlled, for example an air handler. These application diagrams incorporate a wide variety of indications including characteristics of the application; fan, coils, etc. and identification of control points and sequences. A typical approach is to use graphic icons to represent this information. These icons might also include a clock for time of day schedules, and specific icons for proportional, integral, derivative (PID) loops among many others. Control packages which allow the user to write custom loops and assign icons of various types are highly desirable to the sophisticated commissioner.

Controller based pallet programming is not as user friendly as an application based system. This is because they begin with control hardware as the common denominator rather than the application. In essence however they use the same basic components, icons for sequences, points, etc. and control diagram depictions. As with code entry systems, palette programming libraries are extremely effective tools for system commissioners.

The generic commissioning topics mentioned above provide added convenience for system users. The "File Save" feature gives users a tools for maintaining a copy of system files on a PC or disk media. This is extremely helpful if a system must be reprogrammed due to control modification or controller malfunction. "File Retrieve" accesses stored files from the appropriate media, for example a hard disk. Though transparent to the user these files may be compiled or encoded to conserve disk space. In those cases, the retrieve function will convert the files to a user readable format.

"File Copy and Exchange" functions are extremely useful for multiple system users as well as contractors and system integrators. These features allow the programmer to copy and modify an existing file in the process of initializing a new controller. The file exchange feature allows users to transfer data files between two Central OIs. Again this feature is useful to a contractor with more than one branch. System data files may be transferred from one site to the other for on line monitoring from both sites.

"File Exchange" may also be applied with a compare feature for Central OIs that do not employ Systematic File Update. File Compare ensures that the Central OI and controller have the same system file. Systematic File Update is a coordination feature which maintains the latest system file in several Central OIs, and is discussed under Remote Communication. Controller Upload and Controller Download are the features which allow a Central OI to transfer system files to the controller. Each of these is critical because upload and modify features are essential for future controller modification.

Setpoint/Parameter Changes

The above discussion of commissioning covered most of the concerns regarding parameter changes. Yet there is a difference between commissioning parameters, and day to day operational changes such as temperature setpoints and time of day schedules. Parameters of this type should be available to users with lower level passwords, yet they should also be monitored to avoid abuse. A good example is time of day override, from the normal schedule, to allow occupied temperature and perhaps lighting. Making such a feature available to a security guard or a tenant may be perfectly acceptable. This gives the systems manager more freedom while keeping tenants content.

Building System Operations: Monitoring

System monitoring may be one of the most valuable tools available to an end user. Through this feature the system provides a window to the entire facility, or controlled environment. Monitoring offers data in a wide variety of forms: current system status, trend logs, operator logs and override logs to name a few.

Monitoring of system status should be available to all level users. There are obviously some screens that the lower level passwords would not be able to access, but in general all those with access should be given full system monitoring privileges. This discussion will depart almost completely from commissioning systems, because programming styles should not affect monitoring. To some degree it is still true that Data Field Entry systems may utilize menu driven monitoring packages, while Code Entry systems may require command entry.

Command entry means that a user may have to learn enough of the programming language to request data and review parameters. These requirements changed rapidly with the trend towards completely separate Commissioning Central OIs and Monitoring Central OIs. In many cases today, the monitoring package will bear very little resemblance to the programming software. The term User Friendly is a major driver shaping the Central OIs and their appearance to the ultimate system owner. Further variations may occur as well because some systems today are beginning to allow the commissioner to develop custom screens for end users.

The most basic Central OI monitoring requirement is current status. In this mode the user should be able to review individual or groups of points. Less sophisticated systems want to review groups by specific type such as: all digital inputs or all points on controller #1. More friendly systems allow users to generate custom groups of points which have meaning as a group. For example: Outside Air Temperature, Discharge Air and Static Pressure For Air handler #4 (AH 4), Space temperature for "X" number of zones served by AH 4 and output status on Fan, Cooling and Heating stages on AH 4. With this data a user can learn the exact status of AH 4, and make operational decisions or take any action required.

An important monitoring feature at the site is trend or history logging, and the Central OI must have access to all data available in a local control system. Users may log any input points, analog temperatures, etc., and capture current values in a report. The user may select an interval to capture data, for example once an hour, along with a roll over feature which simply writes over entry #1 after X entries. Another option is to select a start and end date for logging. This information gives the user a memory in the facility to use in assessing overall performance or reconstructing a sequence of events, like events leading to a low temperature alarm.

It is also crucial for the monitored information be available to the user in a Data Interchange Format (DIF), or some other universal data structure. Having data in this format that allows transport to a report generating package. With more off the shelf technology becoming common in Central OI applications, universal data may be transported to other programs real time. For example, information may be transferred from the Central OI to a spreadsheet program that will be able to display data tables or charts and graphs real time. These types of features are very useful to building owners and managers.

Trending may also be used to diagnose chronic equipment control or comfort problems. The user choose critical points to monitor, sets very short sampling intervals and is able to record conditions in the space or elsewhere for later analysis, much like a strip chart. Trend logs are generally limited because of the memory intensive nature of saving large amounts of data. Therefore user's may have to be sparing in deciding what inputs to log and how often to store information.

The Central OI must be able to dial in to the site, at programmed intervals if necessary, and access trends and other data before it is written over or lost. This feature will be discussed in greater detail under remote communication.

With the exception of information that is saved, Operator logs and Override logs are basically the same as described under trend logs. Operator logs track Central OI interface, and Override logs track occurrences of building operation which are outside the defined parameters for time of day, setpoint, etc. programmed during system commissioning. In the case of DDC systems with an on site Central OI, the remote OI can upload local operator logs along with trends and other data.

The Central OI can also upload operator logs from systems with on site Central OIs in use by customers. Operator logs are important to manage the effective use of a system. Remember, the system must be monitored for security and integrity. This feature allows the user to connect an action or sequence of events, by time and date, to the operator who was logged onto the Central OI. In some cases this data may also be used by the Central OI to time, date and user stamp certain operator interface procedures for later use.

Time stamping can be used for many purposes, such as tracking system file downloads. This provides an end user or building owner with a record to determine whether the contractor has made unauthorized system file changes. The operator log also allows identification of a user who leaves the terminal with a password log on enabled. The local interface offers the same benefits and more. With some systems it is possible to log dial in sessions from any Central OI, and get the same data for any Central OI polling the system. All of this information is valuable in assessing the way a system is used, and the frequency that various users access the system.

Override logs are extremely useful to monitor departures from normal system operation. Control Overrides can be very costly due to added energy cost, and it is not uncommon for building occupants to abuse this feature. This feature is usually abused when systems are left in override after the building is vacated, and controlled loads remain energized. In some cases, building owners have been able to turn the override capability into a profit center. Property managers, for example, often charge tenants a user fee for HVAC use after normal operating hours. This practice, called after hours tenant billing, has become common in some areas. Overrides offer flexibility and convenience to system users as well. They also may be monitored to indicate whether changes to operating schedules and temperatures are warranted due to excessive override. As with trending this is valuable local system data which must be uploaded by the Central OI and not lost.

Building System Operation: Failure and Alarm Logging

The final set of core features are failure and alarm logs. These features provide critical information to the system operator or the outside world. Timely response is made more possible through telephone dial out features, and reports on alarms or failures that have occurred.

A system failure typically means that control hardware, such as a controller, sensor or measuring device has malfunctioned. Reporting these failures is essential to smooth equipment operation, as device failure may result equipment downtime. Downtime is less common now because of DDCs' high reliability, however at the same time fail safe wiring configurations are rarely employed today. Fail-safe or redundant control systems were commonly provided in the past but generally seen as unnecessary today.

Fail-safe systems provided a second layer of control apparatus, to be enabled upon failure of the DDC system. Again, these systems are less common, thus increasing the need to report system failure alarms immediately. Obviously a major system failure that results in complete shutdown would a serious problem. In fact this condition is one of the strongest arguments for distributed control.

Distributed Control consists of intelligent micro-processor based DDC controllers installed throughout a facility. This approach severely restricts the damage which can result from failure of an individual controller. However, many of these systems rely on data from other controllers on the network. This requires an additional layer of failure monitoring on the network components of the system. Monitoring of data collisions and overall communication status is critical to distributed automation systems. Ultimately, a failure alarm should be generated when any component of the automation system fails to perform within normal operating conditions.

Upon initiation of a failure condition, the system designer will specify a series of events to take place. In most cases a manufacturer will write an application specific sequence during design to be enabled upon failure, and provide an alternate means of control for distributed controllers. The worst case is that a controller will be shut down. With some systems the commissioner is given a choice of sequences to enable on failure.

The most sophisticated approach is to offer a commissioner the power to program a custom event initiated sequence to enable on alarm. This will allow the programmer to anticipate the effect of a failure on controlled equipment and effect a custom control action. Upon any failure, the first decision to be made is whether control should be suspended. Second, consideration must be given to distributed systems which allow the use of global or shared data. This means that controller "A" relies on information from controller "B" to make control decisions. Provisions must be made for alternate control routines if controller B or the specific point needed go into failure. A Local failure report is often reported through an annunciator light, etc. or a failure message to an operator machine interface, alarm printer or Central OI.

The final step generally taken when a system alarm occurs is to initiate an alarm dial out to another location. Attempts will normally be made to dial out to a Central OI at one of three phone numbers. More recently other failure report formats have included dial out to a phone paging systems or voice synthesis alarms. The voice synthesis technique, discussed under Central OI User Interface, may also include a feature which allows the user to step through some system interrogation, or force points to an on/off state using the keypad on a touchtone telephone.

The second topic under this heading is the Alarms. Alarms differ from failures in that these are totally user defined. Examples of alarms might be:
- High or low temperatures,
- controlled equipment failure,
- electrical power failure (upon restore) and
- Equipment protection alarms such as compressor or heat fail,

Alarm conditions are provided and defined through the local system software. Certain user levels can identify alarm limits or conditions to monitor. Further the user may establish a time duration before such alarm will be enabled. Upon enable of an alarm condition, the system will carry out a series of activities based upon the criticality defined for the condition.

The most critical alarms will follow the same set of procedures defined for a failure condition. For less critical alarms, the system will generally follow all the same steps, except dial out. Finally, as noted under failures, event or alarm initiated programming is desirable. This will allow a defined action to be enabled upon alarm, and is extremely useful to ensure that dependent loads are shut down. Shut down of a cooling tower upon failure of the condenser pump is a good example of this type of programming.

Alarm functionality is one of the most desirable and valuable features available to end users. A dial out alarm can start a sequence of events including dial up and interrogation of the local system. This interrogation can provide data to be used in determining if service call is required and who to send. It may also be possible to identify the scope of the task, tools and parts required along with other data to enhance the productivity of service personnel. Timely report of alarms in conjunction with the steps outlined above can save downtime, a commodity that is extremely valuable. Downtime may result in an inconvenience due to adverse space temperature conditions, or cripple data communication due to a mainframe computer shutdown in a data processing center. In either case these interruptions in normal operation translate to lost dollars, which makes Central OI alarming a critical feature.

REMOTE COMMUNICATION

The ability to communicate with building systems from a remote location is one of the fundamental features of any DDC system. It is the process of exchanging data between a Central OI and a remote site which may also contain a Central OI or simply a communication module. It has also been a major concern in the industry to address the need for data to be available in a industry standard format. Even more basic is the desire of most users want to employ a PC based Central OI, both hardware and software, to communicate with multiple manufacturers equipment, and conduct other activities as noted under enhanced features.

Communication implies alarm dial out as noted under the previous section as well as a variety of data access features available through dial in. The critical issues to be covered under this category are listed below.
- Remotely Accomplish All Local Features
- Modem Technology
- Auto Dial polling
- Auto Dial mass update
- Systemized Data Files

An obvious, yet unquestioned requirement for remote system communication is that all features provided by a local Central OI or controller operator interface must be available via remote. This includes the full range of Building System Operations features with both new and existing systems.

Modem technology is a complex topic better left to another text. However modems have some impacts upon automation system communication that must be discussed. The speed of communication may be most critical, and this is in fact a combined issue between the modem manufacturers and the telephone service companies. High speed modems are only as good as the quality of distribution lines that they access. Nevertheless, high speed modems may become a necessity as systems maintain larger databases and must share great amounts of data.

Other issues affecting our systems include new technology, such as FM modems which in some cases are being considered as a viable alternative to trenching between buildings in a complex. Such modems exchange data via radio frequency transmissions and offer cost saving benefits to installers. Careful evaluation is essential to ensure that critical alarms will not be interrupted or missed by the Central OI due to radio or electrical interference. Of greatest importance in the control industry is that we leverage the advances made by experts in this business. It is not desirable for control manufacturers to develop modems, this is best left to experts in that industry. Yet this industry should carefully monitor the advances taking place and implement them as soon as practically possible.

Auto dialing was discussed briefly under Building System Operation. This is an indispensable feature for automation systems, and is used for two functions: polling for data access and mass update for simple programming changes. Before these functions are discussed, it is important to describe Auto dialing. This feature offers great benefit to all users, particularly those with multiple locations. Among these may be sites that are geographically removed from the Central OI by a few miles or by thousands of miles. The user is able to select one or more sites to dial up automatically. Sites may be arranged in any group that fits the users organizational needs.

Auto dialing is accomplished by the Central OI in an automatic mode, typically during an unoccupied time. This allows the system to be interrogated for trends or other data, as well as to undergo limited parameter modifications during less critical times of the day. It also frees the user to employ the Central OI for other critical tasks during peak hours of the day. The auto dial routine will call each site identified, carry out a defined set of tasks, sign off and then call the next site.

Two auto dial functions are important to discuss: polling and mass update In the auto dial polling mode, a Central OI will call successive systems and access predefined reports. These reports may be specific to the auto dial process or any of the trend, alarm or data logs available in the local system. The benefit of this process is to acquire system data in a timely fashion, and before limited data memory in the system may necessitate writing over the data. Auto dial polling also provides the user with a periodic check up capability for any system. It will be possible to request alarm reports which did not warrant an alarm dial out from the local system, and to conduct routine analysis of system effectiveness.

Auto dial mass update is a feature which makes use of access to multiple facilities for modification of simple program parameters. These changes may be carried out on a global basis, and apply to any control software type. Obviously Data Field Entry type systems lend themselves extremely well to mass update, users may change of a variety of system setpoints and parameters. Mass update can also be done on a limited basis with other types of systems if defined variables are accessible. In this case the system refers to setpoints or key control data as variables. The Central OI simply accesses the system and changes the value of a given variable. A simple example of mass update, would be day light savings time changes. Assume that a particular system requires manual modification of these time changes twice a year. The auto dial mass update function would allow the user to call multiple sites in an automatic mode on the appropriate date and change the time. This feature would be a tremendous productivity improvement for the end user.

The final topic under remote communication is Systematic File Update. This topic involves the task of managing system files to ensure that multiple Central OIs, that communicate with the same system, have access to the latest version of a system program. This topic could be covered under several different headings, however the issue becomes most evident when multiple users communicate with a local system from other locations, and Plug and Play is not applied.

Systematic File Update ensures that all users maintain a current copy of the controller/system program. This means that changes to a specific controller, ie: to add a new sensor point, are updated on the Central OI operated by the end user, contractor and any other party authorized to have system access. Problems arising from mismatched controller program files can be far reaching for end users who rely on contractors for service and support, have multiple offices with system access and may allow local system interaction.

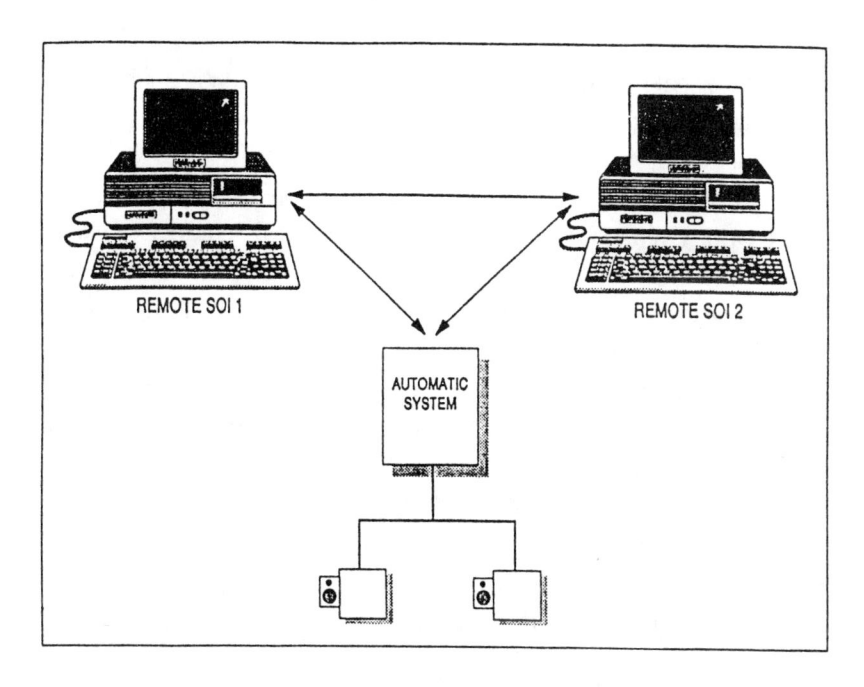

DDC Operator Interface: Handheld and Central   313

Figure 13-3 Systematic File Update

The concept of Systematic File Update is extremely desirable to end users, however it's important to allow the function to be manual or automatic. By now the reader may have noted from the tone of this section, that this is not a feature that exists today with any know system. The concept is based on the authors experience as an end user and contractor managing DDC systems. In much the same way as Auto Polling, it would be desirable to transfer data to other Central OIs during off hours. Some users would also wish to update only after a significant individual change or number of changes had been made to the system. It would also be desirable to limit the number of authorized users. This would reduce the number of changes that are made to the system, and at the same time simplify the update process.

In the interest of further simplifying Systematic File Updates users would be encouraged to consider coordinating these efforts from the controlled site, if feasible. Obviously this would be dependent on the size of the local system, and not feasible where a local Central OI is not installed. For such jobs coordination from a Corporate Energy Office or

Building owners Central OI would be necessary. On larger jobs where one or more Central OI was to be provided this could be an excellent tool for managing interface.

On very large jobs, it could be desirable to provide an on-site local area network for multiple Central OIs to ensure that authorized changes made at any Central OI are held in a Server. In this way the server or network master for local Central OIs could be the central device to hold changes, and act as the update device to any other authorized users. The reader should note that this could only occur if Central OI software was network compatible. There is a risk of major software corruption with Central OI software that is not designed with appropriate safeguards for network operations. In essence, there are multiple users on the network but only one user at a time may access any given file. This eliminates several changes being made simultaneously to the same file, and the resulting data file errors at the Central OI.

CENTRAL OI ENHANCED FEATURES

A number trends have been spawned as a result of expanded Personal computer use within control systems. Along with the growing awareness of features that the Central OI offered to systems, a proliferation of Business software has resulted in a demand for more data. The rapid pace of development in the computer industry has led users to expect the same type of developments with specialized packages for the control industry, such as an Central OI. Some of these requests have been satisfied through the continuing trend toward standardization and use of off-the-shelf technology with Central OIs. Using such standards as Microsoft Windows[6] as a foundation for a Central OI package makes a wealth of enhanced software features available through other compatible software packages.

Expanded familiarity with PC technology and trends has also stimulated ideas for enhancing the Central OI. Concepts to improve both the quality and productivity of system management and reporting using the PC are developed daily. The "Central OI Enhanced features" discussed here typify these types of trends. They may be described as non-conventional Central OI capabilities, but it may be more accurate to refer to them as integrated business and management features.

DDC Operator Interface: Handheld and Central

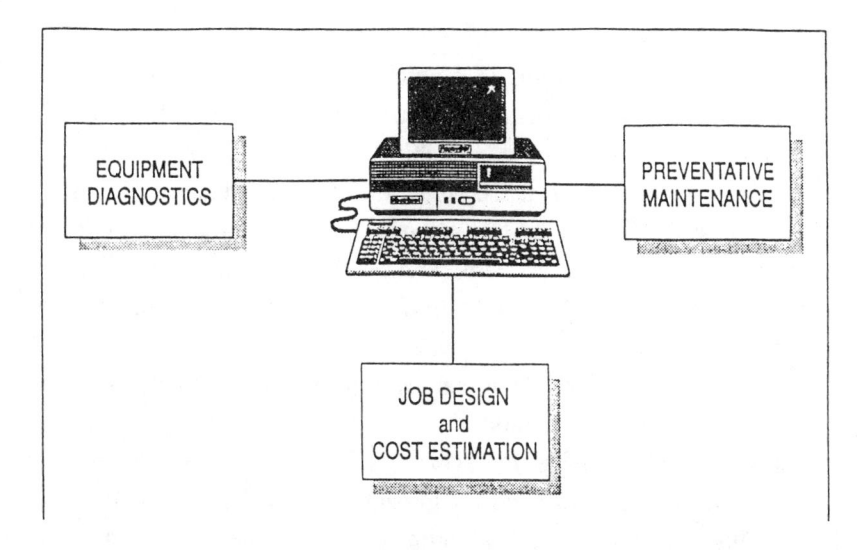

Figure 13-4 Enhanced Central OI Features

The Central OI has been targeted for a host of enhanced features in control industry media, but only three key items will be covered in this text.

- Job Design and Cost Estimation Tools
- Equipment Diagnostics and
- Preventative maintenance.

These are important because of the impact they offer to system benefits. Also these types of tools would be special product developments because they are specific to our industry. This means that they would not likely be available through off the shelf technology like a spreadsheet, etc.

Job Design

There are two areas of activity that are important to note under this topic: Architectural Design and System Design. Architectural design systems are termed Computer Aided Facility Management/Computer Aided Design CAFM/CAD products. These are germane to this discussion because they are currently in use and they point out some of the exciting possibilities for developing System Design tools.

Architectural Design

CAFM/CAD systems are used with both new and remodel architectural projects. They can operate in networked and multi-user/multi-tasking applications to expand the availability of data to other systems. It is important to mention that in the context of enhanced features, a system is defined as: PC Hardware, PC Software, Custom microprocessor based products such as a DDC System, Security System, etc. These systems integrate space design along with architectural, electrical, plumbing and mechanical systems.

This section will provide an example of the process used with these systems for architectural remodel. Access to building data is extremely useful in analyzing square footage needs, and developing an ongoing database of area, furniture and equipment required for various building users. This type of data can significantly speed the time required to design or remodel office areas. Kenneth Ledeen of Sigma Design, Inc. discussed a John Deere case study on CAFM/CAD systems in a recent Building Operating and Management article.[7]

John Deere's 1.1 million square foot, Waterloo, Iowa foundry, made use of a CAFM/CAD system to analyze space modifications, for a series of renovations, with the maximum cost effectiveness and efficiency. This system has allowed them to manage large projects and avoid the inaccuracies and difficulties associated with iterative revisions on architectural drawings. Combining an existing database of building information, along with computing power of the design system, has allowed John Deere to speed the review process as revisions occur. Highly accurate 2 or 3 dimensional drawings can also be developed quickly.

The benefit of computing power allows the system to check facility and equipment dimensions and verify measurements of new additions to ensure an accurate bill of materials. With the existing database it is also possible to generate two prints for each remodel job, one depicting the facility currently and one showing the modifications. Further, ID tags may be assigned to new and existing furniture and equipment which is invaluable for tracking inventory. As will be discussed later, this data may also be shared with other management systems which maintain accurate records on fixed assets.

DDC Operator Interface: Handheld and Central

Using the same capability with electrical, mechanical and control drawings can significantly enhance the job design process. Focusing in particular on controls, to the authors knowledge, systems of this type for control job design are not available. However, given the benefits they could offer it may not be long before such developments occurs. It would be ideal, for example, to use existing CAD files in development of job drawings, thus speeding the development of submittals, and adding control equipment to the owners final software inventory of components. Job drawings could easily be developed to integrate the building CAD files with Control templates. Such templates would be on file in a PC and include control components, interface to networks, Central OIs and ancillary equipment. Integrating the job design tool with a spreadsheet could fully automate estimation with equipment pricing database and calculations for margins, return on investment data and other factors. A word processor would also allow generation of proposals and submittals.

Examples of additional software systems that are available today and could be expanded for control job design include: packages which allow 3 dimensional piping layouts with symbols for connections and fittings and an R.S. Means packages. R.S. Means Company offers integrated estimating and scheduling software with direct access to over 40,000 lines of Means cost data. This system also address such issues as: change orders, bids and cost overruns. Another package from the Computer-Aided Design Group expands architectural design to include complete cabling layouts for computer and communication networks.

An interesting movement underway in the facility management industry is attempting to integrate data available with the multitude of design packages. Reading about the efforts underway in that industry indicates the similarities between various computer based industries. These users, like controls users, are dealing with development of a standard for maintaining data between software packages. Currently there is ongoing research underway at Stanford University's Center for Integrated Facility Engineering.

In a recent article in Building Operating Management titled "Integrated Building Data Base,[8]" Curtis Peltz outlined the process and the benefits associated with this software standardization. This research is oriented towards establishing file standards between the multitude of CAD systems currently in use. Today one owner may have buildings that were designed on these various non-compatible CAD systems. That individual would be unable to get access to the data from package A through another package. Standards would allow building owners to gain the types of benefits John Deere is reaping in Waterloo for all existing buildings. As such trends continue to develop, it is not inconceivable that, in the future, submittals will be made via floppy disk or modem interface.

An integrated building database standard would make it possible for owners to save time and cost in redesign, while at the same time maintaining highly accurate files on each building. In much the same way, Job Design Tools for control systems would enable contractors to enhance job profitability through a number of features being applied with CAD systems. Effective job layout would allow for smooth system start up and quality system documentation for the owner.

Equipment Diagnostics

The second major area to be covered under Central OI enhanced features is system equipment diagnostics. This is an evolving feature that allows the Central OI to help reduce service costs. Controller diagnostics have been available with existing devices for several years. These functions initiate either manual or automatic self test routines to provide a verification of controller operations. In some cases these routines will diagnose a component on a pass/fail basis. For more complex devices, it may also possible to test such components as inputs and outputs independently.

Equipment diagnostics, unlike controller specific routines, assume operation of the automation system to allow diagnosis of the controlled apparatus. These sequences vary widely. At the simplest level routines are well suited to application specific distributed controllers. For example, the microprocessor power in a distributed air handler control can also be employed to identify failure conditions and diagnose causes through monitoring controller points. As an example, the air handler control identifies a discharge air low limit alarm. Assuming that the required inputs are available, a simple routine verifies: hot water temp at the coil, outdoor, return and mixed air damper position and fan status. Based on the results, a determination of probable alarm cause can be alarmed to a local or remote system.

Application specific device are capable of identifying a majority of the alarm conditions, and can initiate routines to determine alarm causes. This data can be extremely beneficial to the service contractor. Conventional systems often identified such data in simple formats such as a two or three character code that the service person can find in a look up table. New systems general have a Central OI or more powerful handheld interface on site to provide more data in full english interface. Equipment diagnostics provide a full range of tests to pinpoint exact service necessary, tools and parts required along with other pertinent data, thus preparing service personnel prior to dispatching them to a site. In the above example, diagnostics might indicate a faulty outdoor air damper motor causing the damper to be stuck open.

More sophisticated equipment diagnostics can be implemented through high powered application specific controllers or via interfacing a Central OI or a PC with the control system. The PC would most likely be a temporary part of the system, and allow an operator to diagnose the entire system of controllers. Many advances have been made in this area, and some users are beginning to use "expert systems" to mimic the steps a person would take in resolving an HVAC or control system problem. These systems are programmed by human experts with extensive application experience, the process is called knowledge engineering.

Providing knowledge engineered routines in a high powered control device or network member would allow the same type of diagnostics outlined above. In addition, this type of functionality can be applied to multiple controllers, and also to integrated control features. With a PC based system, more extensive routines can be provided to aide in system start up, shake down and tuning for air balance, etc. Expert system approaches to diagnostics for an entire DDC system could significantly reduce contractor call backs associated with comfort problems. In much the same way, applying expert diagnostics to existing systems with comfort problems would be an effective service tool. The Expert system could be installed temporarily at the site in either case.

Full scale integration of equipment diagnostics with other Central OI features is a product for the future. However, knowledge based systems are already being developed to address job design tasks as outlined previously, and to do HVAC diagnostics. An interim step in use by some manufacturers is to employ knowledge based systems with telephone support. These system are programmed with all known data to answer question that are asked about various products. With regard to controller integrated knowledge based systems, there are several related activities being researched along with these diagnostics including indoor air quality. It is a reality today that efficiency and comfort issues share the spotlight with indoor environmental concerns.

The diversity of the tasks and systems described above lends more credence to the movement for standardized PC software. This also highlights yet another area of concern for control industry protocol standardization, as discussed earlier in this book. The Central OI would be an essential component of integration between the various software packages and databases. It is clear though that additional research and development are required to bring many of these features to fruition.

The benefits of these systems are obvious for the building owner with multiple facilities or very large and complex buildings employing a variety of systems. Higher powered, lower cost microprocessors may also make it feasible to combine control, knowledge based systems for diagnostics and other functions in a building supervisory panel designed to simplify owner management, and to allow temporary or permanent installation.

Preventative Maintenance

A key component of any system is preventative maintenance (PM). Preventative maintenance expands the types of equipment benefits available through diagnostics to regular service. A host of packages have been developed and marketed over the past 7 to 10 years, and many advancements are being made in this area. In reviewing any mechanical procedure, these packages can provide a building owner or maintenance manager with:
- an outline of the tasks to be performed,
- when to conduct the activity,

- identification of trades needed to do the job (for enhanced scheduling) and
- a list parts and materials required.

The benefits of PM have been documented over the years. A good program can extend equipment life, and minimize both the cost and lost income associated with unexpected equipment failures and downtime. A computer based PM program adds new benefits to:
- enhance scheduling of tasks through automated work orders,
- enhance task efficiency by providing a detailed list of parts, tools etc. required,
- maintain accurate maintenance records,
- manage subcontracts and invoicing,
- provide valuable report data for management planning,
- automate inventory and parts purchasing systems and
- provides complete audit trails on expenditures.

All the computerized PM features discussed contribute to ensuring that repairs can always be made on a timely basis and are very useful in the budget planning process. There is a slight downside to computerized PM, and that is the time requirements for maintaining Preventative Maintenance computer package databases. These systems are only as accurate and valuable as the database of information used to develop schedules, so this is an essential task

A key added benefit, which has only been available with some packages, is integration with a local area network (LAN) or DDC system. There are systems on the market today which allow PCs on a LAN throughout a large facility to access the same database and reports. By integrating a DDC system, the PM package could schedule work based on actual runtime, rather than on a calendar basis, manual readings taken by a trades person or on projected use scenarios. The speed of response to emergency maintenance could also be enhanced through DDC systems interrogation of the mechanical equipment to ascertain the trade and parts required for repairs. Few if any systems currently available offer this type of feature for all the obvious integration reasons previously mentioned.

CENTRAL OI AND PROTOCOLS

A Central PC interface may be used for a number of functions: programming, monitoring, alarming, local and remote communication and even control. Yet accomplishing many of these functions through a PC interface with controllers, and other systems, must return to protocols and the rules for communication between the system members. Protocols, as discussed in the networking chapter, allow sharing data, exchange of commands and managing information flow between controllers and Central OIs. In the simplest terms, data must be available in a form that the PC can make use of in order to provide any of the features discussed previously.

One of the first complaints made by users was that Central OIs used proprietary data protocols and made it impossible for them to use that data. Again using off the shelf technology as the based for today's Central OIs, and making data available in industry standard data file format has helped to address these problems. However, it remain possible in only rare cases to use one Central OI to communication with multiple manufacturers systems.

Users have been clear in requesting interface capability for multiple systems through one Central OI. Bear in mind that this issue impacts onsite applications and remote monitoring. Onsite interface becomes an issue for many applications, consider a college campus with multiple systems installed in various buildings. To use one interface for all of the systems installed would require a standard communication protocol. At the same time, remote interface with multiple systems involves the same issues.

It becomes evident through these discussions of enhanced Central OI features that the issue of integration is quite significant. A standard protocol for interface with Central OI software data files to allow greater efficiency in building management is desirable. The extensive benefits this integration could bring to a multi-location or large complex user, as well as, to a single system user more than justify added work in this area. However, the controls industry must address more immediate concerns first, that means addressing a standard protocol for controller interface.

Universal Front Ends

This chapter can not be closed without a brief mention of universal Central OI software packages. These types of products have been available in the industry for several years, through a number of independent software houses. They employ computer code internal to their operation which allows them to communicate with controllers from more than one manufacturer. The idea of integrating enhanced features has not been addressed as yet by universal software, but it may be closer to reality through this vehicle that by control manufacturer developed software. In essence these are Central OIs, primarily focused on local and remote monitoring functions, that can aide the multiple system own in performing DDC management tasks.

This chapter has provided extensive coverage of Central Operator Interface systems. The importance of these packages to distributed systems is unquestioned. This technology is critical to every DDC system, whether as an on site component of the system or for remote interface via modem. As will be discussed under managing DDC systems for success, this capability is invaluable to any system owner. Further, this component of the overall DDC system is anything but static, in terms of new features and functions.

The focus of this chapter however has been the critical features to be carried out by an Central OI today. Yet users must continually evaluate new roles for the Central OI with their systems. In that light, note that on site Central OIs are being applied in a greater percentage of jobs. In addition, the continuing enhancements to these packages make it critical for users to pay special attention to Central OI research and development.

FOOTNOTES

1. Microsoft Disk Operating System (MS-DOS) is a registered trademark of Microsoft Corporation, Redmond, Wa.
2. IBM is a registered trademark of International Business Machines Corporation, Poughkeepsie, N.Y.
3. Operating System 2 or OS2 is a registered trademark of Microsoft Corporation, Redmond, Wa.
4. Apple MacIntosh is a registered Trademark of Apple Corporation, San Francisco, Ca.
5. LOTUS 123 is a registered Trademark of Lotus Development Corporation.
6. Microsoft Windows is a registered trademark of Mircosoft Corporation, Redmond, Wa.
7. "CAD/CAFM", Kenneth Ledeen, Building Operating and Management Magazine, Milwaukee, Wi., October 1989
8. "Integrated Building Database", Curtis Peltz, Building Operating and Management Magazine, Milwaukee, Wi., July 1989

Chapter 14

Financing Energy-Saving DDC and Other Retrofits

By Rita Tatum

DDC systems and energy management retrofits are excellent ideas, correct? Throughout this book we have talked about energy cost savings, increased comfort, productivity and many other benefits. In spite of these benefits, the capital investment required to finance these projects is often the major obstacle to executing a project.

In the 1970s and early 80s these projects were easier to justify. Energy costs were projected to increase exponentially over the coming decades, and America was committed to reducing oil imports. With the "oil glut" in the early 80s, and the reduction in energy prices, the incentive and the interest in energy management waned somewhat. Projects investments became harder to secure, and that makes financing an extremely important topic for this book.

There has been resurgence of energy management interest in the late 80s and early 90s. This interest accompanied concern for the envi-

---

Rita Tatum is Contributing Editor with Building Operating Management magazine. This chapter is reprinted with the permission of Building Operating Management. It appeared as an article in the March 1992 edition.

ronment, and was a positive influence in justifying more investments. Yet it is extremely difficult in many cases to quantify environmental benefits from energy projects to an owners satisfaction. As a result, DDC system investments are competing on equal ground, at best, with every other capital project for funding.

Energy projects may even be at a disadvantage if the other capital projects appear to be directly aligned with the companies' mission. This is the case in the private sector, and even more pronounced in public facilities that are suffering reduced budgets. Schools are particularly good examples, because in many states the children of baby boomers are entering school and requiring significant new construction. This combined with increased costs for teachers, books, etc., diverts fiscal attention from dealing with energy use in existing buildings.

These issues, along with a prolonged economic downturn and slow recovery, make capital resources and the topic of financing critical. This chapter provides a good clear definition of the more common alternatives to direct purchase of equipment. Direct purchase is always the best alternative if resources can be secured. No financing or administrative charges are incurred and the owner begins to benefit through cost savings from the day installation is complete. The reality is that this approach is not always available, and financing alternatives do exist. It is far better to install the system and have your return somewhat reduced than to take no action and continue to miss opportunities.

We know that capital improvement dollars are scarce for DDC systems and other energy-savings retrofits. Yet this does not eliminate the potential for taking action to implement energy management. Lease-purchase agreements, performance contracting, and utility demand-side management and rebate programs offer the building owner and manager several viable financing options.

LEASE-PURCHASE AGREEMENTS

For HVAC units, DDC systems and some other projects which are capital-intensive purchases, building owners and managers may wish to consider a lease-with-option-to-buy plan. Though there are many lease forms available today, manufacturers generally offer capital and/or operating leases. Either lease can run up to five years and contain an option to purchase the system at the end of the lease term.

The five-year lease term can be important in some areas, where utilities provide energy-saving rebate programs. For example, Pacific Gas & Electric (PG&E) co. offers financial incentives for high-efficiency energy equipment that will be in service for five or more years.

"If the building manager has a five-year lease, then those incentives may be available," explains Denise Rushing, director of PG&E's commercial, industrial and agricultural retrofit program.

PG&E figures the new equipment's efficiency against federal and California's Title 24 requirements. "The new efficiency is subtracted from the equivalent federal or state standard," says Rushing. "On average, we find customers are saving $200 to $300 per kilowatt."

On January 1, PG&E began offering direct rebates for energy-saving equipment through its commercial retrofit program. To qualify, the high-efficiency equipment must exceed state and federal standards.

Option to purchase may be offered

Offered by some manufacturers, a capital lease, sometimes called a finance lease, often includes an option to purchase the equipment at a price below its market value at the end of the lease term. The final price is pre-determined in the lease agreement and can range from $1 to a percentage of the system's purchase price.

On balance sheets, building owners often list capital leases as conditional sales or security agreements because they often exercise their option to purchase the equipment at the end of the lease. Operating or "true" leases allow the lessee to purchase the equipment at its market value, renew the lease or return the equipment. Iowa also relies on lease purchase plans for the energy-conservation efforts in about 400 buildings scattered across the state. Payments are structures to that the actual energy savings should retire both the principal and the 7.34 percent average interest rate within 15 years.

The program is predicated on an initial cost of $2.75 per million Btus for natural gas and the 1987 rate for electricity at each location.

An inflation factor of 4 percent per year was used in calculating the lease-payment plans.

Equal monthly payment options

Many of the manufacturer's concessions available in purchase deals also are available in leasing arrangements. Both equipment distributors and financial institutions that offer equipment purchase loans often offer alternatives to the equal monthly payment schedule. For example, there are skip-payment options, high-low and low-high plans, and sum-of-the-digits methods.

Using a skip-payment plan, the company may choose to skip a payment during tight cash flow periods. Both sum-of-the-digits and high-low plans reduce interest charges by paying the lender back more rapidly. Under these plans, monthly payments are higher during the early months or years and lower in later periods.

Such plans work well when current finances are strong, but future prospects are uncertain. For example, if a number of 5-year tenant leases are up for review in two years, sum-of-the-digits or high-low plans may offer a good option for building owners and managers.

The low-high plan keeps payments low for the first year or two and then raises them. This plan increases finance charges, but allows the building owner to acquire equipment that can save energy costs today. Many times, the energy dollars saved by upgrading more than compensate for the increased finance charges.

Performance contracting

Another financing option gaining increasing acceptance from building owners and managers actually began in the institutional and governmental markets about a decade ago. During the 1980s, a number of energy service companies (ESCOs) began filling a niche for school systems, who often were bound to accept the lowest bid on projects. However, the lowest first cost often meant higher life-cycle costs, particularly for energy intensive equipment.

Offered through ESCOs, performance contracting is a debt-service obligation, often a form of lease-purchase agreement or a shared-savings contract. "Basically, if the ESCO doesn't achieve a guaranteed savings, then the contractor doesn't get paid," explains Shirley Hansen of Hansen Associates in Annapolis, Maryland. Hansen is the author of Performance Contracting for Energy and Environmental Systems, published by Fairmont Press.

Currently, Hansen estimates there are about 20 ESCOs that offer their services on a nationwide basis, with many more available at the local and regional levels. According to Hansen, performance contracts do not follow any one formula. In fact, they are tailored to the individual customer, though some bear a resemblance to more traditional lease-purchase agreements.

ESCOs look over the building to find energy-saving opportunities that are cost effective within a reasonable payback period. They bring a proposal back to the building owner outlining what they can accomplish. Provided the owner agrees, the ESCO then installs – often at no initial cost, though some may charge a modest construction cost fee – those building components. Maintenance, monitoring and other details are ironed out with the building owner. Then a schedule, addenda and a master contract are drawn up.

College avoids upfront capital costs

One notable performance contracting example is Adrian College in Michigan. The performance contractor arranged for third-party financing for the entire $320,000 energy-saving retrofit, so that no upfront capital was required from the college.

In return for the performance guarantee, the college paid a portion of its excess energy savings to the contractor. During the first two years, that portion was 30 percent, declining to 25 percent for the next two years. For years five and six, the percentage dropped to 20 and the ESCO received a 15 percent share during 1990, the final year of the agreement. During the first three years, the program's improvements saved the college more than $500,000 in energy costs.

A decade of reasonably priced energy weeded out many ESCOs. Those national ESCOs left today are reputable, according to Hansen. "Those still out there have proven track records," Hansen observes.

"So if you're trying to evaluate one ESCO vs. another ESCO in performance contracting, you probably should decide based on your local area's service."

Utility programs

Among utilities, the buzzwords for the 1990s in energy conservation are three big words: demand-side management (DSM). Essentially, DSM encompasses all electric utility services that impact how customers use electricity.

The DSM concept is two-fold. The first strategy of DSM is to even out the demand for electricity so that existing power stations are operating at efficient capacities from dawn to dawn, rather than skyrocketing during business hours only to dive-bomb six feet under for the graveyard shift.

The second strategy is to deter the need for new electricity capacities. According to the Edison Electric Institute (EEI), DSM programs have grown from 134 in 1977 to nearly 1,300 today. DSM programs have deferred more than 21,000 megawatts (MWs) already; EEI projects by the year 2000 more than 45,000 MWs of capacity will be diverted with DSM programs.

The terms utilities use to describe DSM include peak clipping, strategic conservation, valley filling, load shifting, strategic load growth, and flexible load shaping, which may include interruptible service or curtailment periods for commercial customers.

Peak clipping refers to reducing customer demand during peak electricity use periods, often by using some form of energy management system. Strategic conservation often is rewarded by utilities with rebate programs. It may include building energy audits, weatherization, high-efficiency motors, Energy Management or "DDC" systems and HVAC systems, and numerous others.

Valley filling increases electricity demand during off-peak periods, allowing the utility to use its power generating equipment more effectively. Load shifting is similar to valley filling, because it too uses power during off-peak periods. Both valley filling and load shifting programs often are used by utility representatives to help sell thermal storage systems.

Strategic load growth

Another DSM program that resembles these two is strategic load growth, which encourages demand during utility-selected seasons or specific times of the day. Flexible load shaping adjusts the load according to operating needs and may result in interruptible or curtailment rates for manufacturing customers.

Large customers who participate in Consolidated Edison's curtailable electric service program can reduce their summer electricity bills without buying new equipment. During each of New York's four summer months, these customers agree to reduce electric demand by at least 200 kilowatts on request. More than 100 organizations participated in this program during 1990.

Duquesne Light Company in Pittsburgh and Georgia Power also have interruptible economic development rates that operate similarly. The Pittsburgh utility offers energy audits and works individually with customers on how to use energy most efficiently.

Utility-backed programs

Georgia Power Company, part of the Southern Company, offers Southern's Good Cents building program for commercial customers with HVAC and lighting rebates, along with energy audits. Georgia Power also has developed an indoor lighting efficiency program, according to Paul Bowers, vice president of marketing, "which we are proposing to go statewide this year." The program was filed with Georgia's utility commission in January.

Besides rebates, some utilities offer low- or no-interest equipment loans; financing, leasing and installation assistance, and assured payback programs. For example, Wisconsin Electric Power Company encourages third-party arrangements with ESCOs.

According to Mark Jacobson, Wisconsin Electric's new product development specialist, the utility currently has four ESCOs – Kenetech

Energy Management Co., Johnson Controls, Viron and HEC Energy Services – participating in its Smart Money program.

Wisconsin Electric Power Company also offers rebates up to 50 percent of the project cost and loans with multiple rates and terms (from 0- to 16- percent interest) for three to seven years. These programs are available to building owners and managers who install energy-efficient lighting, HVAC systems, window glazing, high-efficiency motors or building automation systems.

Commercial and industrial customers account for more than 70 percent of Con Edison's annual electricity sales. As a result, this utility offers numerous Enlightened Energy programs, including energy audits and ApplePower rebates for efficient lighting, steam air conditioning, gas air conditioning, high-efficiency electric air conditioning, cool storage and high-efficiency motors.

At least one utility, Houston Lighting & Power (HL&P) Company, offers its renovating customers a commercial cool storage program. To encourage the use of cool storage technology, HL&P provides qualifying building owners with a $300 cash incentive for each kilowatt reduction in peak demand.

In addition, building owners are eligible for the utility's cool storage billing rate, which redefines the on-peak demand as noon to 7 p.m. Monday through Friday for the entire year. This billing rate also eliminates the annual on-peak minimum demand charge or "ratchet", allowing building managers to use their cool storage systems effectively year-round to reduce their demand charges.

Building values increased

Alan Ahrens, HL&P supervisor of load management, says that "because of lower operating costs, a cool storage system will pay itself off in about three years or less and additional savings can be expected for the life of the equipment.

"For the commercial property developer or owner who's interested in a long-term investment, cool storage systems are the best way to go. In other parts of the country, building owners have seen their buildings

Financing Energy-Saving DDC and Other Retrofits 333

increase in value and marketability as a result of cool storage," says Ahrens.

Currently, the utility has three cool storage systems on-line, including a pilot system at its H. O. Clarke Service Center. Three more systems will begin operations this year. In the Dallas/Fort Worth area, Texas Utilities has more than 135 cool storage systems in operation.

Manufacturers who team up with utilities can save even more. For instance, Kraft General Foods (KFG) and Boston Edison formed an Energy-Efficiency Partnership that will reduce the ice cream manufacturer's costs dramatically. "Through this project, KFG was able to decrease their electricity by one-third, from 7.5 cents per gallon of ice cream to 5 cents per gallon," observes Patrick F. Doyle, manager of Boston Edison's energy management department.

The partnership allowed the ice cream manufacturer to upgrade most of its electrical energy-consuming capital equipment and obtain substantial rebates for the energy saved. The rebates from Boston Edison will return more than 85 percent of KGS's $3 million investment. In addition, more than 6 million kilowatt hours – one-third of the plant's total energy usage – will be saved each year. The project includes refrigeration and defrosting equipment replacement, variable lighting installation, and monitoring equipment implementation.

Direct Digital Control and Utility programs

A survey of DSM programs published in Energy User News, July 1992 showed that a third of the utilities surveyed offer rebates for controls. The programs offer two separate rebates for controls, one for lighting and a second for HVAC. The technologies that may be installed for the rebate include timeclocks, lighting controls or Energy Management Systems, referred to in this book as Distributed DDC. End user surveys have rated control systems as the second most popular retrofit in commercial building, and the rebates make these projects even more attractive.

The subject of commercial Demand Side Management (DSM) technologies was also covered in a guidebook published by the Western Area Power Administration. This is a federal agency that oversees a wide variety of public utility programs, such as Bonneville Power. In that guidebook, the administration provided evaluations of technologies that utilities may incorporate into rebates, and other types of DSM programs.

The Commercial Technologies guidebook described Energy Management, or DDC, Systems and their functions. It also presented rough cost estimates, and the results of a survey done by the Electric Power Research Institute (EPRI). EPRI surveyed 200 EMS users, and found that the actual savings from these systems were 15% of the building energy use. Based on the cost estimates established, EPRI also showed simple payback for 3 different buildings with varying sizes and control complexity. Again their survey showed that paybacks ranged between 1.5 and 3 years. These results provide further credibility for control system projects, and explain the continuing interest in DDC systems for Utility DSM programs.

An enviable position

Commonwealth Edison Company in Chicago is in an enviable position because its baseload capacity through 2000 is already in place. Nevertheless, Commonwealth Edison offers its Least Cost Planning, which includes a load reduction cooperative program, modeled on Boston's Northeast Energy Cooperative. "Ours is piloted to reduce 20 MWs," explains Jerome Hill, director of Least-Cost-Planning. "Currently, we are testing the cost effectiveness of this concept by operating a pilot program for three years beginning this summer."

Selected from the 200 largest consumers of electricity, the first six customers now are getting a detailed walk through. If the final program works similar to ones in Boston and California, a group of businesses will cooperate to curtail or reduce their electricity consumption to a prescribed limit when the utility requests it. Compensation by a special electricity rate is performance-based, with the worst performance during any curtailment period being the base for all electricity charges.

Ken Pientka, Commonwealth Edison's technical programs administrator, says his section is analyzing different segments of the marketplace for natural gas and electricity consumption. Commonwealth Edison did this for the hospital industry, charting those that were in the upper third for energy use and costs. Opportunities for energy savings were presented to these customers, along with typical capital outlays and payback periods, a followup seminar, and methods of implementation.

During the late 1980s, utilities began aggressively offering commercial rebate programs to meet DSM objectives. The rebates help defray high-efficiency requirements' initial first costs. Some utilities pay 30 to 50 percent of the installed cost, while others base their rebate programs on the peak-kilowatt-demand savings achieved by the new equipment. But these vary across the country, depending on the individual utility's DSM needs.

Chapter 15

Plan and Specification of Graphic Central Operator Interfaces For DDC/Building Automation Systems

PREFACE

One of the most complex DDC or Building Automation System topics remains the Specification. Rapid changes in technology combined with decreasing costs and more system retrofits have had major impacts on the traditional control specification. In fact specifying systems may in some cases be as much an art as it is a science. Given that complexity this book includes two chapters on this topic.

This first chapter was written by Harris Bynum, one of the most knowledgeable controls engineers in the Industry. Though it is titled Specifying Graphic Displays, the chapter is provided first due to the valuable insight Harris brings to this topic. Rather than taking a micro-view of Graphic Display as one section of System Specifications, Harris presents a holistic approach to the topic. With exceptional clarity, he shows the ideal outline for an overall specification, and then the impact that Graphics have on appropriate sections.

Prior to presenting Harris' chapter however, it is important to consider the industry changes noted above. How is the concept of specifications changing and what is their role in the control industry. To assume that every system installed in the United States during a given year is "specified" would be a mistake.

The nature of both construction and energy management as industries, along with the economy and its impact on new building starts, has changed dramatically in recent years. At one time there was a significant marketplace involving new construction and perhaps 60 to 70 percent of the automation systems installed were in new facilities. Today as many as 50 to 60 percent of the systems go into existing building retrofits, and Plan Rooms in U.S. cities are not the scene of bustling activity of previous years. Why this trend, one reason is that in recent years only about 70,000 buildings have been built in the U.S. That number sounds large, but it is quite small when viewed as a function of the built environment which includes over 4.5 million existing buildings. Yet to say the new construction or control system specifications are a thing of the past is also incorrect.

Another reason for the retrofit trend is that Building owners discovered the many DDC benefits, beyond energy and cost savings. As a result they are installing these systems in existing buildings across the country, and the world, with greater frequency. In fact DDC systems are rapidly becoming as common in commercial buildings of all sizes as they have been on Industrial processes for more than two decades.

Increased DDC implementation has been further spurred by the reduced costs for these systems today. All of this means that more and more systems are retrofits, and do not get installed through the typical new construction process. This simple fact has resulted in a major shift in control system sales and delivery. Rather than choosing an installer based upon a bid situation, many jobs are now negotiated. It is more likely for a user and the vendor of an existing system to design an expansion or upgrade of existing equipment outside the typical bid process. In many cases that user may work with a turnkey Contractor to fill the role of the consultant, installer and system commissioner.

In spite of these continuing changes in the controls industry however, many systems continue to be installed in new construction or with

Plan and Specification of Graphic Central Operator Interfaces For DDC/Building Automation Systems

major remodels to existing facilities. Beyond that it remains critical for building owners, even with a turnkey contractor, to clearly outline their system requirements. As a result, this information is as critical as ever. Because of these needs Harris Bynum's chapter on specifying graphics with a building operator perspective is must reading. The next chapter provides another alternative for DDC specification, and looks in more detail at each individual section.

There is always a risk of confusion when presenting two different perspectives on a topic like specifications. Yet in the end this was deemed the best approach because a designer must view many approaches to identify the best technique for them. These chapters on Specifying DDC systems should be reviewed along with other data in determining the best approach for a particular specification. Consider these chapters as background. It is in no way recommended that these alternatives be viewed as anything more than examples.

Plan and Specification Documentation for a DDC/BMSystem

By Harris Bynum

INTRODUCTION

One need only attend most ASHRAE symposia on the subject of DDC to see the desire for greater knowledge, training and education on this whirlwind technology. The good news is that in most cases DDC is doing the precise, stable, reliable task promised and specified.

The author detects a serious desire among designers of direct digital control (DDC) systems for a better understanding of this technology, too often perceived as black box magic. As HVAC system designers learn more about the structuring of DDC programs, the nature of communication protocols, and the capabilities of 1990 central operator stations, they will design more synergistic and useful control and management applications more consistent with the power of the processors specified. Through the specification of more and more powerful graphic displays, a much better understanding of DDC and the HVAC system will prevail, and Building Management System (BMS) users will be more effective. In addition to suggestions of better DDC/BMS "System" applications, the author also offers, from a vendors perspective, constructive comments related to the balance of the DDC/BMS system specification.

Reviewing almost 2000 control and automation system designs over 25 years shows a sound evolution from pneumatics to electronic

---

Harris Bynum, Atlanta, Ga., is National Manager of Automation Systems, Technical Support Department of Honeywell Commercial Buildings Group.

Plan and Specification of Graphic Central Operator Interfaces For DDC/Building Automation Systems

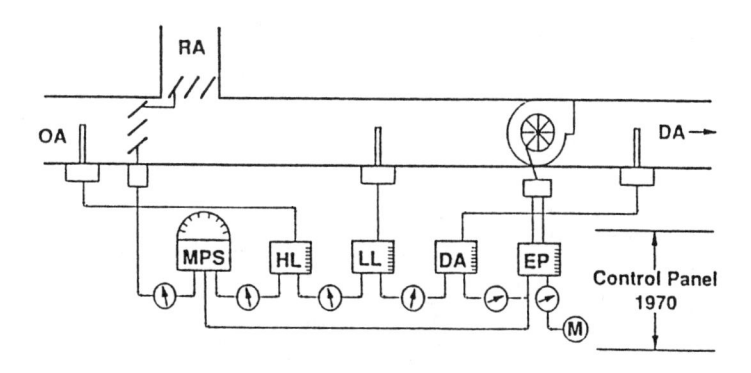

Figure 15-1 Pneumatic Control System

automation to digital energy management. But as the energy management system design process broadened onto DDC, something fell through the crack. As one of North America's largest users of DDC phrased it, "the problem with DDC is that you cannot see the program".

The complex pneumatic control strategy was easy to observe through a series of gauges as noted in Figure 15-1, but this "picture" is not available on DDC jobs. Oddly enough, the only reason for this is that it is not specified. This chapter, therefore, addresses the designer/specifier's DDC documentation in the interest of education. "Documentation" in this context means the specification document itself. Equally as important, however, are the software graphics which make up the system owners living documentation of the ultimate product that is delivered. As a result this paper will focus on how to define more clearly the requirements of vendors and contractors installing the system, as well as those of the ultimate user of the system for graphic operator displays.

To further illustrate the new requirements for graphic specification, consider Figure 15-2. It shows a graphic portrayal of a simple hot water convertor with outside air reset, four "points", and a picture, as it has been specified for many years.

Figure 15-3 shows the same converter, as it could be portrayed on the screen of a DDC/BMS, but can you see the program? Here it's

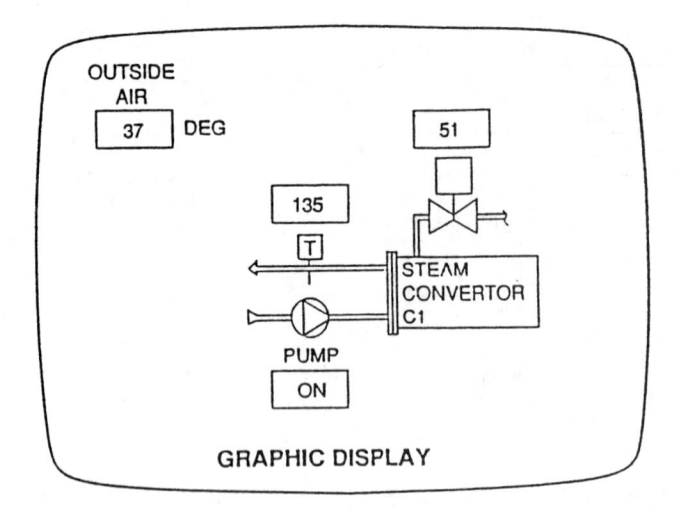

Figure 15-2 Graphic Display of Hot Water Converter with "Points"

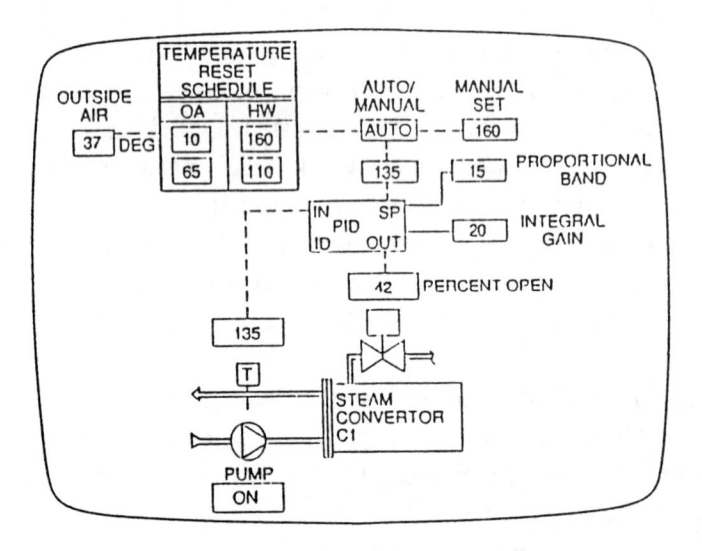

Figure 15-3 Graphic Display of Hot Water Converter with Points and Program Elements

apparent that the PID (Proportional, Integral, Derivative) setpoint (SP) is varied by changes in outside air temperature unless the AUTO/MANUAL switch (software switch) is placed in the manual mode. Not only can the four physical points and the program be dynamically monitored, but the reset schedule, proportional band, integral gain, auto/manual switch, and manual mode set point can be modified directly from the graphic in the comfort and convenience of the BMS office. The "program" need not be black-box magic.

If graphics similar to Figure 15-3 are required, the control and automation design methodology used for the last 25 years of providing a points list is not a good tool. This is because many of the points do not have a recognizable name, and because the graphic is also a picture of a program in addition to a picture of a piece of HVAC equipment.

Achieving the goal of user friendly graphic interface diagrams require a clearly documented specification. Unfortunately, the evolution of DDC/BMS specifications through several technology era's has occasionally produced disorder within the document. In the interest of clarity and consistency and of accomplishing our goal, the next sections offer general direction on the order and contents of the document.

SPECIFICATION ORDER

Orderly design documents are the first prerequisite for an orderly DDC/BMS contract; one with a minimum of confusion and misunderstanding. The Construction Specifications Institute (CSI) specification basic order format of "General-Products-Execution" will work well if the designer will spend a day developing an orderly index, and then treat the index sacredly. All parties concerned; designers, owner/users, and vendors will greatly appreciate a well ordered and published index.

Following is a suggested orderly index:
    Part I General
        Scope/System Description
        General Conditions
        Bidder Qualifications
        Quality
        Bid Procedures (Approved vendors, proposals, etc.)
        Abbreviations

    Submittal Requirements
    As-Built Documentation and Manuals
    Warranty and Services
Part II Products (What the DDC/BMS is)
    System Architecture
    Central System Hardware
    Test, Portable, Diagnostic Equipment
    Remote Controller Hardware
    Remote Devices
        Sensors
        Valves
        Air Compressors
        Etc.
    Central (Operator station) Software
    Remote Controller Software
    Spare Parts
Part III Execution (Specific implementation requirements)
    Scope
    Sequence of Control
    Data, Control and Graphics Summary
    Installation
        Mechanical, Pneumatics
        Electrical
        Data Entry
    Validation
        Commissioning Procedures
        Demonstration Procedures
        System Reports
    Text/Diagnostic Equipment
    Training

SPECIFICATION CONTENTS

The index should be carefully followed. Detailed functional (software) definitions should be avoided under "hardware" paragraphs, and duplicity (which often evolves into ambiguity and inconsistency) should not be used.

Although "grandfather" type clauses ("even if this specification is wrong or incomplete, the vendor shall provide a right and complete system at no additional cost") often reflect weakness, one such clause is considered by the traditional BMS/DDC vendors to be responsible: "The requirements noted in this part, Products, are the minimum that will be accepted and nothing in this part relieves the BMS/DDC contractor from complete conformance to all requirements of Part III Execution." Basically, this says the BMS/DDC vendor must provide all control and monitoring specified even if the vendor's version of the system specified under Products will not perform as required. If controllers are specified to be at least eight bit processor types with 64K of RAM, and a particular vendor needs a 16 bit processor with 256K of RAM to perform the specified execution tasks in a timely manner, the larger controller is required. Software architectures and efficiencies vary greatly between vendors.

Part I – General

A main suggestion for the General section is to keep it general, or more specifically, do not get into detail regarding product or execution issues.

Part II – Products

Although the real purpose of this paper is to present suggestions for Part III Execution, a few comments relating to Part II are in order.

The products area of DDC/BMS specifications have been the most disorderly. The products portion of a competitive specification, while not very time consuming, is the most difficult, if not impossible to develop. A safe (protective of the owner and consultant), fair (to several competitors), and competitive product specification is often still accompanied by strong bidder qualifications or a preapproval requirement.

Conscientious consultants often try the LCD (Least Common Denominator) approach wherein they select two or three responsible vendors and try to remove everything from the specification that precludes any of them from complying. Strong "or" clauses are often much safer than removal of a spec item, such as "...EEPROM, or RAM with lithium or automatically rechargeable batteries selected for 72 hour backup power in the even of commercial power loss."

Skilled consultants are usually very sensitive about vendors proposing special additions and modifications to their "standard" system in order to comply with product specifications. The issues are many:

1. How will the vendor support the special developments over the next ten years?
2. Can the vendor incorporate the developments into the system with the allotted time?
3. What impact will the developments have upon the performance of the standard system?
4. Will the developments affect the systems code approvals?

Many consultants, therefore, see their job as one of critical evaluation of a vendor's product and system rather than one of designing a system to be developed. In either case, the skill level of the consultant must be considerably greater than that required for control systems of the fifties and sixties. Keeping up with the continuing evolution of software growth and partitioning, protocols, networks, controllers and microcontrollers of major vendors takes a very significant portion of a consultant's time.

Another suggestion that could simplify the product specification, submittal, and evaluation is to edit the design documents carefully for irrelevant requirements. An example is the detailed requirement for controllers to have "peer-to-peer" communication features allowing controllers to communicate with each other independent of the general purpose central computer when no specific controller-to-controller communications are required to accomplish the functions detailed in the "Execution" section. Another similar instance is a requirement for a duty cycle program when duty cycling is neither listed as an execution requirement nor is it applicable to the HVAC system designed (such as a VAV system).

It is recognized that often certain features specified are intended for specific or possible future use and will not be reflected in the Execution section except under the validation portion, within which demonstration of the feature should be required.

The words "capability" and "capable" should be avoided when describing computer software. Requiring a chiller optimization program to be provided is much stronger phrasing than requiring a controller to be capable of optimizing the performance of a chiller. All computers have tremendous inherent capability, but implementation of that capability can take extensive software efforts.

Part III – Execution

As noted earlier, execution is the primary specification section of concern in this chapter, and in particular the topic of "Data, Control and Graphic Summary." In general, the "Execution" portion of the design documents should tell the vendor what to do with the system products and devices. The "Products" section described the components. Here the specification outlines how they will produce precise environmental conditions in an energy efficient manner, and present complex digital network of points and programs in a simple and meaningful manner (graphic, text based or otherwise). Features detailed under the "Products" section should be used in the "Execution" section. Consultants usually consider the Execution section as sacred, i.e., if a vendor's system cannot perform a specified execution function, the system is not acceptable. A suggested index for the major elements of this section was shown previously.

Though this chapter highlights graphic interface and display, it is critical to set the style for further discussion. so the scope and sequence of control sections are each discussed briefly below. These discussions allow for more specific illustrations of how to specify graphics by clarifying the applications depicted later.

Scope

The "Scope" is a brief statement of execution requirements, such as "provide a fully programmed and implemented direct digital control and building management system as specified including pneumatic and electrical installation, all data and graphic entry and construction, validation, training, documentation and warranty."

As with all specification documents, DDC/BMS requirements are also noted on design drawings. Most of the requirements of a

DDC/BMS can be defined within the standard page sized specification with certain references on mechanical and electrical drawings:
- Location of filters, coils, dampers, equipment, etc.
- Location of space environment sensors
- Location of motor starters
- Sources of DDC/BMS power
- Location of plant room digital controllers

On larger jobs (where page size is too small), a BMS network drawing should be provided showing as a minimum:
- Operator stations and printers or System Master Panel/Peer Controller Managing Remote Communication
- Schematic of primary network controllers with primary network cable layout (not to scale)
- Secondary/primary network interfaces

Recognizing that if a digital controller fails, control of HVAC equipment fails, designers should define the maximum amount of HVAC equipment that can be connected to each single primary controller. Secondary networks should connect to related primary network controllers, i.e. a secondary network of VAV box controllers should be connected to the controller of the respective VAV unit to assure efficient control strategy communications.

Sequence of Control

In many cases, the sequence of control noted in specifications today is very close to that used for pneumatic control with energy management. But digital is different; different in repetitive cost, different in performance, and different in BMS and network capabilities.

Sequencing of Valves and Dampers

Prior to DDC, designers did a good job of specifying where pilot positioners were required for sequencing and where actuator spring range sequencing was acceptable. With DDC, another option, DDC sequencing with multiple analog outputs, is available and should be dictated where it is appropriate.

Where DDC sequencing is selected, normally open valves should be considered. Normally open valves are less costly, simpler, more powerful (for close-off), easy to close when equipment is off, plus they will open upon a pneumatic line or air compressor failure (which may mean rudimentary control of heating or cooling via pump and/or water temperature control.)

Position display of valves and dampers is an extremely important feature for the ultimate system user. One of the most valuable bits of information one needs to analyze HVAC system problems is the position of valves and dampers. Although these "points" are highly recommended, a detailed discussion on the source (software vs. hardware) of the point value is beyond the scope of this paper.

In allowing for graphic display it is important to consider the sequence process. The process of sequence of operation design is quite traditional.

- Sketch HVAC System; Ductwork, Fans, Coils, Dampers, Filters, Humidifiers
- Add Controlled Devices; Control Valves, Damper Actuators
- Add Control Sensors; Temperature, Pressure, Flow Stations, Etc.
- Add Safety Devices; Fire, Smoke, Freeze, Pressure
- Define the control and automation relationships between all of the above

A good sequence then is a verbal and complete description of the interrelationships of all the sketched elements. Good sequences, like good specifications are orderly, i.e.:

- System Start-Stop
    Optimum Start
    After-Hour Operation
    Night High/Low Limit
    Safety Relationships
- System Volumetric Operation
    Supply Fan Pressure/Flow Control
    Return Fan Pressure/Flow Control
- System Temperature Control
    Warm-Up/Cool-Down
    Discharge Air
    Minimum Outside Air
    Economizer

The Data Control (D/C) and Graphics Summary Process

The discussion of specifications has provided a fairly detailed coverage of general issues. Armed with this background it is now possible to consider Graphic Interface and Display. The focus here is on specifying graphics to result in optimum implementation and to offer greater value to the system owner. This value is a direct result of better system understanding, thus allowing benefits of the system to be enhanced. These benefits might be from energy and cost savings, improved productivity of workers due to better environmental control or a host of other reasons.

To achieve the goal of optimum graphic display and interface it is necessary to again consider the scope and sequence processes. During the sequence process a sketch of the application was developed.

At this point, that sketch is appended to include other elements of information that might be useful to a BMS operator. The nature of the "Operator" must be considered prior to proceeding, just as with pneumatic designs for the past 40 years. Pneumatic jobs varied significantly from bare controls, to free standing panels with fluorescent canopies and a multiplicity of pressure, target and indicator gauges and light; an array of multi-position and gradual switches and positioners; a graphic or engraved diagram; and nameplates. Who is the user and what are his requirements? The previous sketch usually has all physical elements that might be of interest and can be cleaned up as the basic "graphic" as noted in Figure 15-4.

Then additional data are added to suit the operation and maintenance requirements of the owner, as noted in Figure 15-5.

First, it is strongly suggested that rather than trying to convert the Figure 15-5 graphic sketch to a traditional I/O point summary matrix, that the graphic sketch be published instead. The reasons are four fold: (1) certain software points do not have standard names, (2) the I/O summary is usually derived from an unpublished graphic sketch, (3) this removes the risk of vendor misinterpretations, and (4) the consultant, rather than the low bidder, determines the solution to issues where one HVAC system has too many points to fit uncluttered on one graphic (as is the case with Figure 15-5). Although Figure 15-5 may constrain the

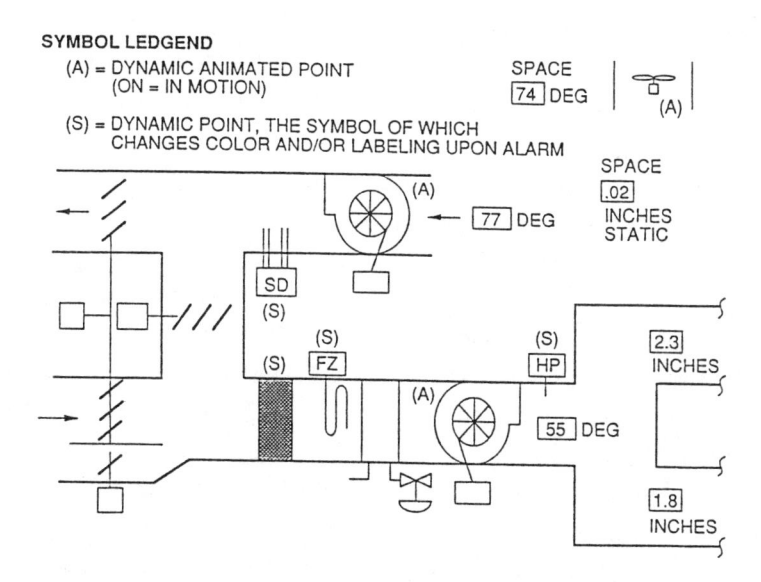

Plan and Specification of Graphic Central Operator Interfaces For DDC/Building Automation Systems

Figure 15-4 System Sketch Modified as a Basic Graphic

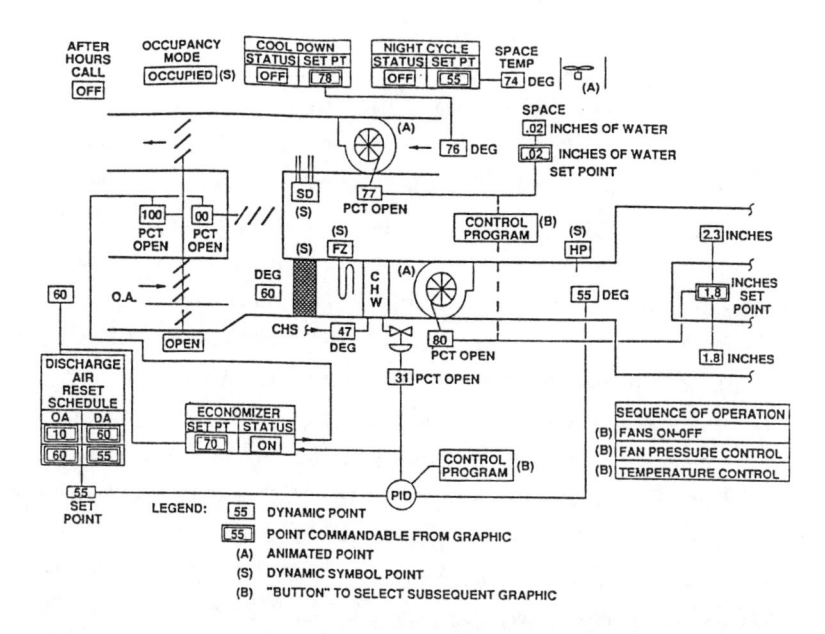

Figure 15-5 System Graphic Including Program and Operator Information

data of interest to a particular client, it is too "busy" to be useful, and should be broken down into two or three graphics. In breaking it down into several graphics, the designer may elect to show points such as outside air temperature and fan status on each graphic (another reason why the designer should not leave graphic design up to the low bidder).

The Figure 15-5 legend (which would not appear on the vendor's installed dynamic graphic) defines five types of dynamic or target areas within the graphic artwork: (1) the dynamic display of a hardware or software point, (2) a dynamic set point which is commandable, (3) an animated status point which, when "on", displays motion, (4) a dynamic symbol that changes tables and/or color as the point state and/or status changes, and (5) a target area "button", the selection of which will initiate another graphic screen. Other than these five sensitive areas, all other text, equipment, etc., is considered to be just static artwork.

Figure 15-5 includes 15 physical points, five computer conditioned actuator position points, 15 software "pseudo" points, five "new graphic" buttons, and two points which are physical points to other remote digital controllers. This is not necessarily a "recommended" graphic for large VAV AHU, but instead shows display options available to specifiers that may be used when needed to satisfy a particular client's needs.

The physical points, except for the mixed air temperature and filter, are required for control. The borrowed (from other DDC panels) point, outside air temperature, is required for control, and the borrowed point, chilled water temperature, is most valuable in diagnosing air handler problems (to keep this borrowed point from being misleading when the chiller pump is off, it may be specified to display 00 when the pump is off).

The pseudo points are either necessary to make control adjustments (setpoint), or to display software information of interest.

The Sequence of Operation "buttons" imply that the sequences are to be presented as display screens linked to their respective HVAC component display. If consultants choose to write the sequences as display screens, character matrices of 60 wide by 19 lines are appropriate.

The Control Program "buttons" suggest that dynamic flow charts of portions of the software may be specified. Figure 15-6 is an example of the technique. It would probably be considered gold plating if required for all loops of all systems; however, most major projects have some

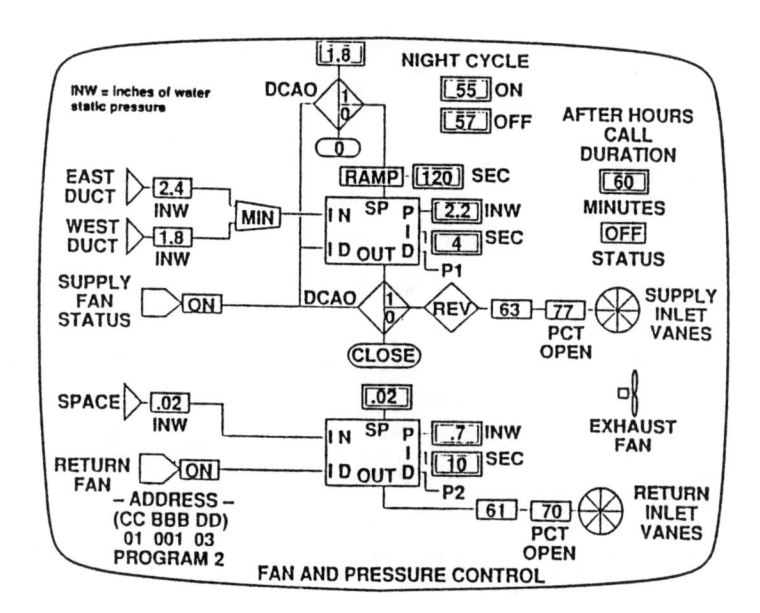

Plan and Specification of Graphic Central Operator Interfaces
For DDC/Building Automation Systems

Figure 15-6 Graphic of a Control Program

critical control loops of concern wherein this type display would be invaluable in loop tuning and evaluation.

The software points above have no appreciable physical attributes; however, they do have costs associated to their definition, processing, and graphic reference. Also, commendable set points, which are now quite easy to modify, should be specified to have upper and lower command limits, beyond which they cannot be easily commanded. Software points usually cost ten to twenty percent as much as a hardware point, but they should require no annual maintenance.

One HVAC Component, One Graphic

The design tradition has been to require a dynamic graphic of each HVAC system component (AHU, convertor, chiller, etc.). A menu or graphic penetration scheme is also usually briefly specified, but not designed.

When designers design a BMS by sketching graphics, including meaningful pseudo points, many other helpful graphics and graphic functions come to mind.

Penetration, Overview, and Composite Graphics

After the graphic-per-component system has been designed, the penetration scheme defining how operators get from an initial sign-on top level graphic down to each component graphic is fairly simple. Each screen typically has four to fifteen penetration targets. Penetration targets can be for any combination of geographical areas (buildings, floors), disciplines (HVAC, Fire Alarm), and special (digital technician, engineering) screens.

Figure 15-7 is an example of part of a penetration design. Each box is a graphic. Graphic 201, if appropriate, shows how the EMS may be used to manage disciplines unrelated to HVAC by people who cannot access HVAC graphics, such as a guard monitoring and controlling exterior lights. Graphic 202, the normal starting graphic for most operators, is Figure 15-8 discussed later. Graphic 203 is a graphic of the physical digital network, which is separated from HVAC graphics and would allow maintenance, diagnostics and configuration management of the digital hardware architectures. This Graphic 203 should also be dynamic, such that displayed controllers that are malfunctioning would be so noted (red, blinking, etc.). This Figure 15-7 need not be a graphic itself,

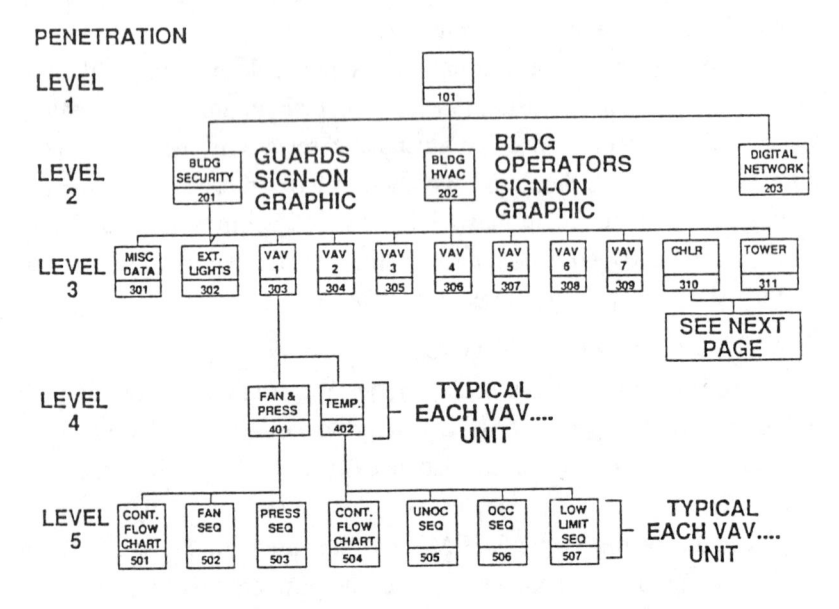

Figure 15-7 Graphic Penetration Design

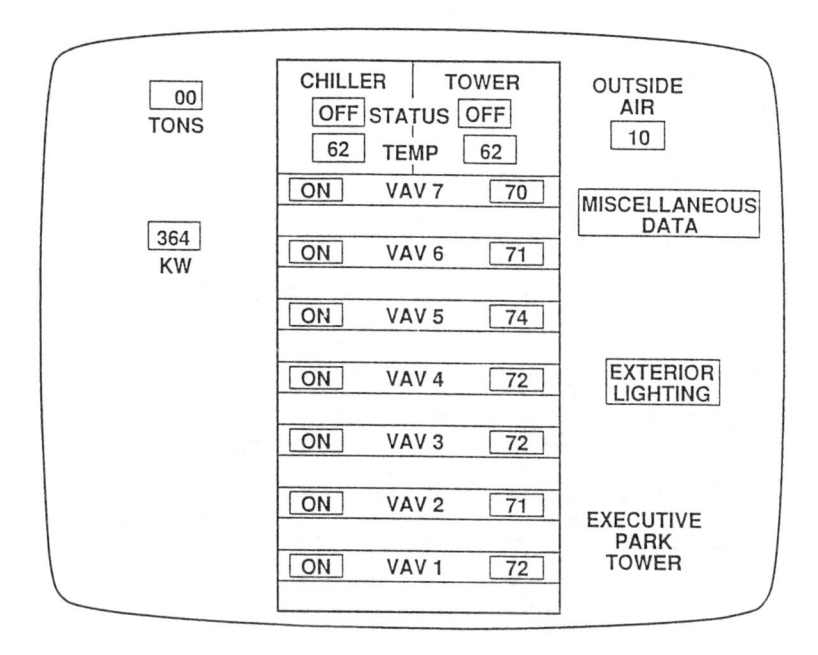

Plan and Specification of Graphic Central Operator Interfaces 355
For DDC/Building Automation Systems

```
          ┌─────────────────────────────────────┐
          │  [00]    CHILLER │ TOWER   OUTSIDE   │
          │  TONS   [OFF] STATUS [OFF]   AIR     │
          │          [62] TEMP [62]    [10]      │
          │         [ON]  VAV 7  [70]            │
          │  [364]                   MISCELLANEOUS│
          │   KW    [ON]  VAV 6  [71]    DATA    │
          │                                      │
          │         [ON]  VAV 5  [74]            │
          │                                      │
          │         [ON]  VAV 4  [72]  EXTERIOR  │
          │                            LIGHTING  │
          │         [ON]  VAV 3  [72]            │
          │                                      │
          │         [ON]  VAV 2  [71]  EXECUTIVE │
          │                              PARK    │
          │         [ON]  VAV 1  [72]   TOWER    │
          └─────────────────────────────────────┘
```

Figure 15-8 Building Overview Graphic

but it would be appropriate in a specification if the designer cares to dictate the penetration scheme.

Designing graphics makes it easy for a designer to conceive and specify overview data on upper level graphics, similar to Figure 15-8, such as instantaneous KW, Tons, and occupancy status of each building on a campus display. Data can add meaning to multi-level penetration graphics.

Other very useful graphics are composites of chiller systems, hot water systems, etc., such as Figure 15-9. Composite screens are invaluable in energy performance and operational analysis.

Global Commands

After preliminary design of control strategies, operational benefits of certain global commands such as those shown in Figure 15-10 may become apparent. To raise the drybulb temperature economizer changeover setpoint of many AHU's on an unusually dry day from 72°F (25°C) to 74°F (26°C) for 15 AHU's could be a single point command,

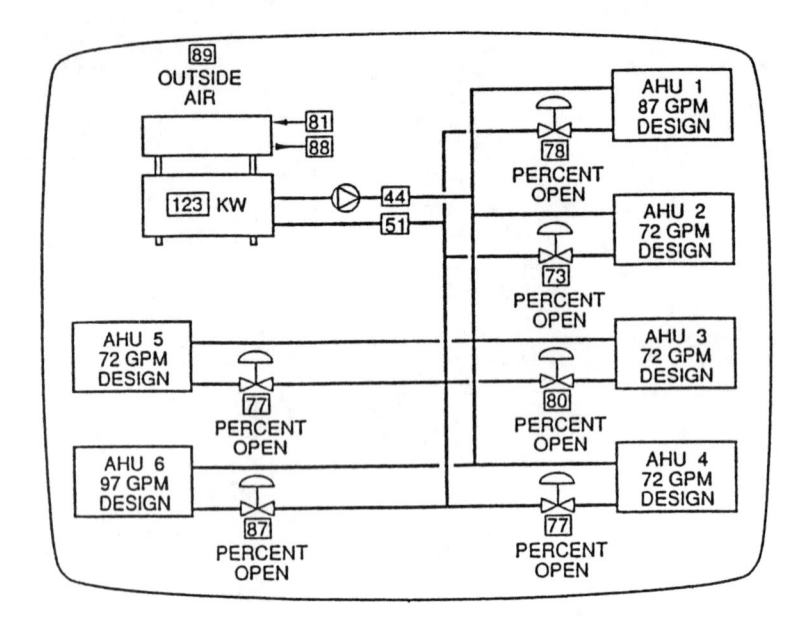

356　　　　　　　　Direct Digital Control: A Guide to Distributed Building Automation

Figure 15-9　Chiller Composite Graphic

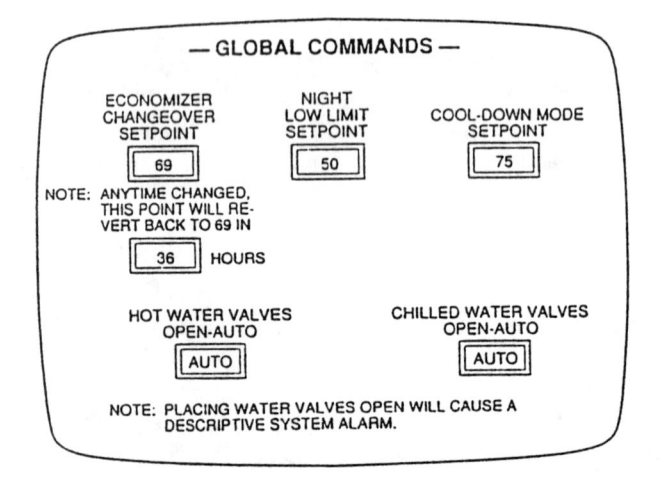

Figure 15-10　Global Command Graphic

which would require the BMS network to coordinate the command among multiple controllers. The night cycle setpoints could similarly be lowered in mild weather. If it is of concern, these "exception" commands may be specified to be automatically rescinded after a day or two. Also, the use of global commands will usually require coordination with the related AHU, etc., "system" graphics and control programs.

Another type global command could be "open" commands to all automatically controlled water valves in each pumping loop to simplify balancing and/or rebalancing. Commands of this nature may be specified to initiate an alarm upon execution, and/or to be automatically rescinded after three or four (adjustable zero to eight) hours. Global command type graphics may be installed at level two (Graphic 204 for instance) if desired to restrict access from normal "operators".

SUMMARY

To summarize, the suggested DDC/BMS design process is:
- Sketch a graphic picture of each piece of equipment or area to be monitored and/or controlled
- Add points (inputs, outputs) required for control to the graphic
- Write control sequences
- Add monitor only points (filters, etc.) to the graphic
- Add desired setpoints to the graphic
- Add other desired software points (night cycle and economizer cycle mode display) to the graphic
- If the graphic looks too cluttered for one screen, break it down into two or more (20 to 25 points max per graphic) (the 20 to 25 points is a practical limit, not an absolute limit)
- Add "buttons" for other related graphics such as "sequence"
- Define these "other" graphics
- Sketch any special, global and engineering graphics required
- Sketch any composite graphics required
- Sketch intermediate multi-level penetration graphics
- Add overview points desired to penetration graphics
- Design graphic penetration tree
- Edit (or write) Part II Products and the balance of Part III to complement the control and graphic scheme above
- Edit (or write) Part I General for project relevance

CONCLUSION

Specifying detailed graphic requirements is simpler and less risky that trying to define state-of-the-art enhanced displays via the point matrix method. Most complex point matrices are made from unpublished sketches anyway.

Sketching graphics seems to lead the designer into conceiving of other helpful or essential overview, global or engineering displays. The BMS truly becomes a system far greater than the sum of it's hardware based point graphics, and operation and management functions become much easier.

A better utilization of the DDC/BMS products via 1990 implementation requirements will drive the industry into many other BMS uses. How many times will the BMS owner try to explain the HVAC system configuration to new operators in the next 10 years? Would a few graphics constructed specifically for this use be helpful? Could the requirements of "Design Documentation Procedures" required by ASHRAE Ventilation Standard 62-1989 be best provided by a BMS of the 90s? (See Figures 15-11 and 15-12.)

Yet in every case, just as with pneumatic panels, the graphics and level of automation required must match the users unique needs. Also, although graphics and displayed software points do not have physical or annual maintenance costs associated with them, some initial cost is involved.

Plan and Specification of Graphic Central Operator Interfaces
For DDC/Building Automation Systems

SYSTEM VENTILATION DESIGN

The Ventilation System of this building was designed under the following assumptions:

A. The use of the spaces is typical office and administrative

B. No unusual contaminants are introduced into the occupied spaces by equipment or processes

C. The fresh outside air introduced into the ventilation system contains no unusual concentrations of harmful contaminants

D. People densities are seven per 1000 square feet for office areas and fifty per 1000 square feet for conference rooms

E. Outside air is introduced at a rate of 20 cubic feet per person

These design procedures and the requirement for this design documentation were outlined in the "ventilation rate procedure" of ASHRAE Standard 62-1989.

– PRINT THIS SCREEN –
Next Page

**Figure 15-11 Example of Graphic Display Regarding
Ventilation Design Procedures**

SYSTEM VENTILATION DESIGN
IMPORTANT

If at any time space use changes (from those assumed on the previous page) occur or if unusual contaminants or unusually strong sources of specific contaminants are introduced into the space, the ventilation system design must be reevaluated to assure proper indoor air quality.

Contact the original consulting engineers

_____
_____

_____

Or the indoor air quality design professional of your choice for reevaluation if required.

Deliver this two page message immediately to:

John Doe – Chief Hygienist
Extension 2436

Print this page
Reset this output

| Reset |

Note: To assure continuous acceptable indoor air quality, this message will be automatically output every April 1 and October 1

**Figure 15-12 Example of Graphic Display Regarding
Ventilation Design Procedures**

Chapter 16

Procurement Practices for High Performance DDC Systems

By Ken Sinclair

PREFACE

In Chapter 15, rapid changes in both control technology and the control industry were discussed. These changes complemented by decreasing technology cost and more system retrofits have had a significant impact on the control industry.

In this chapter, specifying control systems will again be the topic. The approach here is different however, and offers another viewpoint on how to purchase DDC Systems. It is important to state initially that the examples stated and the user requirements outlined here fit a very specific customer profile. The user who is able to make use of these concepts will tend to be very sophisticated, and likely have in house talent dedicated to managing their systems.

Chapter 15 noted many changes in building control system specifications. These were particularly impacted by a growing trend toward retrofit versus new construction installations. In this chapter there will again be mention of retrofit systems, but here we are talking as much about

---

Ken Sinclair is President of Sinclair Energy Services in Sidney, British Columbia, Canada.

upgrading existing computer based systems as traditional electro-mechanical or pneumatic systems. These are all important issues when we look at the process for specifying this equipment, and how it is purchased.

Given the changing industry environment, this chapter will describe a different approach to system specification, the request for proposal (RFP). Consistent with trends toward retrofit and upgrade to systems, some owners are beginning to use the request for proposal to customize requirements for their buildings. Again, it should be obvious that these tend to be sophisticated users, who are capable of clearly delineating their needs for a DDC system. This approach is interesting because it lends itself well to acquisition of systems outside the new construction process. Respondents to the RFP may be manufacturer branches or system integrators. At the same time, since the RFP is more suited to sophisticated users, equipment only may be sold direct to the user who installs the product in-house.

The topic of programming languages or styles has been discussed elsewhere in this book. An additional note is warranted here, because this system is again focused on the sophisticated user or system integrator. The requirement noted is for an Operator Control Language (OCL), see glossary at the end of the chapter for a definition of OCL. Custom programs are clearly indicated through use of an OCL, but bear in mind however that this is a user preference. The full discussion of programming style along with its advantages and disadvantages should be considered carefully before specifying an OCL.

As with the previous chapter, consider the proposals of this author as background. They should in no way be used verbatim or without careful consideration and review of the customers' needs and unique requirements.

Procurement Practices for High Performance Direct Digital Control (DDC) Systems

By Ken Sinclair
HVAC Congress, Anaheim, CA, April '91

INTRODUCTION

This chapter deals with the rapidly changing procurement practices for purchasing and installing computerized Direct Digital Control Systems. An alternative to traditional product specifications is the Request for Proposal (RFP) concept. The RFP approach allows for competition on technical merit and relative cost. This concept is essential to purchase the best value and functionality in the rapidly evolving technology marketplace for DDC systems. The term DDC system for this article is synonymous with Energy Management and Control System, EMCS and Building Automation System (BAS). Mandatory requirements of this required DDC system are defined in the RFP and all vendors presenting proposals must meet these requirements or state their differences. The proposal concept allows the vendor to sell the technical merits of their product in a formal manner for a particular project. Evaluation can be done based on technical performance versus cost on a project basis. Different vendor's systems may better suit different projects. This paper will expand on the present mandatory requirements of a high performance EMCS. Evaluation criteria will be presented as well as cost information of several million dollars of EMCS/BAS purchased through the RFP approach.

A second concept will also be introduced called "The Field Device and Wire Specification." This concept is used when the electronics can be pre-purchased and installed/programmed by the Owner or separate Contractor. This allows competition from trades not traditionally in the

EMCS or BAS industry. This concept greatly increases competition, lowering the cost of installation which may be as high as 60-80% of the total project cost. Note that this statistic is true in larger facilities which are a particular focus of this chapter. Installation as a percent of total job cost varies based on job size and complexity.

Existing traditional Building Automation Systems are more cost effective and easier to replace than most perceive. Some of the systems that were replaced have had maintenance savings alone which have provided better than a three year simple payback. The relatively low cost to upgrade, coupled with almost zero maintenance costs provides an excellent platform to explore the energy saving potential of a high performance EMCS. The Request for Proposal (RFP) environment allows vendors and contractors to best fit their skills and equipment to the project. An incredible opportunity exists on this type of project to reutilize existing wiring, relays, and even sensors if suitable. We have seen vendors even build ribbon cable connectors to match old equipment to eliminate re-termination. Thermistor sensors are now at such a low cost that normally all RTD sensors are replaced. Although most vendors can build software tables to reutilize RTDs, the relative small resistance change per degree, coupled with electronics designed for thermistors does not provide as high an accuracy. In most cases field cabinets and existing control panels are reutilized to re-mount new equipment.

Often the high cost of maintaining the man/machine interface devices (video terminals, printers, etc.) of the older systems contributes heavily to maintenance savings. The use of standard personal computers allows significantly increased functionality at reduced maintenance costs.

REQUEST FOR PROPOSAL (RFP) VERSUS THE SPECIFICATION

The Request for Proposal to procure Energy Management and the Control System (EMCS) and Building Automation System (BAS) has been widely used successfully, in British Columbia for the last eight years. The advantage of the Request for Proposal over traditional specification can be significant.

The traditional approach allows the specification of a product that will perform the required function. Since these products must be well understood and a proven product, the tendency is to specify an obsolete product.

A second major problem that occurs is that to maintain a competitive market, approval must be given to an equal to the specified product.

In reviewing this scenario, no product is equal to any other product, so a compromise must be made. Once the vendor is approved as equal it is difficult to insure that the quality of the original specification is supplied. The vendor is granted equal approval knowing that his product is not the same as the specified product. Most vendors would prefer that they be approved as an equal rather than to be the specified vendor for this reason.

The Request for Proposal approach acknowledges that the product that will perform our demands may not be a known quantity. The Mandatory Requirements section of this document lists the actual requirements for the project. This allows for assessment of what we require and not an assessment of how a product might fit the requirements. This approach has allowed many non-traditional control companies to put together financial and technically attractive proposals. These approaches have been fresh and innovative.

Generally the non-traditional proposals have come from smaller independent control companies with greater beliefs and understanding of computerization and less traditional control knowledge. Obviously some risk is involved when a relatively unknown contractor presents an attractive proposal on a reasonably sized project. This risk has been controlled on retrofit projects by not paying any portion of the contract until the contractor provides a complete working system tested for 7 days and substantial performance is achieved.

In new construction, where the relatively long time of installation would prevent the above approach, contractor holds backs could be reduced to the value of the electronic equipment only, normally less than approximately 20-30% of the total contract. In the case of retrofits, to insure that the contractor completes by a certain date, a completion date is mutually agreed upon. The date is guaranteed by a liquidated damages clause that has the contractor pay the owner the agreed amount per day, if the contract is not completed and/or has not passed the 7 day test. This should not be confused with a penalty clause or bonus clause in which the date is set by the owner.

Liquidated damages is a mutually agreed upon completion date which is decided upon before the award of the contract. The contractor knows that if his completion time is greater than that of his competitor that the assessed liquidated damages outlined in the RFP will be multiplied by the number of days extra and added to his price. The amount of liquidated damages is calculated using the value of the owner's annual savings divided by 365 days, less the financing cost of the project.

MANDATORY REQUIREMENTS

It is a constant source of amazement as to how simple the mandatory requirements are for the DDC system. Generally we are trying to accomplish simple relationships between time, temperature, pressure and humidity, etc., and a number of pieces of equipment. If an Operator Control Language (OCL) is part of the mandatory requirements (see glossary at the end of the chapter for a definition of OCL), the details of these relationships can be easily described.

A copy of the mandatory requirements has been included in the appendix. This document has been used for approximately eight years and is continuously refined by a committee of consultants and owners using it. The document has led to the purchase of over 100 high performance systems. The cost history of some of these projects and others are outlined later in this chapter.

It is hoped that the copy of mandatory requirements included in the appendix will be self-explanatory to those involved in the process of specifying or supplying EMCS/BAS.

Clause 1 Mandatory Requirements include the following sections:

1.1 General overview of contractor responsibility
1.2 DDC system function – details of hardware and operator control language (OCL) mandatory requirements
1.3 Memory – details of hardware memory mandatory requirements
1.4 Processing speed – Details of hardware speed mandatory requirements
1.5 Operator interface – Details of hardware access mandatory requirements

1.6 DDC system inter panel communication – details of hardware communication mandatory requirements
1.7 System expansion – details of expansion mandatory requirements
1.8 Sensor or associated equipment mandatory requirements

The Clause 1 mandatory requirements is of course only part of the total RFP document. Legal and contractual sections of the RFP document tend to be customized for each client. Words in legal and contractual documentation should be reviewed to acknowledge the Request for Proposal approach.

Clause 2 Alternative proposals are also requested.

DETAILS OF PROPOSAL EVALUATION

In evaluating RFPs it became obvious that several vendors were providing extremely useful features that were well beyond the mandatory requirements. Making these features mandatory requirements would significantly reduce the number of vendors able to propose systems. Clause 9 Details of Proposal Evaluation attempted to provide monetary credits for these features and has been included in the appendix. Local vendors have been encouraged to enhance their systems to score full points on the evaluation criteria. Approximately twice a year the evaluation criteria is reviewed and upgraded. Evaluation features can be transferred to the mandatory requirements if sufficient vendors can perform features which would be useful to a particular project. In this manner evolution is encouraged by the evaluation criteria and the minimum level of technology is controlled by the mandatory requirements.

LIFE CYCLE COSTING

Several life cycle costing issues must be included to select best value vendor/contractor. The cost of extending warranty should be analyzed, as well as the cost of replacement parts and new components necessary to extend the system. Prices for all of the above items are requested as part of the RFP. The number of calendar days to complete the project, insured by the liquidated damages, is used in the life cycle cost analysis. The sooner the system is completed the sooner energy savings can begin. Time is money! Almost without fail, the low cost proposal that meets the mandatory requirements, and evaluation criteria, will have the lowest life cycle cost.

INSTALLATION STANDARDS

Installation standards are often controlled by the building owner, or are done to match existing construction, therefore a copy of this portion of the document has not been included in the appendix. A general emerging trend is the following of the good installation practices of the communication industry and not the traditional power wire standards often followed by the control industry. All electrical codes must be met but often significant leeway exists in the installation of room sensors, etc. Conduit is generally only used in mechanical areas.

SUBSTANTIAL PERFORMANCE TEST PROCEDURES

It is important to inform the contractor of what is expected of him before substantial completion will be granted. The concept of a 7 day test has solved many problems in the interpretation by the consultant and vendor as to whether the system is substantially completed. If the system is 100% usable with all mandatory requirements demonstrated, substantial completion will be granted. If there are major barriers preventing it from being a usable system substantial completion will not be granted.

A copy of Clause 4 Substantial Performance Test Procedures has been included in the appendix.

WARRANTY AND EXTENDED WARRANTY

Normal 12 month warranties are required and are tied to Certification of Total Performance to insure total completion is actually achieved. This concept increases the contractors incentive to finish small items that prevent total completion. Several years ago service contracts could represent 15-20% of the total capital cost of a system. The new high performance system typically costs from 2-5% of the total capital cost and to extend the warranty as maintenance is generally not required. These savings are again amplified by the lower capital cost of the new systems.

Extended warranty contracts are requested as part of the RFP to allow the contractor to express confidence in their product. Extended warranty costs are used in life cycle costing to evaluate the cost of owning one system over another.

In the author's experience most clients seldom if ever enter into an extended warranty contract; they generally only repair the system as required. Generally no maintenance is required on the equipment described in the mandatory requirements because it is all solid state with no moving parts. If DDC panels fail they can generally be replaced in less than 10 minutes with no tools. The building owner often purchases spare panels which are kept running, on line, to insure that they are ready for service when required. An exchange policy often exists with the vendor to repair larger panels, but small DDC panels are most economical to replace if they fail. Most sensors, relays, and transducers are of the same nature and replacement when repair is required.

NOMENCLATURE

It has been found that control of system nomenclature (naming convention) is mandatory. If contractors are allowed to name points confusion exists for the life of the project. The RFP document, vendor documentation, and system tagging must use the same nomenclature. The support of the building operator in establishing the naming convention is imperative to insure names used reflect the name used by the on site staff. An example of suggested nomenclature is included in the appendix.

COMPLIANCE CHECK LIST

A check list can be extremely useful in reducing RFP evaluation time as well as clarifying what information is required by the proponents. A copy of Clause 8 Compliance Check List is included in the appendix.

SYSTEM OVERVIEW AND DETAILED POINTS LIST

The RFP must include a detailed point list for all points that are to be connected to the system. Note that this is a marked variation from the recommendation in Chapter 15; therefore, careful consideration of customer needs and the technology applied, graphics vs. text-based monitoring is important. The point list should be the ruling document as to the contractor's expected scope of work. All negotiations should focus around the point sheets. Point sheets should include the exact number and type of points that are to be connected to the system. The nomenclature should also be part of the point lists to insure that all field tagging and documentation reflect the same names. A sample point sheet has been provided in the appendix to demonstrate this approach. The point sheet was copied from a project where an existing BAS is being replaced.

On larger projects a summary of point sheets are required as well as diagrammatic layout showing user access nodes, etc. If panel location is to be controlled this must also be shown.

FIELD DEVICE AND WIRE SPECIFICATION

Once the first major phase of a project has been completed, or for several other reasons, it may be advantageous to maintain control over the computer hardware. The actual computer hardware is generally a small portion of the total project. By purchasing equipment and self installing/or installing and commissioning selected system as part of a separate contract, equipment type and continuity can be maintained. Once the computer hardware is known, sensors, relays, and transducers can be specified with greater accuracy. Insure that a vehicle of reasonable competitive purchase is available to all contractors for all specified devices. Provide several sources of supply for each device, if possible. A few pages have been provided on the standard Field Device and Wire Specifications. It is amazing the simplicity of this specification if computer hardware is pre-decided. In the author's experience, approximately 70-80% of the traditional contract is piping, wiring, sensor relays and transducer valves, dampers, etc. This concept is an excellent method of insuring equipment continuity for large hospitals, universities, colleges, etc.

Often installation is bid by non-traditional control contractors. Examples: electrical contractors, signal wiring contractors, and fire alarm contractors. If the points list and wiring details are clearly given and total end to end expertise is always available for consultation, few problems arise. The close control of the system expertise on a total complex basis provides overall coordination to the project that cannot be achieved with the combination of several stand-alone projects.

EFFECT OF RFP ON PER POINT COST

Between 1983 and 1991 the author kept a record of over 60 RFP including EMCS/BAS retrofits or upgrades. The per point costs have dropped from over $1000 per point traditional systems circa 1983, to less than $60 per point total DDC systems of 1991. Often the difference between the high and low proponents on an RFP are as much as 40-50% and almost without fail the lower price provides newer technology, greater technical features, and a simpler approach to the project.

Four categories of purchased systems, based on actual RFPs, are listed below indicating an approximate range of cost in Canadian dollars per point.

Retrofit of existing central pneumatic control systems to EMCS	$300-$400 per point
Retrofit of existing BAS/EMCS to high performance EMCS	$60-$200 per point
New construction including valves and damper actuators:	
Central system only	$300-$500 per point
Terminal control	$50-$200 per point

Total DDC systems have clearly become the lowest cost system to install in new construction providing incredible features over traditional pneumatics. Care should be taken to provide seamless communication between terminal control and central system control to utilize the full potential of the new systems.

CONCLUSION

The RFP approach provides a controlled method to purchasing the latest DDC features in a competitive atmosphere, and allows real evaluation of technical merit. The approach acts as a catalyst to the industry driving them to provide more features for lower costs. The simple modular approach of new control system suppliers greatly decreases system complexity and the need for special maintenance. The use of a simple Operator Control Language allows an understanding of control strategies by operators and designers without a tremendous understanding of the control industry. Concepts are present in "IF THEN" logic, not obscure control piping and wiring diagrams that also require an understanding of each physical control device. Note that the OCL approach is a flexible language approach and as such is not "simple". The requirement is a thorough understanding of the control language, as well as the application or controlled equipment. With dedicated staff that are knowledgeable in building systems and computer technology, these tools can be very effective. Also new alternatives to "IF, THEN" such as graphic object-oriented programming with control diagram sequences further simplifies the OCL approach.

Consider this approach for your next project. The valuation of the proposals will teach you about the DDC marketplace in your area, and will provide a focus on those products that are most cost effective per feature. You will be surprised at the excellent proposals provided by both traditional and non-traditional control contractors. Advise all proponents as to the best value contractor selected, and the reason why, and how their proposal could be improved for future RFPs. Encourage evolution; do not specify obsolescence.

APPENDIX

Clause 1	Mandatory Requirements
Clause 9	Details of Proposal Evaluation
Clause 4	Substantial Performance Test Procedures
Clause 6	DDC Point naming/nomenclature
Clause 8	Compliance Check List

Sample Point Sheet
Guidespec for Field Device and Wiring Installation for EMCS

CLAUSE 1 MANDATORY REQUIREMENTS

1.1 General

It is the intention of the Corporation to obtain a complete operational Energy Management Control System (EMCS). It is the Contractor's responsibility to provide all items (labor and material) required to complete the installation, including but not limited to, all computer software and hardware, operator input/output devices, remote panels, sensors, control circuit transformers, as required to meet the performance as set forth in the Points Lists and the control sequence description (attach product documents). The Contractor shall provide all wiring, piping, raceways, conduit, installation supervision and labor, including calibration, adjustments and checkouts necessary for a complete and fully operational system. Installation shall be in accordance with the Installation Standard (Clause 3).

The location of all devices shall be reviewed with the Corporation's representative prior to installation.

1.2 DDC System Functions

1.2.1 The DDC system shall have the capacity for timed start/stop on daily schedules, as well as the capability for the owner to develop and run custom application programs. For this, the DDC system shall have a proven Operator Control Language (OCL) which shall be capable of reading the value and/or status of all system points and initiating both digital and analog control actions from any user defined combination of calculations and logical expressions which shall at a minimum include:
1. Addition, subtraction, multiplication and division.
2. Square roots, summations, absolute differences.
3. Logical "and", "Or", "less than" and "greater than".
4. Time delays, in seconds, minutes or hours.
5. Ability to imbed comments in system generated documentation.
6. Ability to use time-of-day and day-of-year in algebraic calculations.

If any of the above functions are accomplished via firmware, the proponent shall provide full detail.

1.2.2 Show precisely how the following strategies will be handled by the Operator Control Language:
1. If real time is between 8:00 AM and 4:00 PM and minimum space temperature is more than 0.5°C below the building objective temperature, then start Pump #1.
2. As soon as Pump #1 has been on for 10 minutes, turn on the boiler.

3. If day of year is before July 2, then factor A is equal to 0.1667 − (July 2 − day of year)/1368, other wise the factor A is equal to 0.1661 − (day of year − July 2)/1368.
4. Select high of space temperature #2, space temperature #3 and space temperature #6.
5. If the real time is greater than 8:00 AM minus a factor, times the difference between setpoint and space temperature and if the real time is less than 3:30 PM and if outside air temperature is less than 10°C, then enable boiler.

1.2.3 Other mandatory monitoring and control features of the DDC system are:
1. Operator defined digital and analog alarms and automatic alarm condition reporting.
2. Auto lockout of alarms when alarmed system is shut down.
3. Direct keyboard override of all digital and analog outputs, with an indication on the display of any point that is operating under keyboard override.
4. Addition, deletion, definition and modification of points and point types from operator keyboard.
5. Trend log graphing of user selected points and times.
6. Run time totalization.

1.2.4 The DDC system shall have the capability to be taken off line in the event of failure or for maintenance and returned to operation without the need for entering any portion of the software program manually. To accomplish this, non-volatile memory or an off-line disk storage device shall be utilized to provide software backup and reload.

On-site backup and verification of the entire system, with full applications software, shall be less than TWO (2) seconds per real point.

1.2.5 The DDC system shall be protected from power line surges and voltage transients. Provide with proposal, a technical description of this protection.

1.2.6 The DDC system shall be provided with automatic protection from any power failure of up to TWENTY FOUR (24) hours duration.

This protection shall at a minimum include continuous real-time clock operation and automatic system restart upon power return. The proposal shall describe the power failure protection system to be employed. System will be tested to confirm rated hours.

1.2.7 The DDC system shall have a 10% expansion capability for each point type on a panel-by-panel basis. The proposal shall describe the system configuration and note spare point capacity, by point type, on a panel-by-panel basis.

1.2.8 Panel replacement shall be possible without any hardware modification. Describe replacement procedure in technical data submitted with proposal.

1.2.9 Any panel malfunction shall not affect the proper operation of a multi-panel system.

1.2.10 Provide with the proposal, information concerning the standalone capability of the proposed panel, giving complete details.

1.2.11 Indicate how points located on one panel can be accessed and utilized in the control language of another panel. Explain any limitations of the above.

1.3 Memory

Each standard panel proposed shall have enough non-volatile or random access memory for all of the following:

1. *Trend Logs* – TWO (2) TL for each input and output point connected to the panel with 100 samples each.
2. *Controllers* – TWO (2) for each analog output point connected to the panel.
3. *Variables* – THREE (3) for each output point connected to the panel. Variables are "virtual points" (as opposed to physical

points) but which have all the attributes of real or physical points.

4. *Operator Control Language (OCL)* – TWENTY (20) syntactically correct lines each with at least 4 operators, for each output point connected to the panel,

or

TEN (10) syntactically correct lines, each with at least four operators, for each output point connected to the panel, if the OCL has the ability to call common routines or use wild card commands.

5. *Descriptor* – ONE (1) for each user definable point, real or virtual, in the panel. In addition, on multi-panel systems, every descriptor in the system must be accessible from a single I/O port.

6. *Time Schedules* – ONE (1) for every 3 output points connected to the panel.

7. *Totalizers* – ONE (1) for each digital point in the panel.

1.4 Processing Speed

1.4.1 *Effective Panel Processing Speed* – Maximum permissible execution time is THREE (3) seconds. Execution time is defined as the time it takes the standalone panel CPU to execute all application software in the panel, from some point in the software back to the same point, assuming full memory usage, as defined in 1.3, while simultaneously responding to operator or terminal display requests and carrying out normal inter-panel communications averaged over a ONE (1)-minute period. This will be done by setting up a counter in each panel and monitoring their counting rate.

Provide with the proposal the estimated execution time for each panel in the system.

1.4.2 *Effective System Processing Speed* – This applies to multi-panel systems only. System processing speed is intended to address inter-panel communications and will be checked by evaluating system display response. This will be done by setting up a display of all panel counters and checking how frequently each counter updated on the refreshed display.

Displays shall refresh within TEN (10) seconds. Every counter shall show an updated value on the display within SIXTY (60) seconds at the previous update appearing. Provide confirmation with proposal that required system processing speed will be achieved.

1.5 Operator Interface to DDC System

1.5.1 *Direct Access* – The system shall be accessed via an I/O device. Provide operational details and system requirements.

1.5.2 *Remote Access* – A telephone modem shall be supplied and installed to enable remote communications. Modem to be 1200/2400 baud equivalent to Bell 212A and be DOC approved. Telephone line installation is the Corporation's responsibility. Explain in detail the modem operation and limitations for remote access into the system over phone lines.

1.5.3 *Access Security* – The system shall have individual operator password access security. The proposal shall indicate levels of password security available.

1.6 DDC System Inter-Panel Communication

1.6.1 Some means shall be provided to ensure communication integrity. Provide detail of your system with proposal.

1.6.2 To prevent damage to the system, each data highway line shall be provided with a means of isolation, either optically or by some other means. Provide detail of your protection system with proposal.

1.7 System Expansion

Submit with proposal, unit prices for all system components, including panels, guaranteed through TWO (2) years past the total performance date for all hardware required to expand and maintain the system. Unit prices are for supply only.

1.8 Sensors and Associated Equipment

The Energy Management Control System shall be supplied with all sensors, relays and associated equipment to fully connect the points listed in the computer hardware points lists (attach product documents). Field point installation shall be performed in a neat and orderly fashion with all components marked or labeled to correspond with the marking or labeling in the As-Built drawings. Use baggage style plastic pockets and cards for mounting labels (supplied by the Corporation).

1.8.1 All sensors and controllers shall be of commercial grade and shall be installed according to the manufacturer's recommendations. Provide full details of all sensors and controllers proposed, including their range and accuracy.

The sensors shall conform to the following table:

Description	Accuracy	Range
• Duct Air Temperature	+/-0.2°C	0 to 40°C
• Space Temperature	+/-0.2°C	0 to 50°C
• Outside Air Temperature	+/-0.2°C	-40 to 40°C
• Water Temperature	+/-0.2°C	0 to 120°C
• Humidity	+/-5%	20 to 80%
• Supply Duct Pressure	+/-10 Pa	0 to 500 Pa
• Building and Return Air Pressure	+/-1 Pa	-25 to 25 Pa
• Current Sensors	+/-2% of full scale	
• Pneumatic Transducer	+/-2 Pa	0 to 500 Pa

NOTES:

1. Provide defined curve and calibration point data.

2. Provide data on Repeatability, Linearity, Deadband and Hysteresis.

CLAUSE 2 ALTERNATIVE PROPOSAL

The system, as described in the Mandatory Section, is the minimum requirement. Proponents are encouraged to submit alternate proposals for the system which may provide enhanced performance and/or reduced cost.

CLAUSE 9 DETAILS OF PROPOSAL EVALUATION

Proposals that comply with the Mandatory Requirements, Clause 1 of the EMCS Specifications, will be evaluated based on the following criteria:

1. The price quoted for the mandatory items, the "Amount of Proposal", will form the basis of the evaluation. Evaluation Items 2 through 6 will each involve calculations to arrive at the capital cost credit or a capital cost penalty. The credits and penalties will be subtracted from or added to the Amount of Proposal. The proposal with the lowest adjusted capital cost will be awarded the contract.

2. If one or more of the alternative items quoted pursuant to Clause 2 of the EMCS Performance Specifications provide an acceptable payback, the estimated net benefits of these optional items over a FOUR (4)-year period will be deducted from the Amount of Proposal.

3. Replacement parts costs for all components must be provided with the proposal. An amount equal to FIVE (5) percent of the cost to supply replacements for all components in the system (excluding wiring) will be added to the Amount of Proposal.

4. A price for an Extended Warranty, as described in Clause 5 in the EMCS Specifications, must be provided with the proposal. The present value of FOUR (4) years of this Extended Warranty will be added to the Amount of Proposal.

5. A credit will be given for systems that inherently provide the following software features which are in excess of Mandatory Requirements:

    a) *Full Screen Editor* – The Mandatory Requirements specify a very basic level of editing. A credit will be given for full screen editing capabilities. Maximum credit will be THREE (3) percent of the Amount of Proposal.

b) *Graphing of Trend Log Data* – The Mandatory Requirements specify that historical data be retrievable in a tabular form. A credit will be given for systems which can produce easily readable graphs from such data. Maximum credit will be THREE (3) percent of the Amount of Proposal.

c) *Custom Menus* – The ability to create custom menus linked to specific operator ID's will be given a credit. Maximum credit will be ONE (1) percent of the Amount of Proposal.

d) *Custom Screen Display Format* – The ability for the operator to create custom displays will be given a credit. Important criteria are: the number of displays available in each panel; flexibility in selecting points from other panels; the ability to use the same point in more than one display; flexibility in display test capabilities; and the ability to locate each point anywhere on the display screen. Maximum credit will be TWO (2) percent of the Amount of Proposal.

e) *Operator Control Language (OCL)* – The OCL in some systems includes powerful global operands or commands or graphic programming styles which simplify the writing of application software. Maximum credit will be TWO (2) percent of the Amount of Proposal.

f) *Graphics* – Some systems have dynamic graphics capabilities for displaying menus, data, schematics, etc., as an integral part of the system's capabilities. Availability of such capabilities will be given a credit. Important criteria are: ease of graphics development; degree of integration with the operating system (eg. ability to issue commands from graphics screens); screen resolution; and ability to link graphics screens to one another. Maximum credit will be THREE (3) percent of the Amount of Proposal.

g) *Time Schedule* – Some systems have screen displays of weekly and annual schedules. This facilitates developing or editing time schedules. Systems which provide screen displays, screen editors and annual schedules in a calendar for-

mat will be given a credit. Maximum credit will be THREE (3) percent of the Amount of Proposal.
6. A penalty will be levied against system for the following:
   a) *Command Feedback* – The ability for an operator to be able to issue a command and then see the results of that command on the screen display with a minimum of keystrokes is very important. The number of keystrokes required will be evaluated for each system and the penalty will be proportional to the number of keystrokes required. Maximum penalty will be THREE (3) percent of the Amount of Proposal.
   b) The proposed construction schedule (i.e. the number of days required to complete the project) will be considered in the Proposal Evaluation. Penalty will be based upon the value of daily liquidated damages times the extra number of days over the shortest period.

CLAUSE 4 SUBSTANTIAL PERFORMANCE TEST PROCEDURES

4.1 Overview

Before the SEVEN (7)-day acceptance test may begin, the EMCS must be completely operational including the following:
1. Every point shall be checked end to end to ensure accuracy and integrity of systems. Each point will appear on one of the sheets in Clause 7 and be signed off by both persons involved in the commissioning procedure.
2. Basic control strategies shall be written in Operator Control Language (OCL).
3. Time schedules shall be built and in control, replacing time-controlled equipment.
4. Displays shall be built for each logical air handling system, boiler systems, chiller, etc.
5. Each control loop measured variable, controlled variable and set point if calculated; shall be placed on a FIFTEEN (15)- minute

continuous trend for at least TWENTY-FOUR (24) hours to prove stability of loop.
6. Each space sensor shall be placed on a THREE (3)-hour trend for One Hundred (100) samples.
7. Runtime totalizer shall be set on all digital outputs.
8. Load/save of panel programs must be demonstrated.
9. If included, sample dynamic graphics shall be built as demonstration of graphic capabilities.
10. All features of system shall be exercised.
11. Operator shall be briefed on operation of system.
12. A trend on one panel shall be set up for a point from another panel. This point shall also be trended in its own panel for the same intervals. Comparison of the two trends will indicate if any communication problems are occurring during the SEVEN (7)-day test.
13. Related DI's and DO's shall be connected to show alarm condition.

4.2 During SEVEN (7)-Day Test

4.2.1 Fire alarm will be tested to ensure correct action of all fire and smoke sequences.

4.2.2 Power failure for building will be simulated, system recovery monitored.

4.2.3 Control strategy upgrade shall be started which will exercise most features of the system.

4.2.4 Demonstration of modem operation will be required.

4.2.5 Demonstration of hardware low limits and damper interlocks will be required.

4.2.6 Spot checks of points end-to-end integrity will be carried out. If several problems are identified, a complete reconfirmation of system integrity will be required by Contractor.

4.2.7 Printer shall be left on for complete SEVEN (7)-day test. All printouts will be kept for review at completion of test. An alarm printer shall be supplied by the Contractor if none is supplied in this contract.

4.3 Documentation

The following documentation must also be in place before completion of SEVEN (7)-day test and Substantial Acceptance is granted:

1. Panel layout sheets complete with point name, point address and wire identification number. One copy attached to each respective panel door.
2. All points tagged with point name, point address and panel number.
3. Points Check-Out data sheets showing final set points values during calibrations.
4. As-Built control drawings showing interface with existing controls.
5. As-Built ladder wiring diagrams showing all hardware interlocks.
6. Complete Operators Manual.
7. Apparatus and Maintenance Manual for all sensors, transducers, solid state relays, etc.
8. Reduced floor plans showing sensors, terminals and panel locations.
9. Electrical approval certificate.
10. All of the above information, with the exception of #2 (point tags) shall be bound and presented in TWO (2) manuals to be left on site.

Once the basic above requirements are met and all other features of the system are complete and acceptable, Substantial Completion shall be granted. A deficiency list shall be prepared and holdbacks applied. All deficiencies shall be corrected prior to Total Performance. Warranty shall start from the date of Total Performance of the work.

DDC POINT NAMING CONVENTIONS/NOMENCLATURE

In general, the number of characters is limited. Therefore, a shorthand system should be developed, and the name should be in segments separated by an underscore, period, etc. The guidelines below should be considered.

1. Identify the system or building name and if desired node number in the point name (ie: "ECR: for El Camino Real").
2. Identify the portion of the building served (See Example #1); for example "C101" for Classroom 101.
3. If desired, identify the type of equipment controlled (See Example #2). For example "SZ" for single zone or "RTU" for rooftop unit.
4. Consider identifying the control equipment, for example "BDDC" for Building Direct Digital Controller.
5. Identify the point type and number along with the sensed or controlled unit such as: Temp., AI3, etc., (ie: DA TMP for Discharge Air temperatures).
6. As desired identify an equipment number and be sure it matches all other systems used by maintenance or operations staff.

Using the above examples and observing the rule of 18 characters maximum, some point names might be:

ECR_C101_SZ_DA_TMP
ECR_C101_SPACE_TMP
ECR_RTU1_FAN_DO1

Observing these simple rules and customing them to any application will simplify diagnostics for DDC Management staff.

EXAMPLE #1
DEFINITIONS – POINT NAMES

The following are example definitions of H.V.A.C. controlled equipment for retail stores by areas.

SHR	–	Showroom	JLY	–	Jewelry
SLF	–	Self Service	CPU	–	Customer Pick Up
SPT	–	Sporting Goods	SGT	–	Sight & Sound
TOY	–	Toys	STR	–	Stockroom
RCV	–	Receiving	EXP	–	Expediting
ENT	–	Entry Vestibule	EXT	–	Exit Vestibule
OFF	–	Office	BRK	–	Breakroom
JLK	–	Jewelry Office			

There are four (4) basic codes to the assignment strategy:
1. PREFIX = Area or load controlled.
2. Place of Hardware: Control Equipment
3. Physical location (front, center, back, left or right).
4. Nomenclature for controlled equipment (Local Unit I.D. #)

Example:
```
JLYETF1
1 2 34
```

1. JLY = Jewelry
2. ET = Electronic Thermostat Module
3. F = Front
4. 1 = H.V.A.C. Unit No. 1

The following are example definitions of Lighting/Auxiliary loads prefix options:

EMPLGTS	–	Employee Lights
CUSTLGTS	–	Customer Lights
EXTSIGN	–	Exterior Lights
ELWTHTR	–	Electric Water Heater
PKLTCANP	–	Parking Lot/Canopy Lights
JLYGTS	–	Jewelry Lights
EXSCLTS	–	Exterior Security Lights

Example: 50% EMPLGTS = 50% Employee Lighting

EXAMPLE 2
CLAUSE 6 NOMENCLATURE
SUGGESTED NOMENCLATURE FOR EMCS POINTS

1. GENERAL

SYSTEM		SUB-SYSTEM		FINAL ELEMENT	
MZ	– Multi-Zone	MA	– Mixed Air	V	– Valve
DD	– Dual Duct	HC	– Heating Coil	D	– Damper
VAV	– Variable Air Volume	CC	– Cooling Coil	P	– Pump
CV	– Constant Volume	HD	– Hot Deck	C	– Converter
SZ	– Single Zone	CD	– Cold Deck	B	– Burner
HP	– Heat Pump	SA	– Supply Air	T	– Temperature
SF	– Supply Fan	RA	– Return Air	S	– Setpoint
RF	– Return Fan	OA	– Outdoor Air		(Analog Pt)
EF	– Exhaust Fan	BW	– Boiler Water	S	– Status
Blr	– Boiler (Single)	HW	– Hot Water		(Digital Pt)
B#	– Boiler (Multiple)	CH	– Cold Water	L	– Low
CHLR	– Chiller (Single)	R	– Room	H	– High
C#	– Chiller (Multiple)	D	– Duct		
S, ST	– Steam				

Any of the above acronyms could be display names

2. NON–SPECIALIZED

MIN	– Minimum	DHW	– Domestic Hot Water
MAX	– Maximum	EBB	– Electric Baseboard
AVG	– Average	RAD	– Radiation
WUM	– Warm-up Mode	WS	– Weekly Schedule
CDM	– Cold-Day Mode	CO or CTLR	– Controller
DAY	– Day	W	– Washroom
NIGHT	– Night	K	– Kitchen
OCC	– Occupied	P	– Primary
UNOCC	– UNOCCUPIED	S	– Secondary
		T	– Trend

3. TEMPERATURE PREDICTOR

PHT	– Projected High Temp	HLT	– Hour of Low Temp
PLT	– Projected Low Temp	CD	– Cold Cay
DHT	– Day's High Temp	WD	– Warm Day
DLT	– Day's Low Temp	BO	– Building Objective
HHT	– Hour of High Temp		

CLAUSE 8 COMPLIANCE CHECK LIST

EMCS – REQUEST FOR PROPOSAL (RFP) COMPLIANCE CHECK LIST

RFP CLAUSE	DESCRIPTION	COMPLIANCE	REMARKS
**************************** **Instructions to Proponents** ****************************			
Clause 4	Completion of Proposal and acceptance form		
4.4(a)	Previous Experience		
4.4(b)1	EMCS Block Diagram		
4.4(b)2	Hardware Description		
4.4(b)3	Control Language		
4.4(b)4	Construction Schedule		
4.4(b)5	List of Subcontractors		
4.4(b)6	Compliance Checklist		
************* **Performance Specification Mandatory Requirements** *************			
Clause 1	Completion of Proposal and acceptance form		
1.2	DDC System Functions		
1.2.1.1	Logical Expressions +, *, −, /		
1.2.1.2	SQR @ SUM, ABS (——)		
1.2.1.3	Logical and, or, < , >		
1.2.1.4	Time Delay		
1.2.1.5	IMBED Comments		
1.2.1.6	Time of Day & Year		
1.2.2	OCL Examples		
1.2.2.1			
1.2.2.2			
1.2.2.3			
1.2.2.4			
1.2.2.5			

EMCS – REQUEST FOR PROPOSAL (RFP) COMPLIANCE CHECK LIST

RFP CLAUSE	DESCRIPTION	COMPLIANCE	REMARKS
1.2.3	Other Mandatory Reqmts.		
1.2.3.1			
1.2.3.2			
1.2.3.3			
1.2.3.4			
1.2.3.5			
1.2.3.6			
1.2.4	Data Storage/Reload/Backup		
1.2.5	Power Line Surge		
1.2.6	Power Backup (min. 24 hrs)		
1.2.7	System Expansion Capability		
1.2.8	System Modularity		
1.2.9	Point Status on System Failure		
1.2.10	Remote Standalone		
1.2.11	Panel-to-panel Comm		
1.3	Memory		
1.3.1	Trend Logs		
1.3.2	Controllers		
1.3.3	Variables		
1.3.4	OCL		
1.3.5	Descriptor		
1.3.6	Time Schedules		
1.3.7	Totalizers		
1.5	Operator Interface		
1.5.1	Direct Access		
1.5.2	Remote Access		
1.5.3	Access Security		

EMCS – REQUEST FOR PROPOSAL (RFP) COMPLIANCE CHECK LIST

RFP CLAUSE	DESCRIPTION	COMPLIANCE	REMARKS
1.6	Interpanel Communication	_____	_____
1.6.1	Communication Integrity	_____	_____
1.6.2	Data Highways Protection	_____	_____
1.7	System Expansion Unit Prices	_____	_____
1.8	Sensors & Assoc. Equipm't	_____	_____
1.8.1	Sensors/Controllers	_____	_____
	Duct Air Temperature	_____	_____
	Space Air Temperature	_____	_____
	Outside Air Temperature	_____	_____
	Water Temperature	_____	_____
	Humidity	_____	_____
	Supply Duct Pressure	_____	_____
	Building and Return Air Pressure	_____	_____
	Current Sensor	_____	_____
	Pneumatic Transducers	_____	_____
Clause 2	Alternative Proposals	_____	_____
Clause 5	Warranty		
5.2	Extended Warranty	_____	_____

REQUEST FOR PROPOSALS SAMPLE POINT SHEET

DESCRIPTION	TAG#	LOCATION	DI	DO	AI	AO
Phase 3 Administration						
Supply fan F304	SF304.S	West mech rm	X			
(34000 cfm)	SF304.C	West mech rm		X Relay		
Supply air temp	SF304SAT	West mech rm			N DTS	
Supply air volume control	SF304SAV	West mech rm				N EI ASD
Supply air static pressure	SF304SP	West mech rm			N PS	
Cooling coil	SF304CC	West mech rm				N EPT
Freeze stat	SF304FRZ	West mech rm	X			
Mixed air damper	SF304MAD	West mech rm				N EPT
Humidity valve	SF302HUM	West mech rm				N EPT
Space temperature	RT1CAFE	1 Cafeteria			N RTS	
Space temperature	RT2CLAIM	2 Claims UT1			N RTS	
Space temperature	RT3EMP	Empl-Relat			N RTS	
Space humidity	RHEMP	Empl-Relat			N RHS	
Return fan F317	RF317.S	West mech rm	X			
(34000 cfm)	RF317.C	West mech rm		X Relay		
Return air volume control	RF317RAV	West mech rm				N EI ASD
Return air temp	RF317RAT	West mech rm			X DTS	
Return air humidity	RF317RAH	West mech rm			N DHS	
Building pressure	BPWEST	West mech rm			N PS	
		Total 18	3	2	9	4

DI – Digital IN DO – Digital Out AI – Analog In AO – Analog Out
APS – Air differential pressure switch PS – Pressure sensor
CRS – Control relay status CS – Current status
WPS – Water differential pressure switch CR – Control relay
DTS – Duct temperature sensor CT – Current transformer
RTS – Room temperature sensor AUX – Auxiliary
WTS – Water temperature sensor RHS – Room humidity sensor
DHS – Duct humidity sensor transducer EPT – Electronic/pneumatic
OAS – Outdoor air sensor EI – Electronic interface
X – Existing * – Optional
N – New

GUIDESPEC FOR FIELD DEVICE AND WIRING INSTALLATION FOR EMCS

PART 1 – GENERAL

1.1. Work Included
.1 All sensors, relays, contractors, operators and other field devices required for the control system.
.2 Operators for dampers and control valves.
.3 All wiring between field devices and DDC controllers or DDC standalone panels, DDC controller network wiring, and DDC standalone panel network wiring.
.4 120 volt AC wiring and connections to all DDC controllers and standalone panels, and to all other components requiring line voltage power.
.5 Commissioning of work.
.6 Instructions to the Owner.

1.2. Related Work   *NOTE – Insert appropriate list of references.*

1.3 Quality Assurance   *NOTE – Insert your office standard wording here.*

1.4 Submittals   *NOTE – Insert appropriate office standard wording here.*

1.5 Division of Responsibility
.1 This Division is responsible for:
  .1 Provision of all sensors, actuators, relays, contactors, power supplies, conduit, wiring, tubing, and other field devices.
  .2 Mounting of a steel mounting tub and associated termination strips, supplied by Owner, for each DDC standalone panel.
  .3 Mounting of all required DDC controllers, supplied by Owner, on or adjacent to the specific components to be controlled by same.

.4 All required wiring, including terminations, and conduit for field input/output devices to either DDC controllers or to termination strips in DDC standalone panel mounting tubs.

.5 All required network wiring and conduit between DDC controllers and a DDC standalone panel, and between DDC standalone panels.

.6 All required 120 volt AC wiring and conduit, including connections, for power supplies, interlocks and other specified functions.

.7 Commissioning of all work installed. This includes continuity checks on all wiring installed, and calibration of all devices and components provided, including a record of all checks and calibrations. It also includes checking and documenting the functional performance of all equipment and components installed and cooperating as the Owner verifies that performance.

.8 Provision of shop drawings, operations and maintenance information, as-built drawings and any other documentation required under this specification.

.2 The Owner is responsible for:

.1 Supply of all required DDC controllers, and turning them over to the Contractor for mounting on or adjacent to equipment being controlled.

.2 Supply of a steel mounting tub and terminations strips for each DDC standalone panel, and turning them over to the Contractor for mounting.

.3 Supply and installation, within the steel mounting tubs, of all required DDC standalone panels, including wiring between termination strips and the panel wiring connection points.

.4 All required software programming.

.5 A telephone line, telephone modem and installation.

.6 Verification of correct installation and functioning of all the Contractor's Work.

1.6 Coordination

.1 Provide coordination between the Owner, the Engineer, the Contractor, the Commissioning Agent, and others as necessary.

Cooperate and attend as required to complete the installation and commissioning procedures as specified in this Section and in other Sections of this specification.

1.7 Control system Layout

.1 Each DDC input and output point is identified on a schematic diagram of the relevant HVAC system, and also in the points lists included as part of this Section of the specification. Review the schematics and points lists to obtain information on panel number, point descriptor, input/output number, field device and wiring detail references. When the project is complete, update this information as required to reflect the as-built condition and include in project documentation submitted to the Owner.

.2 The schematic diagrams show the relative physical location of each field input/output device.

.3 The points lists identify each specific input/output device, its required wiring and connection details, and other relevant installation information.

1.8 Definitions and Abbreviations

.1 The following are abbreviations, with related definitions, used throughout this Section of the specifications.

.1 DDC	Direct Digital Control
.2 EMCS	Energy Management Control System
.3 DI	Digital Input
.4 DO	Digital Output
.5 AI	Analog Input
.6 AO	Analog Output
.7 HVAC	Heating, Ventilating, and Air Conditioning
.8 MCC	Motor Control Center

.2 The following is the symbol legend used on the points lists:

.1 WPS	Water Differential Pressure Switch
.2 DSP	Duct Static Pressure
.3 CRS	Control Relay Status Input
.4 WTS	Water Temperature Sensor
.5 RTS	Room Temperature Sensor

.6 DTS	Duct Temperature Sensor
.7 OAS	Outdoor Air Temperature Sensor
.8 RHS	Room Humidity Sensor
.9 DHS	Duct Humidity Sensor
.10 EPT	Electric/Pneumatic Transducer
.11 KWT	Kilowatt Transducer
.12 AUX	Auxiliary
.13 PS	Pressure Switch
.14 MS	Momentary Contact Switch
.15 MC	Maintained Contact Switch
.16 ES	End Switch
.17 CS	Control Relay Output
.18 SV	Solenoid Valve
.19 CT	Current Transformer
.20 VT	Voltage Transformer
.21 PI	Pulse Counter Input
.22 FZ	Freezestat
.23 DA	Damper Actuator
.24 VA	Valve Actuator
.25 N	New
.26 X	Existing

PART 2 – PRODUCTS

2.1 Metric Identification

.1 All indicators, eg. for temperatures, pressures, etc. shall read in metric units.

2.2 Shop Drawings

.1 Shop drawings submitted shall show full details of range, span, accuracy, signal characteristics, repeatability, linearity, hysteresis, dead band, and drift as applicable.

2.3 Alternate Equipment

.1 Follow the procedures for requests for approval of alternate equipment or components elsewhere in this Division. Requests shall provide complete information on the product being submitted, including all the information required for shop drawings and, for analog devices, defined curve/calibration points.

2.4 Field Devices

NOTE: If there need to be more than one type of any specific category of field device, add additional device.

GLOSSARY OF TERMS USED IN THIS PAPER

EMCS	Energy Management Control System
High Performance	A system that can provide a win win win scenario for saving energy, improving environmental conditions, and improving client comfort.
BAS	Building Automation System
RFP	Request for Proposal
Mandatory Requirements	The minimum functionality that must be present in the proposed system.
RTD	Resistance temperature sensing device – linear small change in resistance/degree of temperature
Thermistor	Resistance temperature sensing device – non linear large change in resistance/degree of temperature
CRT	Cathode ray tube (display screen)
PRT	Printer
DDC	Direct digital control
PC	Personal computer
MS-DOS	Microsoft Disk Operating System
OCL	Operator Control Language
Proponent	Contractor/vendor presenting proposal

Chapter 17

Managing DDC Systems for Results

This may be the most important chapter in the book! The benefits of micro-processor based automation or DDC Systems were demonstrated without question over the past 2 decades. Initially this book noted that the benefits were confined to reducing energy use and reaping the corresponding savings. Market research as recently as 1989 still indicated that these are among the top reasons that building owners install DDC today. However, there have been a host of additional benefits provided by these systems to further contribute to their popularity with users. Among these benefits are occupant comfort, complying with indoor air quality requirements, providing integrated fire and life safety and other functions such as access control for an office building.

These benefits can be achieved and even enhanced by users who aggressively manage DDC Systems. At the most basic level, system management is essential because tenant comfort, smooth building operation and cost savings are highly dependent upon regular attention to details. These details include regular system verification to ensure operation and system finetuning for enhanced efficiency. With this level of effort, DDC Systems will consistently meet or exceed projected results.

The DDC system picture is not completely without blemish however. Over the years more than one control system, particularly microprocessor based ones, have failed to meet projected benefits. Many reasons can lead to such an outcome, but perhaps first on any list is the quality of DDC System management.

This chapter is not in any way intended to place blame on any group for systems that miss the mark. Rather it is focused on a process to ensure results. The two major headings for discussing this process are:
- DDC System Management Functions and
- Management Tools.

Management Functions include the critical areas of concern regarding system performance, and the corresponding tasks that must be completed. Management Tools are those technologies that aide in ensuring DDC System Results. These include both components of the DDC system and non-control related tools that are important to managing systems.

The process of Managing DDC systems for results can also be termed "DDC Optimization". Throughout this book, the term DDC System has been synonymous with Building Automation (Control) System or Energy Management System. In addition, these systems are Distributed in control of widely discrete pieces of equipment including HVAC, lighting, etc. This concepts are all important as the topic of optimization is broached because they introduce greater potential for benefit and new capabilities, along with new opportunities for shortfalls in performance.

Regarding DDC Systems there are two key characteristics to note immediately,
1) they do use distributed system intelligence, and
2) many systems are installed remote from system operators.

As a result, one of the most critical components of DDC Optimization is communication. Locally, communication is necessary for network interaction between all of the distributed controllers and perhaps and on site Central Operator Interface. Remotely, sites without local operators or who require interface from remote experts, installers or management personnel must have "dial in" and "dial out" features. To accomplish the latter, one of the Management Tools discussed will be a Personal Computer based operator interface.

For any project to produce the benefits that were projected initially, a process must be established to optimize performance. Obviously, the results of this process justify continuation, and perhaps expansion of an overall DDC program as well. Outlining the DDC optimization process is the purpose of this chapter. The process involves proven techniques that have been developed and proven by the author, and others in the industry.

DDC SYSTEM MANAGEMENT FUNCTIONS

Catch phrases come and go in our industry, and the business world, one in particular comes to mind "information era". A DDC contractor once referred to this concept somewhat less delicately "data diarrhea." Whether a catch phrase or not though, this reality is at the heart of the DDC optimization challenge.

Since the early 1970s there have been Energy Managers throughout corporate america. These tend to be highly sophisticated technical managers, and often they are faced with the formidable task of achieving top managements goals for reduced operating costs without mashing the toes of operations staff who live with the companies mission every moment of every day. Optimizing DDC System performance tends to be just such a task, and the Energy Manager is generally the individual who must implement the process.

Energy Manager is a tough job, and these individuals are faced with significant technical and logistical concerns in many areas, not just DDC system management. In spite of these other demands however, it is imperative for corporations and managers to reaffirm the importance of their DDC programs. The benefits noted above are real and unquestionably support the mission of the organization. Continued investment in this technology is a wise use of corporate resources. Beyond this however, implementation of DDC is not always a conscious decision. More and more HVAC equipment comes from the factory with pre-integrated DDC controls, and these must also be maintained and optimized. In these times of corporate belt tightening, why not optimize these systems, and harvest additional benefits to the company. At the same time this establishes the groundwork for expanded implementation of the DDC technology.

Ultimately for DDC optimization requires an effective program designed to: 1) validate system operation, and 2) to access raw data and make computations in the process of finetuning system performance. This chapter discusses concepts and optimization techniques to ensure efficient DDC System operation. This operation may also extend beyond conventional energy management because often automation includes energy and building automation programs. As a result effective DDC Management requires diligence and time-intensive oversight.

DDC management functions that will be discussed are:

(1) Validating DDC Systems including:
– Hardware Maintenance
– Security Issues
– Performance and
– Management Reporting, and

(2) System Fine Tuning.

It is also important to consider the maintenance of controlled equipment in the evaluation of DDC systems, though this topic will not be covered in this book.

VALIDATING DDC SYSTEMS

Under system validation the focus is on ensuring that system remains in operation as designed. An important tool is the front-end system for validating automation performance, and make certain that control performance and savings projections materialize. The most basic task is hardware maintenance. It is essential to verify that systems are operational and that regular maintenance is conducted.

Hardware Maintenance

Maintenance requirements are typically quite limited with electronic equipment. As a result, users often forego long-term preventative maintenance programs to address this issue.

It is important to consider that all components of the automation system installation are not electronic. Normally, at least a portion of the project is comprised of off-the-shelf electrical and electro-mechanical gear, such as low voltage relays and transformers of various sizes. With further implementation of distributed DDC there are fewer such components yet system interfaces and many HVAC safeties and interlocks, etc. are still discreet ancillary components.

In new construction, for example, about 40% of the Variable Air Volume systems installed are applied with pre-integrated DDC controllers on the air handlers and boxes. As a result there are still a large number of transducer and pneumatic interfaces found in new buildings. This is because such interfaces are the most common form of system retrofits.

Redundancy, the old concept of duplicating controls to ensure "fail safe" operation was common in older systems, and is still applied in some systems today. This practice is particularly true in retrofits, and where owners are unfamiliar with the technology. It is also an issue where there are DDC to pneumatic interfaces, because systems are often designed with fail-safe features. It is crucial to ensure that an Automation system failure does not cause the equipment to revert to fail-safe mode, resulting in missed savings and possible control problems.

The primary focus of hardware maintenance is to ensure that DDC System savings are not jeopardized by a minor component or equipment failure. Electronic equipment has earned a well deserved reputation over the years for good reliable operation, but that does not mean that it should be neglected. So a regular hardware maintenance and inspection program must be developed along with an operating budget for necessary repairs.

Security Issues

Security in this case is the process of avoiding system abuses and missed cost savings. Security issues of greatest concern are those system inefficiencies which are introduced via "unauthorized" and "uneducated" interface with the system, and this discussion will be confined to those topics. Such problems are limited to those exchanges which occur at a local hand held operator interface or through a Personal Computer based front-end. With respect to local panel or hand held interface exchanges, unauthorized refers to any interaction that occurs through unsanctioned channels.

It is commonly known that manufacturers provide for three password levels of system interface or access: "read only," "read and program," and "read, program and change passwords." Controller level security assumes that the individuals conducting the interchange do not have terminal or site interface passwords.

Classic examples of a system's savings deterrents from unauthorized interface are: pulling the plug, fuse, etc.; jumpering or re-circuiting wires around the system; or any number of other creative system stoppers. Often the severity of this problem is determined by the amount of direct local presence an energy or system manager has on site.

It is assumed that most users deal with uneducated exchanges on site from time to time during or prior to system training of new employees. This may also be caused by employees who are unfamiliar with the automation equipment. These are a primary sources of lost savings particularly if regular maintenance does not make them visible.

Problems stemming from uneducated interaction generally arise through local system operator machine interface features. Other include override or bypass equipment that is added external to the system, such as momentary switches or telephone enable features, and is abused.

It is also possible to have an uneducated interface with system software. This happens when site personnel are provided with a local frontend and a password. That allows them to make changes that effect system integrity. Further discussion related to this concept will be addressed under performance and programming styles.

The topic of Personal Computer (PC) central operator interfaces for local and remote communication introduces a number of interesting variables into the discussion of unauthorized exchanges. Normally these PC interfaces are password protected, thus limiting unauthorized interface. Unauthorized interface can be severely limited by careful maintenance of password integrity.

Bear in mind that unauthorized interfaces might also occur through such problems as computer viruses. This type of problem is possible with any type of PC system. Uneducated interface is often a greater problem from remote communication. There are several considerations under this heading which will be discussed under system performance, as it is also necessary to consider programming access and style issues. The primary point is that interchanges should be monitored, and that individuals responsible for this activity should be trained, supervised and follow an established procedure to minimize problems stemming from this cause.

Performance

System performance is a broad category, and is affected by many variables. To simplify the discussion, assume that the actions recommended thus far have been implemented. This will allow the discussion to be limited to control software. Note that this does not include the PC or front-end software in any way.

The first important topics under control software are access and style. Consider that a segment of the equipment available on the market is application specific or parameter based.

Application specific control software allows limited user interface. Many mini-computer, micro-computer and Distributed Direct Digital Control systems employ proprietary programming languages which are used to program the control sequences. Users may customize operations with these controllers through programming parameters and setpoints. In such cases manufacturers often limit that interface to: "Read Only" and Program Parameter" passwords. These will allow a user to access data and change setpoints, but not to modify the actual control algorithms, etc. As noted programs are not changed, though it is common for users to program parameters in a way that controls many different types of equipment, including some that was not originally intended.

The concept of control programming is not to be confused with the manufacturer's source code. Source code is the manufacturer's proprietary program instruction and is often called firmware.

System control software of interest here is the user level programming language. Three types of control software will be discussed: "Equation Driven" or "Object Oriented Graphic Sequences" and "Parameter/Setpoint Entry". Of key concern is that programmable control software is the focus, and it is not limited to equation driven language based devices. New devices on the market with Expanded Parameter/Setpoint type programming offer a broad range of equipment combinations and sequence options. Therefore control software may be viewed in two general categories: flexible and structured.

As a clarification, all languages are structured, but the Automation System firmware provides a language for user interface which will vary in format and style. Structured styles, mentioned above, provide very specific formats which require a response to a prompt. Such systems are often Microsoft Windows based, but require fairly simple point and click selection of control parameters and setpoints for a control application.

Application Specific control software typically employs error trapping which limits responses to a particular prompt to answers within a specific range. This may appear to refer more to the front-end system then the control. The key point though is that the contractor or end-user has access to parameters for sequences or setpoints only rather than to the actual control language.

Flexible control software varies in degrees but shares critical characteristics. Primarily this approach allows control loops to be programmed using logic statements or graphic control diagrams. Typically logic sequence languages require statements to be written into programs or equations to accomplish each specific task. The graphic approach involves the selection of HVAC components or control icons, such as a symbol for "clock" or "PID loop".

These systems offer two approaches for programming to the more sophisticated user. but some manufactures have employed techniques to simplify systems through graphics.

The flexible programming approach, sometimes called Operator Control Languages", requires step by step control sequences to be written in software code. A shortcut to simplify this technique is to include library file strategies for the programmer to draw upon or link together. These may be used discretely or incorporated into an individual program for a piece of equipment or application.

Another technique to further simplify flexible programming is graphic sequences through diagrams style programming. The key is that regardless of the technique, flexible programming allows the controller to meet any control sequence. The essential requirements are a knowledgeable programmer who understands the application and the control language, and a detailed testing sequence to ensure that the program controls as intended.

Each of the categories outlined under performance applies to both flexible and structured program styles, but the extent of the affect will vary.

As with global system issues, software performance factors will be examined in two general categories: Validation and Finetuning.

Validation is a check on software integrity which involves: System and Synchronization and Maintaining Program Integrity to avoid software problems through Programming Errors and Software Inconsistencies. Finetuning is a process of improving efficiency through adaptive programming modifications based on examining historical data.

System Synchronization means updating all related user files to include changes in parameters that are stored permanently on disk. This should be a part of a general software maintenance, yet it becomes complicated when more than one front end is employed. This is even more difficult when geographic distance is applied, and individuals who make changes to system parameters and strategies are not diligent about communicating those changes to other users.

Further complication can occur if the program residing in the Automation System is lost, due to extended power failure, etc., and a new program must be downloaded. This situation can result in the need to retrace every step taken in the process of fine tuning and modifying the original program, a time-intensive process. Equally important to the Energy Manager, is the fact that this may erode the facility user's faith in Automation systems. Some of the more sophisticated front-end systems on the market are automating this feature, and even go so far as to initiate communication with multiple remote front ends to synchronize front-end data files.

Program integrity is an issue which is caused by software inconsistencies of several types. Basically this problem occurs when there are errors made in programming, the initial program is modified incorrectly or a parameter input to the program is inconsistent with related control strategies. These concerns can be acute where flexible program styles are employed, and there are no internal error trapping routines to guide the programmer.

Dependent upon the data checking features used, this can also happen when programs are transmitted via modem. Interference or noise introduced to the line during transmission can result in a minor transposition of characters which is not picked up by the front end, and render the control strategy ineffective. Program integrity can best be ensured through a well conceived and implemented system acceptance. This includes exercise of all system modules, programs and components under various load and operating conditions during the system start-up phase.

Management Reporting

Reporting is an essential function to DDC System management. Managers must produce reports on the fast track while maintaining a high level of quality. There are two general categories of reports: Operating Reports and Management Reports.

Operating reports usually contain day to day information that is of importance to control, or energy management, and maintenance staff only. These typically provide hard data for use in making decisions on regular operations, evaluating program performance and justifying new projects. Such reports would also provide a full range of data for use in a preventative maintenance program.

Management Reports are intended to review major milestones and provide bottom line results. These reports will often employ charts, tables and other graphic displays of the information. Historically these have been presented on paper to management, but with the advent of real time data transfer between System interface software and application packages PC presentations are becoming more common. It is now possible to have the energy manager bring a color monitor into a management meeting and show real time information in graphic charts and diagrams throughout the presentation. This supports the primary function of management reporting which is to present program results to upper management is a concise and state of the art manner.

System Finetuning

System Finetuning is a key DDC management process which ensures that the system meets projected benefits, and in many cases exceeds original projections. It should be clear at this point that all of the preceding activities must be successfully implemented before finetuning can occur. Finetuning as a function differs from the activities discussed thus far in that it addresses modifications to software, and targets system enhancements.

The system validation issues discussed above are oriented towards maintaining the system in an optimum working environment. Validation also ensures the integrity of original system programming.

Finetuning, as the term implies, is a process of evaluating the existing system design, software control loops and parameters to determine whether control may be implemented more effectively and efficiently. Once an ongoing effort has been applied to system validation, the finetuning activity follows as a matter of course.

Unfortunately many users are totally consumed by or encounter difficulty in completing the validation process. This condition might be called "fire fighting", and users who find themselves in this mode may never complete validation. Due to the fire fighting mentality, or for other reasons, there are all many system owners who do not allow time for finetuning. Therefore many DDC systems continue to be managed with a crisis mentality, and finetuning is not done.

This chapter assumes that system validation is actually an essential finetuning activity. This validation process ensures that the system achieves the desired control performance and savings. In contrast to a crisis program, validation enhances performance over a system which is installed and not maintained.

With a well validated system, Finetuning becomes a specialized task which is applied to specific aspects of the control system. The user finetunes a distributed zone controller, a cooling control strategy or an occupancy schedule to optimize that specific automation system function. Finetuning requires that specific processes be pinpointed for examination, and that each of the steps outlined below is carried out.

- *Targeting use areas* – Interrogate the system to determine current operating characteristics and opportunities for control enhancements. This process may also include tenant or staff interviews to determine changes to the scope of use and occupancy for particular spaces.

    Targeting of energy intensive or wasteful processes is one type of opportunity. Other options may include the use of more sophisticated control strategy, or a fully distributed controller, for a point(s) that were previously controlled. For example, lighting in a retail store may be controlled on a time-of-day basis. To fine-

tune this use area the company could monitor daily store entrance and egress through door switches and a building security system. This would allow more precise control by energizing lights at the exact time of arrival, and de-energizing immediately upon departure.
- *Evaluation* – Establish trend logs to monitor use of the targeting areas. Conduct daily or weekly reviews of system trend longs to verify opportunities. Review logs and historical information to supplement current data and allow for stronger analysis.
- *Benefit Analysis* – Next the user must assemble the data and make benefit/savings calculations. This is generally necessary to ensure that the proposed modification will enhance system performance. A personal computer and spreadsheet software are the most effective tools available to complete this task. Note that this is called benefit analysis. It is extremely desirable to use system that allow easy conversion routines from Automation system data file directly to a particular spreadsheet for analysis.

An important note is that it is possible to make a control change which does not save dollars, but does provide benefits. The presence of a DDC System allows for greater flexibility and convenience which may be of value to the occupants. For example a commercial office building may allow tenants to dial up the control system and warm up the building before they go to work on a winter weekend. This may use more energy than in the past when the tenant had to arrive, turn a timer switch and wait 30 minutes for the building to warm up. However, the convenience and benefit to the occupants may attract new tenants to the building.
- *Trial Implementation* – Another benefit of DDC systems is that they allow changes in control strategies to be effected on a test basis and carefully evaluated. If the strategy change is to be used in more than one facility, this test run should be done with a local facility system that can be carefully monitored. An alternative is to simulate the test prior to control of actual equipment. System changes of this type may be done quickly, and trend longs are again used to evaluate the results.

DDC Systems aide in testing because they can also speed up time or simulate conditions that may not appear until a later date, for example weather. Building occupants should be informed of these system changes as well, to build good will. Finally, calculations can be compared with actual results. Managers can then make intelligent decisions on to implement this modification permanently or to re-evaluate immediately or at a later date.

- *Re-evaluation/Field Implementation* – This step may involve minor corrections to the strategy or simply targeting a field test site. If field testing is required the comments made above still apply, and the site should be monitored regularly to ensure smooth operation.
- *Final Review* – When testing has been completed, a final review should take place. The purpose of this is to plan a strategy for global implementation of the finetuned control strategy. Among the considerations which must be addressed are: applications which can benefit, transition to the new strategy, program updates, timing of implementation, informing tenants and plans for a 90-day review.

The finetuning process should be conducted periodically on all controlled equipment, distributed controllers and loads in all DDC systems. Benefits from this can far exceed energy savings. One of the key factors tenants cite for leaving an office space is comfort. In fact property managers have documented that tenants will begin to search for new offices after the third complaint about space comfort. The same comfort issues apply to customers in a retail store or patrons of a local restaurant. Finally, do not overlook the fact that space comfort has a definite impact on productivity. Even if the facility is an owner occupied office building there are many reasons to ensure optimized comfort and efficiency.

Finetuning ensures optimum system performance which can significantly enhance comfort, thereby adding solid benefit to the Automation system. In addition, the automation system can provide information for preventative maintenance which, if used for service scheduling, will extend equipment life, enhance operational efficiency and contribute to long-term occupant comfort due to tighter control and less downtime.

This keeps building owners happy and, of course, to the contractor, it means fewer callbacks for better job profit.

DDC MANAGEMENT TOOLS

This chapter has covered in considerable detail the process of DDC System Management. There should be no question remaining that this process can be critical to the viability of a system investment. Implementing this process, however requires tools. These tools are in fact a combination of activities that are carried out by managers and the equipment needed to accomplish the activities. There are four primary tools.
- Communication
- Database Management
- Analysis and
- Reporting.

This chapter does not discuss these tools in detail, but attempts to illustrate how they affect the effective operation of DDC Systems. These tools will be covered briefly as they relate to the issue at hand.
- *Communication* in this context is defined as the process of human interface with the DDC system firmware. It should not go unmentioned that in the world of Distributed DDC systems, this implies local area control networks, front end communication software. Though these topics are included, they should be completely transparent to the user.

Communication for the user is a tool to interrogate systems, carry out the validation and finetuning task discussed above and ensure optimum system performance. The basic requirement to execute these tasks is communication both locally and remotely. A host of issues will be discussed throughout this book including networks, front ends and operator machine interfaces. The ability to successfully manage systems relies on these components to offer a real time status window for current system operation.

Yet our industry is not static, and we are seeing requirements for more features than those outlined here. Expanded features are becoming extremely important: After-hour tenant billing packages to monitor and invoice, preventative maintenance packages and integrated graphics and system design aids to name a few. This communication may not be limited to automation system related networks and front ends, but will likely include Local Area Network interface to other PC or control systems for data exchange. So communication implies much more than control system interface, and our plans for the future must consider these expanded requirements for processing power, data storage and exchange of data to industry standard formats. This is critical, but in spite of these requirements we must maintain our focus on communication of the most critical and basic element of DDC System management.

Database management is an important concept touched on in the discussion of communication, but what is really necessary? Managers want access to data from their systems for documentation, effective system management and alternative profit center options like tenant billing.

One of the greatest benefits that end users gained from DDC systems was unforeseen, and that is information. This feature alone has contributed significant savings to building owners, only a portion of which are in the form of energy savings. The information from these systems may be used to evaluate investments, bill tenants, maintain the building, monitor building occupancy and augment fire and security systems.

For all the reasons above, the information resident in the DDC system is valuable, but unusable without data files structured to be compatible with other systems. Therefore system information must be converted to any number of formats, Data Interchange Format (DIF) being most common, for use with a host of applications programs. Such application programs might include conventional spreadsheets or database managers, or could be customized PC software for features like tenant after-hours billing.

Discussion of DDC Management tasks, in particular finetuning, highlights the need for full access to system information. A variety of tasks are performed such as sorting data by category, etc., placing it in arrays, summarizing and preparing data for analysis. As a result DIF formats are a short-term solution; ultimately it is essential to the performance aspects of DDC management that managers be able to build interactive databases. This means that all information from the system may be arrayed on a real time basis in spread sheets and database management systems to track and monitor a host of conditions. Underlying all the discussions in this book will be the issue of information and how to acquire it quickly for a variety of uses. One of the key uses is analysis, the process of combining information in a meaningful way.

Analysis is only possible when data is available in a useful format. The topic of analysis here involves conducting engineering calculations or other operations on system information. This is the primary building block for system performance enhancements and in particular fine tuning. The primary function of analysis is to report on the data in a meaningful way for tracking the building system and documenting results.

Expanded use of analysis as a tool is also allowing some managers to turn their buildings and their systems into profit centers. This is because "pseudo" or calculated points have become common with DDC systems due to the growing need for complex algorithms. In addition to externally calculated data the use of these points allows the user to upload information in a partially or fully analyzed format, thus saving time and simplifying the process to be carried out by the manager.

Pseudo points are not physical devices connected to the system, but the product of calculating data from real points and/or constant values or other data. The issue here is that a user can derive valuable information or save time using this feature. For example, a user may track after-hours use through a real point (runtime on a digital input), and every 30 days calculate an invoice that may include an hourly use charge in dollars. Runtime data from the controller may be arithmetically combined from more than one real point. This simple calculation would save labor that the user would typically expend polling the system, etc. So, the pseudo points may be calculated at the controller or in a host system or front end. Expanded use of networking and global point or data-sharing features enhances the value of pseudo points, and system analysis.

Management Reporting – Regardless of the types of analysis carried out by the user, one of the most important jobs is reporting on the data. Reports may be specific or general, and may involve a presentation to document system performance or an invoice for after hours use.

Reporting is designed to take data in a variety of formats and array it for viewing. The reports may be simple tabular listings of information, two- or three-dimensional graphics, screen or hard copy representations. Typically, the key to the analysis step is to massage the information in such a way that it can be presented clearly, and in summary form. However, the benefit of some reports is detailed historical data for diagnostics. This is a non-control function, yet it is very important to the overall success of a system. Many ancillary features of this type are being integrated with systems to provide these tools directly for DDC management.

Successful implementation of an Automation System requires an effective management effort to address each of the issues discussed in this chapter. Lack of DDC management is one of the key reasons that some systems do not meet their performance objectives.

The manager should therefore consider the DDC management process as critical to long term system success. DDC Management functions are carried out through a combination of manual and automated steps, and with the aide of the tools discussed here. Remember, this may be the most important chapter in this book! You must get down to basics, and carefully monitor and maintain a DDC System just like any other piece of vital business equipment.

Chapter 18

DDC System Operator Training

By Jack Meredith, P.E.

PREFACE

Operator training is essential to a successful DDC System. The activities outlined in the last chapter can not be carried out effectively by an untrained operator.

In this chapter Jack Meredith of British Columbia Buildings Corporation, a major end user in Canada outlines a proposed training program. The approach here is to outline the key components for such a training, and not offer an actual program. One of the greatest difficulties for building owners trying to implement energy management programs, including automation systems, is availability of trained staff.

There are very few programs yet in tact for training such specialized personnel. As a result, many end users like Mr. Meredith, have been forced to develop their own in house training programs. Another example of the crying need for this type of training is the fact the many of the local trade union have also begun requesting this type of training

---

Jack Meredith P.E. is Director of Energy and Environment with the British Columbia Buildings Corporation. This chapter was reprinted with permission from the Association of Energy Engineers. It was originally published in the proceeding of the HVAC Congress, Boston, Massachusetts in March of 1992.

for their members. In many cases the local union will go to a control manufacturer and request training. Another source for such training is a trade organization such as the Association of Energy Engineers. Ultimately both of these approaches result in high quality staff training.

There are some problems with generic training through local unions and trade associations. First they must assume that all attendees come to the session at roughly the same skill level, and the ability to customize material for individual needs is limited. Second and perhaps more important is that these programs can not offer practical, hands on training to employees in their own work environment. As a result, attendees must make transfer information to their current situation to the best of their ability.

Mr. Meredith proposes a much more customized approach. This is a comprehensive in house training program complete with software tools for use in carrying out the various building operator duties. Note that these tools are not described in detail, yet it appears that they have been developed around off the shelf Personal Computer (PC) application software. The reader may feel free to develop similar tools in house or contact Mr. Meredith for more detailed information on these tools.

It is critical for the reader to note that the underlying premise in this chapter is first that staff must be dedicated to the task of building operations. These staff members must have a clear understanding of the tasks required, tools to conduct those tasks and a means to measure progress toward their goals. Given a commitment to these basic precepts, the next step is to establish the program for system operations and clearly delineate the tasks required. Finally it will be necessary to dedicate staff to this purpose and develop an effective training program with both classroom and on the job training components. Given that data, let's look at how Mr. Meredith proposes that this training be conducted.

Training High Performance Building Operators (HIPEROPS)

By Jack Meredith – Director, Energy and Environment
British Columbia Buildings Corporation

INTRODUCTION

The idea of a "HIPEROP" High Performance EMCS Operator was conceived as part of the aggressive British Columbia Buildings Corporation's Energy Program. This program has reduced energy consumption by 55% over the last 12 years, saving 60 million dollars in a time before utility companies aggressively supported the demand side management of energy. In striving for building operation excellence, it was found that a "Win-Win-Win" environment could be developed if building operators were introduced to technology and were made to be an important link in the accountability network. The concept turned the building operator from an overhead maintenance item to a self-sustaining profit center. The self-esteem of the building operators was greatly increased also because they could participate in the improvement of the efficiency and comfort of their complex(es).

The HIPEROPS training should be centered around the use of personal computers by the building operator. The training should provide nine computer programs which aid the building operator in understanding energy interactions in buildings and building systems. The computer programs are:

- **MRR** – Meter Reading and Rate Program
- **SEA** – Simplified Energy Accounting Program
- **ZAP** – Zone Audit Program (Heating/Cooling Load Program)
- **BEST** – Building Energy Simulation Training Program
- **THUMB** – Rule of Thumbs in HVAC

- **SYSSIM** – System Simulator Program (for HVAC Systems)
- **PLP** – Part Load Performance Program
- **EMCS** – Energy Management Control System – Mandatory Requirements and Specification
- **OCL** – Operator Control Language Program (Custom Strategies for EMCS)

Why the software approach to HIPEROP? Good building operators must embrace technology. Most large buildings cannot effectively be operated without the use of computer technology. Entry level for a HIPEROP (High Performance EMCS Operator) is personal computer working knowledge so, rather than write a book, this series of HIPEROP text, calculations and demonstration programs was identified. Simple programs can be designed for teaching concepts, but can also be used to solve real building programs and to start the transformation of your building operator(s) into HIPEROPS.

These programs may be developed primarily for building operators, but they are also useful for design professionals, contractors, managers and all those who deal with the HIPEROP building operator. The evaluation of a new building design by a HIPEROP (High Performance EMCS Operator) with a High Performance computerized Energy Management Control System often sends contractors and consultants back to learning mode.

The HIPEROP training must be designed to significantly increase the operational skills of good building operators. Once operators complete this training, they are able to transform their building(s) into high performance buildings achieving excellent client comfort with minimal energy consumption and environmental impact.

A good portion of this training deals with accountability, plus with establishing an accountability network. Building operators cannot transform ordinary buildings into high performance buildings in isolation. They require the support and financial commitment of management. A commitment must be made by their management that will empower their operator to strive for excellence. If management is not prepared to make this commitment and provide necessary support to implement projects with a reasonable payback, there will be minimal value to operators training.

DDC System Operator Training

The training would not be intended for novice operators. To get maximum advantage, operators should have a minimum of five years of actual operating experience in building systems. An understanding of MS DOS and experience programming Building Automation Systems is also highly desirable. A general knowledge of air systems, heat flow, controls, electrical, boilers and computers is also recommended. Although BOMA Buildings Operators Training would not be a mandatory prerequisite, equivalent knowledge is assumed. Operators taking the training should be self-motivated and have a strong interest in improving their skills.

This type of training is about relationships and should be designed to increase operator awareness of various relationships that must be well understood for a person to become a high performance building operator.

The *Introduction* would explain what a high performance building operator is and what the relationship is to successfully operating a comfortable energy-efficient building.

The *Accountability Network* would explain the importance of having a correct relationship with all of the people who can make a difference and emphasize the operator relationship with accountability.

Energy Accounting would use a computerized energy accounting program (SEA) to explain methods of documenting the relationship between energy use, cost and relative building performance. If you cannot measure it, you cannot manage it!

Client Comfort would explain the delicate but important relationship operators must have with clients and building occupants.

Dynamics of Building Energy Flows would explain the relationship between internal lighting, appliance, people loads, building envelope loss and gains. The strong relationship of solar load on a room load is demonstrated. The relationships are demonstrated using the Zone Analysis Program (ZAP).

Computerized Building Energy Simulation would show a relationship of rolling up hourly heating cooling load calculations and lighting KW to annual figures using the BEST Program.

Comparison of Actual Consumption (SEA) to Energy Simulation (BEST) would explore the relationship of the calculated annual total energy usage to the actual usage. The relative difference is explained as the energy consumption of the mechanical system and is shown using the "ZAP" input and BEST simulation.

Identifying Energy Conservation Opportunities would use your understanding of the foregoing relationships to save energy and matches the BEST simulation to SEA actual.

How Systems Handle Building Loads would explain the relationship between various systems, terminal reheat, variable volume, dual duct using the SYSSIM System Simulation Program. The relationship of mass is also demonstrated as part of this program.

Presentation of Existing OCL Modules would be presented using OCL Simulator Program. The relationship of using OCL to effective control systems should be emphasized.

HIPEROP was originally conceived as a training course to be put on in groups of 10 to 12. Based upon local experience, it has evolved to be a network of High Performance Operators who are able to share ideas, technologies and experiences with one another to transform their buildings into high performance buildings achieving excellent client comfort with minimal energy consumption and environmental impact. British Columbia Buildings Corporation's desire is to be able to act as a facilitator in networking as many HIPEROPS together as possible. As such, the HIPEROP training can now become a HIPEROP Users Group, a self-sustaining entity which is funded through individual and corporation memberships.

Chapter 19

DDC Industrial Case Study: Energy Conservation is a Process

By Paul Krippner, P.E.
Alan Stewart, P.E.

PREFACE

This case study is drawn from the Industrial sector and one of the worlds leading medical products manufacturers. ETHICON is one of the Johnson and Johnson family of companies and sets an excellent example for all industrial facilities. This study shows an excellent real world example of how many of the techniques in this book can be used successfully.

It is particularly important to note that this chapter offers a macro-view of energy conservation and management. The author firmly believes that Direct Digital Control systems are a corner stone of any energy management program. These systems result in energy cost reduc-

---

Paul Krippner, P.E. is Chief Engineer with ETHICON, Incorporated at the Cornelia, Georgia plant which is the focus of this case study.

Alan Stewart, P.E. is an Engineer with Honeywell, Incorporated in the Atlanta branch of the supplier that entered into the partnership with ETHICON.

This chapter was reprinted with permission from the Association of Energy Engineers. It was originally presented and distributed as a technical paper at the World Energy Engineering Congress, Atlanta, Georgia in October of 1991.

tions, improve building comfort and offer a host of operational benefits. Yet, as this case study points out, DDC systems should not be viewed in isolation from the rest of the facility.

A DDC system is only one of the possible energy management projects in a facility. Once installed, however it is also a valuable tool to evaluate additional projects. It allows owners to monitor energy use, and to make intelligent decisions about other energy management projects to implement and their relative priority. Mr. Krippner and Mr. Stewart do an excellent job of showing the overall system benefits, and documenting many of the techniques covered elsewhere in this text. The credibility of this case study is enhanced because these authors represent real companies who are making significant progress in managing this facility efficiently.

ETHICON's energy management program is exemplary, and the Cornelia facility documents the success of that program. Therefore the philosophy and practices of the managers running that program are of great interest to us, and deserve our congratulations.

There are a number of technologies discussed in this paper, and a brief note on these is warranted. The DDC System is referred to here as a Distributed DDC (DDDC) or Building Automation System. As noted earlier in this book, these terms are synonymous for our purposes because they all employ distributed intelligent control modules.

Another important concept discussed here is a customer/supplier partnership. At first glance this may appear to contradict traditional DDC system acquisition practice, perhaps it does. The approach certainly is contrary to competitive bid and specification techniques discussed in Chapters 15 and 16. However, in the prefaces to those chapters it was noted that negotiated system sales have become much more common. Actually the concept of a partnership is not so very new. In the mid to late 1970s a number of companies entered the marketplace offering both financing and turnkey energy management service. Over the years this practice has been referred to as shared savings, performance contracting and Energy Services.

The concept of shared savings nearly always involves the type of partnership discussed in this chapter. This is because the capital provided to initiate the project is an investment that the shared savings company must ensure is returned. This is a perfectly natural business proposition, and often results in benefits exceeding projections on these projects. In ETHICON's case the approach is similar, though financing is provided by the customer. The similarity is the fact that both the customer and the supplier see mutual benefits in partnership. The customer receives quality services that contribute directly to the success of his energy management program. The supplier also benefits from the professional reward of a successful project. The ongoing revenue from future projects further benefits the supplier, as does the testimonial of a satisfied customer.

Energy Conservation Is A Process

*By Paul Krippner, P.E., Chief Engineer, ETHICON, Inc.;
Alan Stewart, P.E., Honeywell, Inc.*

Energy conservation is a Process. It should not be considered a project that has a beginning and an end. It is a Process which goes on and on and grows and grows. ETHICON, Inc. has approached energy conservation at its Cornelia, Georgia, facility with the philosophy that energy conservation is a Process. The result has been very successful.

The term "successful" is subjective and can be measured by several standards. One way to measure the success of an activity, such as energy conservation, is to look at the bottom line savings. After 28 months, ETHICON, Cornelia, has accumulative savings of over $489,000.00. If bottom line savings are the goal, then this has been a successful process. The real success story at ETHICON, however, is not just the savings. The real success story is the process which made the savings possible and how a partnership between ETHICON, Inc. and the Honeywell Corporation made the whole process work.

ETHICON realized that a dynamic process would require new ideas, new opportunities constantly being developed. ETHICON, like most companies, did not have the resources to devote full time attention to energy savings activities. In addition, ETHICON did not have on their staff the professional expertise necessary to seek out and resolve complex energy problems. To overcome these inhibitors, ETHICON chose a new approach and developed a cooperative working relationship, or partnership, with one supplier. This relationship resulted in mutual ownership in the success of the project, and even extends to joint authoring of chapters, such as this, to document its success.

ETHICON entered into a formal contract in 1989 with the supplier for the installation of an energy management system. The "partnership"

however, was not created through purchase order or legal contracts. It was a relationship that grew out of ETHICON's wish to have a single source supplier for all their energy and building system-related activites.

The partnership was a "quality" match between ETHICON and the supplier. ETHICON is one of the Johnson & Johnson family of companies. Johnson & Johnson and the supplier have chosen the concepts of The Quality Improvement Process (Philip Crosby Associates, Inc.) as the basis for manufacturing management and meeting customer requirements.

Having a single source partnership meant that ETHICON would not use competitive bids to assure that they were getting the best value or to assure that they were seeing all the alternatives and options. It also meant that a purchase order would not have to be issued every time they needed some professional assistance with an energy problem. ETHICON states that the partnership between ETHICON, Cornelia, and the Atlanta branch of the supplier is "like having a branch of our engineering department in Atlanta that we can call on when we need some help."

Working together, ETHICON and the supplier developed an energy conservation process consisting of:
- Energy Audit
- Energy Management or DDC System
- HVAC, Building System Modification
- Maintenance
- Energy-efficient Lighting
- Attack on the Sacred Cow–the Manufacturing Process
- Energy Awareness Program

ENERGY AUDIT

The basis for any energy conservation process is the energy audit. Most energy audits involve the installation cost and predicted savings associated with an Energy Management System. While an Energy Management System is important to the process, a truly effective audit will also identify and quantify sources of major energy consumption. This in-depth analysis helps establish priorities and is the planning tool for the conservation process. The audit becomes the basis for energy savings calculations and measurements.

An effective energy audit must go beyond just identifying energy savings opportunities, it should also recommend modifications to building systems and to the manufacturing process which would optimize energy consumption.

The energy audit is the tool used to initiate the energy conservation process. The audit continues to be an important tool as the process develops. Each month the audit must be updated to include the changes in manufacturing schedules, any equipment added, or equipment removed. It should also include any energy savings activities which were completed during the month and any planned for future months.

ENERGY MANAGEMENT SYSTEM

Most plants that still depend upon individual electric or pneumatic thermostats for area control will benefit from the installation of an Energy Management System. As previously noted, the terms Energy Management System and Direct Digital Control System are synonomous in this context.

In the past, all Energy Management Systems have not been successful and potential users can learn from that prior experience. Many companies installed Energy Management Systems with the belief that they would require minimum maintenance and minimum attention. They believed that the computer-based system meant automated, hands-off control that would function indefinitely. As long as the person who installed the system remained involved, or as long as someone was emotionally tied to the system, it continued to operate reasonably well. Unfortunately, in most companies, priorities change and new assignments are made. As a result, no one was fully committed to maintain the Energy Management System. As problems occurred, it was easier to bypass or disconnect control lines than it was to fix them. After two or three years, the Energy Management System became useless and was just a "pile of junk setting in the corner."

ETHICON wanted to assure that their Energy Management System continued to operate long after those involved with the original installation had moved to new assignments. They wanted to assure perpetual attention, maintenance and upgrading of the system. To assure that this would happen, an on-site energy specialist was hired through the supplier.

DDC Industrial Case Study: Energy Conservation is a Process   427

The energy specialist, in addition to full-time maintenance of the Energy Management System, has the responsibility of continuously seeking new energy conservation opportunities. The cost of the energy specialist was included in original contract costs and ROI calculations. Remember that Chapter 17 outlines the activities of that specialist in greater detail.

ETHICON feels that one significant reason for the success of their energy conservation process is the full-time on-site energy specialist. Other companies, following the ETHICON lead, have contracted for on-site assistance. Some contracted for full-time assistance and some for a part-time energy specialist to service their plants two or three days each week.

Companies that recognize the benefits and the limitations of an Energy Management System have had good success. Too often, however, companies will consider energy conservation as a project and, having installed an Energy Management System, will stop their efforts there. This is where their energy savings will also stop. Fig. 19-1 is a graph of power consumption at ETHICON, Cornelia, Georgia, in 1989. The center line, the zero line, is 1988 consumption. In 1988, the supplier con-

KWH CONSUMPTION
PERCENT CHANGE COMPARED TO 1988

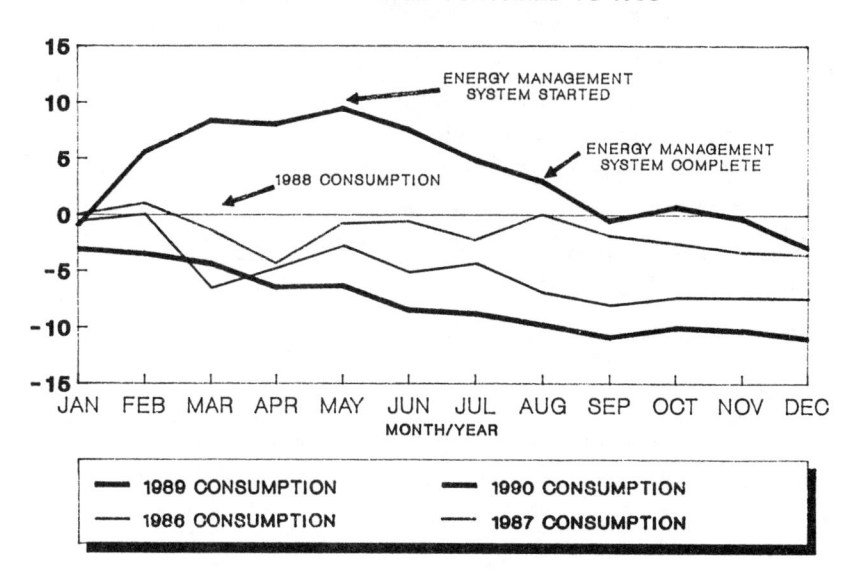

Figure 19-1 KWH Consumption

ducted ETHICON's energy audit and it is the basis for savings measurements. The graph shows that energy consumption in 1989 was increasing dramatically. In May, the Energy Management System was installed and individual areas of the plant were placed, one at a time, under control of the system. The result was a noticeable downward trend in power consumption. The downward trend continued until August when the Energy Management System installation was completed. If ETHICON had considered energy conservation as a project that ended here, then their savings would also have ended here. Considering energy to be a process, however, ETHICON and the supplier continued to develop energy-saving opportunities. As a result, by the end of 1989 energy consumption in Cornelia was below that of 1988.

HVAC MECHANICAL SYSTEM MODIFICATION

The particular HVAC system modifications for Ethicon are important here as an example of the more general approach to good engineering practice. Extensive computer modeling of the following systems was completed during the design analysis: the building envelope, chilled water generation systems, steam generation systems, air-side delivery systems, and process energy consumption systems. The result of this modeling showed we could eliminate two energy inefficient chilled water generating systems. Further, we were able to convert the constant volume air handling systems with terminal reheat to variable air volume with terminal reheat. We were able to eliminate 220 tons of inefficient cooling plant and place 150,000 CFM of air side capacity on variable volume. The DDC system became a valuable tool for use in identifying major mechanical retrofit. Detailed techniques for conducting such analysis are beyond the scope of this chapter, yet there are many good resource books available for further research.

MAINTENANCE

The importance of a good maintenance program cannot be overemphasized. The efficiency, reliability and continued effective operation of building HVAC systems is directly related to the maintenance of the equipment.

A basic preventative maintenance program should include periodic schedules for cleaning, lubrication, coolant level checks, filter and belt changing, bearing inspection, motor and control checks. Annual internal maintenance should be performed on chillers and boilers. As noted in Chapter 17 these programs should include periodic evaluation of the DDC System and its components.

An energy-conscious preventive maintenance program should also include regular inspection of pipe and duct insulation, door and window seals, and for proper operation of fresh air dampers. It should also include calibration of thermostats and steam system maintenance.

Steam systems components, such as steam traps, pressure reducing stations and steam valves are often not included in a periodic maintenance program. These steam system components are installed in locations which are difficult to access and are often neglected. These components, if not properly maintained, are a significant source of energy loss.

Preventive maintenance is the basis for a good maintenance program and was adequate in the past. In the future, proactive predictive maintenance must be used if maintenance is to become world class. Ultrasound and thermo testing of steam traps, vibration analysis of bearings, oil analysis of chillers, eddy current testing of chiller tubes and the use of expert systems to trouble-shoot and diagnose chiller and boiler operation are some examples of predictive maintenance techniques available for building system use.

Most plants perform their own preventive maintenance and contract annual boiler and chiller maintenance with service companies. ETHICON found it cost-effective to consolidate all maintenance activity under one contract with the supplier. Professional service companies, such as the supplier, are able to provide specialized repair and maintenance and have purchasing leverage with suppliers. These service companies can also provide predictive maintenance on building systems. This means that their customers do not have to purchase expensive predictive maintenance instruments.

Maintenance is an important part of the total energy conservation process. In addition, maintenance itself is a process. A process which must expand and improve.

ENERGY-EFFICIENT LIGHTING

Energy-efficient lighting was included in the energy conservation process at ETHICON. With so many types of energy-efficient lights available, and so many factors to consider, ETHICON again called on the supplier for assistance. The problem was not as simple as selecting one or two types of fixtures or lamps. ETHICON was retrofitting an existing building and the replacement of 3,000 existing light fixtures would be cost prohibitive. In some areas, energy-saving lamps could be used; however, in testing and inspection areas, reduced voltage bulbs would not be acceptable. In administration areas and corridors, motion sensors would be used, in the manufacturing areas timers would be more effective. Because of all the options which must be considered, the supplier had to develop a lighting model to assist in selecting the most cost-effective lighting.

In most manufacturing areas, standard fluorescent fixtures were retrofitted with F.R.E.D., Frequency Regulating Electronic Devices. The use of the F.R.E.D. units did not require the removal of existing fixtures and did not require the use of the more expensive low wattage bulbs. In inspection and close tolerance areas, electronic ballasts and Octron lamps were used. All fixtures were cleaned and new lamps were installed in all fixtures.

In any manufacturing plant no single type of lighting will meet the various requirements. It is recommended that a lighting consultant be used to make recommendations. The lighting consultant should be an energy conservation specialist who will select the best lighting components from the many brands on the market.

In some places, a lighting program which would require a significant financial investment would not be possible. In these cases, there are some low implementation cost opportunities. Overall lighting efficiency may be improved by a good lighting maintenance program that includes cleaning of fixtures and lenses and group relamping. Group relamping, versus replacing a lamp as it burns out, is often overlooked as a highly efficient, labor-saving maintenance technique which provides improved lighting and reduces costs. Reducing the amount of lighting in corridors, storage and non-production areas and the use of motion detectors to turn off lights will provide energy savings.

ATTACKING THE SACRED COW

Energy conservation is usually considered the exclusive activity of the plant engineers. Most energy saving efforts, therefore, address building HVAC systems or plant lighting. Too often, plant energy savings efforts stop with the physical plant. To address energy savings opportunities that affect the manufacturing process is considered taboo. It is easy to use the excuse that the product is too critical to change the process, just for energy reasons. The manufacturing process becomes sacred – The Sacred Cow.

At ETHICON, the Energy Management Systems provided a tool to identify and evaluate energy consumption of various processes. ETHICON and along with the suppliers engineers used this information to propose modifications to existing production operations. The supplier also reviews all new processes for energy savings opportunities before the new processes are installed.

If energy conservation is to be a continuing process, it must include modifications to the manufacturing process. These modifications will require cooperation with production and process engineers and could require an element of risk-taking.

The credibility of energy conservation grows as the process grows. Success with the energy management system, with mechanical system changes, etc., develops confidence in the process. This makes it easier to promote energy savings activities which involve changes to the manufacturing process.

ENERGY AWARENESS PROGRAM

There are three very important reasons for saving energy: conservation, environment and cost savings. First, energy must be conserved if this nation is to have energy for future generations. Second, each kilowatt of energy that is saved prevents fossil fuel from being burned to generate that kilowatt of energy. Each time fossil fuel is burned, it produces nitrous oxide, carbon dioxide and sulphur dioxide. These are contributors to acid rain and global warming. Finally, energy conservation produces cost savings which will go directly to reducing plant operating costs. Three very important reasons which receive very little publicity.

Energy conservation activites go unnoticed in most plants. Other energy-conserving efforts, such as reduced lighting levels, lower room temperatures in the winter, and higher temperatures in the summer, are unusually unpopular with plant employees. As a result, energy conservation efforts receive resistance from plant employees and plant management.

If energy conservation is to be a continuous process that receives employee and management support, there must be a planned awareness program. ETHICON uses several methods to publicise energy savings. Throughout the facility, energy awareness signs are displayed in manufacturing and administration areas and in corridors. The signs describe the energy savings accomplishments in that area. The effects of the energy process on costs and energy consumption are also diaplayed on graphs and charts. The Energy Management system is located in a prominent central office area that is decorated with plaques, charts and graphs that exhibit pride and the importance of energy conservation.

Each year during Energy Week, ETHICON has an energy day which emphasizes energy conservation. The day's activities include exhibits by utility and fuel companies demonstrating home conservation ideas. A luncheon is provided for employees with a speaker who will discuss and answer questions on energy conservation. A coloring contest is held for children of employees. The local winners are entered in a nationwide contest for all Johnson & Johnson family of companies. These and other activities throughout the year help to emphasize that energy conservation is a process having many benefits.

Promoting energy awareness by communicating the benefits and successes of energy activities is an important part of the energy conservation process. The communication of energy successes should go beyond a single plant or even a single company. If energy conservation is to succeed, nationwide, then plant energy engineers must share their methods and their successes with other engineers. Through presentation, articles and technical papers the successes of individual plants will help direct and encourage energy activities at other locations.

CONCLUSION

Energy conservation is a Process that provides benefits beyond just conserving energy. Energy conservation reduces pollution of the environment and offers excellent cost-savings opportunities.

ETHICON, Inc., Cornelia, Georgia, has approached energy conservation with the full commitment of its employees and management. In an effort to assure that energy conservation receives full-time attention, ETHICON formed a partnership with one supplier. The ETHICON/supplier partnership developed an energy conservation process that has been very successful. To be successful, energy conservation must be considered a Process – a Process that goes on and on and grows and grows.

Chapter 20

DDC Commercial Building Case Study: The Museum of New Mexico and Energy Management

By Brian K. Johnson, P.E.

PREFACE

The case studies in this book have been drawn from typical applications for DDC systems. In fact as you will note, these studies show a wide diversity of system application; perhaps there is not a "typical" DDC job.

Commercial buildings have probably seen the greatest application of DDC systems. The decreasing cost of systems combined with an increase in both functionality and reliability has continued. Given that situation, the author elected not to document a typical case such as an

---

Brian K. Johnson is an Energy Engineer with the Energy Conservation and Management Division of the New Mexico Energy Minerals and Natural Resources Department. Mr. Johnson is a registered Professional Engineer in three states, a Certified Energy Manager and a Certified Lighting Efficiency Professional. This chapter was reprinted with permission from the Association of Energy Engineers. It was originally published in the proceedings of the World Energy Engineering Congress, Atlanta, Georgia in October of 1992.

office building or retail store. Rather, this case involves greater system complexity than many such applications. As a result, this case documents two excellent examples of DDC success. It should be stated up front that it may challenge the readers usual criteria for determining DDC success.

The notion of DDC success has been formed based upon the issues of energy management. Widespread implementation of computer-based control systems did not begin until the energy crises of the 1970s. These systems were initially termed "Energy Management Systems" as in this case, and their primary focus was to save energy and reduce cost. As noted throughout this book though, DDC offers much more than energy savings. This case is a perfect example of two systems which are extremely successful in aiding the organization to meet its mission. In fact, one of the systems is not actually producing energy and cost savings.

How can an Energy Management System be successful without saving energy? Simple, these systems are protecting priceless art collections through sophisticated, accurate and *efficient* environmental control. Efficiency is important, but the benefits of these systems go far beyond energy savings. In this case, the curators and energy managers are learning to unleash the full power of their system to a significant benefit, which is protection of artwork for future generations while conserving natural energy resources.

Chapter 20

DDC Commercial Building Case Study: The Museum of New Mexico and Energy Management

By Brian K. Johnson, P.E.

PREFACE

The case studies in this book have been drawn from typical applications for DDC systems. In fact as you will note, these studies show a wide diversity of system application; perhaps there is not a "typical" DDC job.

Commercial buildings have probably seen the greatest application of DDC systems. The decreasing cost of systems combined with an increase in both functionality and reliability has continued. Given that situation, the author elected not to document a typical case such as an

---

Brian K. Johnson is an Energy Engineer with the Energy Conservation and Management Division of the New Mexico Energy Minerals and Natural Resources Department. Mr. Johnson is a registered Professional Engineer in three states, a Certified Energy Manager and a Certified Lighting Efficiency Professional. This chapter was reprinted with permission from the Association of Energy Engineers. It was originally published in the proceedings of the World Energy Engineering Congress, Atlanta, Georgia in October of 1992.

office building or retail store. Rather, this case involves greater system complexity than many such applications. As a result, this case documents two excellent examples of DDC success. It should be stated up front that it may challenge the readers usual criteria for determining DDC success.

The notion of DDC success has been formed based upon the issues of energy management. Widespread implementation of computer-based control systems did not begin until the energy crises of the 1970s. These systems were initially termed "Energy Management Systems" as in this case, and their primary focus was to save energy and reduce cost. As noted throughout this book though, DDC offers much more than energy savings. This case is a perfect example of two systems which are extremely successful in aiding the organization to meet its mission. In fact, one of the systems is not actually producing energy and cost savings.

How can an Energy Management System be successful without saving energy? Simple, these systems are protecting priceless art collections through sophisticated, accurate and *efficient* environmental control. Efficiency is important, but the benefits of these systems go far beyond energy savings. In this case, the curators and energy managers are learning to unleash the full power of their system to a significant benefit, which is protection of artwork for future generations while conserving natural energy resources.

The Museum of New Mexico and Energy Management

By B. K. Johnson

PREFACE

There are unique indoor conditioning and energy management challenges in museums. In Santa Fe, the Museum of New Mexico (MNM) is located in a unique climate and must stay within utility expenditure limits allocated through the State government budget process, while handling valuable collections with specific environmental requirements. Adequate humidity for indoor exhibitions is the top priority for heating, ventilating, and air-conditioning (HVAC) systems. Direct Digital Control (DDC) or energy management systems (EMS) implemented by the Energy, Minerals and Natural Resources Department (EMNRD) in two MNM exhibition facilities avoid energy costs, but must be maintained regularly. Energy savings goals must yield priority in favor of maintaining proper indoor conditions. Perhaps of even greater importance than cost reduction are the operational benefits of these systems. These facilities house extremely valuable collections which cannot be replaced. These collections must be preserved, and the EMS is a valuable tool is accomplishing that task.

MNM is one of six Divisions within the State of New Mexico's Office of Cultural Affairs (OCA). The mission of OCA is

> *to foster, preserve, and protect current and past expressions of culture and the arts, which are determined to be in the best interests of New Mexico.*

As part of their mission, OCA is well-known for excellence in cultural collections, through MNM. MNM is comprised of the

- Museum of Fine Arts
- Museum of Southwest History

- Museum of International Folk Art
- Laboratory of Anthropology
- Museum of Indian Arts and Culture.

There are eight separate physical facilities that house these operations, including administration. Behind the scenes, there are operational costs that must be managed carefully; the costs of heating, cooling, and lighting the buildings that MNM uses are a part of this. EMNRD has assisted OCA in meeting its mission through the expertise of the Energy Conservation and Management Division (ECMD). ECMD is designated by the Governor as the State Energy Manager agency.

There have been a variety of obstacles that ECMD and MNM have been challenged with as they pursue preservation of collections and control of operational costs. As new situations present themselves, solutions have come forward from the ECMD/MNM collaborative efforts. These experiences, described herein, may assist and encourage other organizations as they take on energy management jointly with a strong mission to accomplish.

The significant issues encountered, in the joint pursuit of energy management and culture, have involved

- HVAC systems in museums
- EMS in museums
- energy implications of operating museums
- communication.

Two MNM buildings, both in Santa Fe, where ECMD has had extensive involvement, are the Museum of Fine Arts and the Museum of Indian Arts and Culture. The buildings are shown in Figures 20-1 and 20-2. With EMNRD funds and ECMD technical assistance, Energy Management Systems were implemented in each of the buildings, followed by EMS maintenance.

Currently, there are new opportunities for energy management that ECMD is pursuing and that can be utilized by MNM. The main avenues through which they will occur are:

- 1991 State Energy Policy
- Green Lights program for State government.

MNM is developing plans for improving its ability to preserve collections as it embarks on expanding into additional facilities. An envi-

DDC Commercial Building Case Study: 439
The Museum of New Mexico and Energy Management

Figure 20-1 Museum of Fine Arts: Santa Fe, New Mexico

Figure 20-2 Museum of Indian Arts and Culture: Santa Fe, New Mexico

ronmental preservation consultant is now retained by MNM. The consultant is recommending improved indoor conditions by making humidity the highest priority in museum operations. Humidity control is a significant issue for MNM, given the dry ambient conditions of the southwest U.S. region, as a whole, and the unique weather patterns of Santa Fe.

STATE AGENCIES' BACKGROUNDS

Museum of New Mexico

New Mexico is often referred to as a tri-cultural state, because of the Native American, Hispanic, and Caucasian peoples that live side-by-side. Unique expressions in art and culture are produced: the mixture of culture and the natural world are frequently cited for the beautiful and diverse works created. MNM is an important part of this natural collaboration of people and land, with ties to a variety of significant endeavors, such as

- current art of the three cultures
- ancient art of indigenous peoples
- musical performance
- anthropology
- archaeology.

The essence of the above is coordinated within the museum buildings and the environment that they provide. Shows, research, education, performances, and exhibitions are constantly being juggled, with the facilities used to the fullest. This is a tremendous administrative challenge. The buildings, themselves, are also a part of the essence, having been produced through the art of architecture, with several of them being of historical value.

A survey of the current status of MNM facilities would lead one to conclude that they should upgrade maintenance and administration of the buildings and their systems. MNM's rule-of-thumb for the proper indoor environment, given New Mexico's dry climate, is 70 degrees F, while 35% relative humidity (RH) should be maintained. MNM has documented that they are not maintaining these conditions, causing an untold amount of damage to cultural collections.

MNM is also increasing their square footage with expansion of existing buildings and construction of new buildings. This includes administrative space as well as museum program space. With expansion seeming to be at an apex now, MNM knows that it is wise to investigate indoor conditioning and energy management.

Energy, Minerals and Natural Resources Department

EMNRD provides consultation services to State agencies in the areas of energy efficiency and renewable resources. This effort is led by ECMD. Among its many capabilities, ECMD has expertise in building systems engineering. Through ECMD, State agencies have access to experience with building systems, which are usually thought of as the three areas of

- envelope
- HVAC systems
- lighting.

Maintaining proper temperatures and humidity conditions ensure that collections are protected, while controlling energy consumption, is possible through a fourth component: EMS. These four components interact, along with human intervention, to provide building spaces with a level of comfort. Systems powered by renewable resources (e.g., passive solar, active solar thermal, photovoltaics) are also integrated into building systems. Passive solar systems for space heating, which are integrated into envelope design, exist, to a certain degree, in the two MNM buildings discussed here.

EMNRD has supported the Office of Cultural Affairs (OCA) with several projects in the past ten years, assisting them to avoid excessive energy costs and to investigate building systems' problems. The simple payback criterion that ECMD uses is ten years or less (i.e., the time in which a project's first cost is paid for in energy cost avoidance accrued through implementation of the project). Examples of ECMD projects for MNM are

- installation of energy management systems
- maintenance of energy management systems
- test and balance of HVAC systems
- energy modeling.

The above listed projects follow in step with the focus of the ECMD vision, which is to be the catalyst for implementing energy efficiency measures in State agencies which control costs.

This vision does not exclude comfort or other space function criteria which are important to the mission of the agency being assisted; energy-efficient operation and comfort are values which can be achieved simultaneously. It is not the function of EMNRD to alter the quality of services of another State agency by restricting or changing usage of building systems. Ideally, energy efficiency measures should be implemented and costs avoided through projects that customers (i.e., museum visitors) do not even notice.

Temperature and humidity conditions which preserve museum collections (e.g., pottery, weavings, paintings) are top priority. Museum collections, which MNM has built a reputation on, require humidity-based environment control for satisfactory preservation. This means that relative humidity conditions for collections are actually of higher priority than conditions for people, which are temperature-based in typical institutional building design.

GOALS AND CHALLENGES

Ideally, proper indoor conditions would be maintained for a wide variety of exhibition types and storage areas in MNM facilities, continuously. The HVAC systems are designed to distribute conditioned air to all spaces in the Museum of Fine Arts and Museum of Indian Arts and Culture buildings. There is not differentiation between the collection-occupying areas of galleries and storage rooms and the people-occupying areas of administrative offices, theaters, lobbies, and gift shops. This means that the HVAC systems should maintain the top-priority temperature and humidity requirements needed for galleries and storage rooms in all building spaces. This is a practical solution and can be programmed into the EMS to coordinate the necessary HVAC control functions. At the same time, progress toward these goals can be monitored by trending data with the EMS. In practice at these two museums, the conditioning goal has been a difficult challenge.

The conditioning goals of MNM are made of a combination of local needs and international requirements for maintaining collections[1]. First, temperature control is relatively easy to manage, because designers and building occupants are used to relating to the temperature-based requirements of conventional buildings. The 65 to 80°F range is fine for people using the building spaces as well as museum collections. However, a constant temperature within the range, 72° F, for example, must be established. Night setback of HVAC systems or any other type of temperature fluctuation, on a daily or weekly basis, is not acceptable. It may be possible to have a seasonal adjustment of the setpoint (e.g., 72° F for winter, 78° for summer). A gradual swing from one setpoint to the other should be planned to occur over a long period; one degree F change per two days may be adequate to museum conservators. Temperature control is not a difficult issue for museums in Santa Fe.

Second, but of top priority, is how to manage humidity control for collections. The typical international requirement widely recommended is 55% RH, plus or minus 5% RH[2]. By maintaining this humidity in museums, one meets an important requirement for accessing the international market of touring collections. MNM has a requirement of 35% RH, plus or minus 5% RH. Their practical reasons for having lower humidity levels are:

- 55% RH has been established by museum conservators from high-humidity climates in the eastern U.S.
- New Mexico has a dry climate
- 75% of MNM's total collection is locally created
- humidity equipment breakdowns would expose collections to damaging humidity changes[1]
- maintaining 55% RH in a dry, cold climate incurs high energy costs and possible condensation damage to building envelopes[2].

With a high percentage of the total collection being created in New Mexico, dry conditions in the collection materials are already established. For international collections, micro-climate containment methods are frequently used when exhibiting and transporting collections to maintain adequate humidity.

To continue striving for the goal of proper conditions for a wide variety of exhibition types, while maintaining collections in storage, MNM has a number of challenges. The challenges are being addressed in a number of ways. New flexible standards for the museum environment will be released soon that are more accommodating to the unique New Mexico situation. As a part of its expansion plans, MNM is considering converting one museum to specifically meet the higher 55% RH requirement. MNM is also consulting with an art preservation specialist to evaluate the current status of their museums and propose specific options for cost-effective solutions1. ECMD is working with MNM by contributing to the consultant's evaluation, continuing its technical assistance (especially concerning EMS issues), and briefing OCA administration on responding to the State Energy Policy and the State's potential Green Lights program.

ENERGY MANAGEMENT

The two buildings being addressed herein are different in design and purpose. It is a challenge for MNM to administer and maintain different types of buildings, while continuing a unified mission of preserving delicate cultural collections within those buildings.

Two Museums

The Museum of Fine Arts is housed in a brick two-story building, with 52,000 square feet. Constructed in 1917, it is one of Santa Fe's many historic buildings. Collections predominantly displayed in the galleries are paintings. A basement is used to store paintings and other collections. A conventional chiller and boiler are located in the basement; ductwork distributes conditioned air throughout the building. Humidity is increased through a hot water coil that creates steam in the ductwork. An EMS was installed in 1985 for $45,000, which resulted in significant reduction of the building's utility bills during the 1985-86 fiscal year period. An ice thermal storage system provides cooling for the St. Francis Auditorium, a musical performance space in the museum. The annual energy bill is approximately $1.10 per square foot.

Pottery, weavings, paintings, silverwork, and other anthropological materials are displayed and stored in the recently-constructed Museum of Indian Arts and Culture. The building has two levels, one being a basement, is 32,000 square feet in area, and incorporates passive solar and daylighting features in the envelope. A conventional boiler and chiller serve administration, a store, a theater, and classrooms; rooftop units provide heating, cooling, and humidity control for three gallery spaces. An EMS was installed with the original HVAC controls for $85,000 in 1986. Humidity is provided by replaceable, water-filled canister units which run on electricity. The annual energy bill is approximately $1.50 per square foot.

Energy Management Systems, Maintenance, and Results

The two MNM buildings described both have EMS or Direct Digital Control (DDC) systems. DDC provides computerized coordination of conventional HVAC control systems. Instead of directing each of the myriad of system valves and dampers to open, close, or modulate through a series of local discrete electric and pneumatic switches, controllers, transmitters, or thermostats, there is sophisticated computer-based logic that controls these systems. The logic resides in software subroutines and effects the same changes, while coordinating operation of all building systems. Computerized coordination of control offers the opportunity for MNM to control space under close tolerances for relative humidity and temperature to protect collections. Closer control of space conditions also means avoided energy costs.

The reader will understand by now the complexity of a museum application. Further, it becomes obvious that evaluating EMS results involves much more than energy and cost savings. Therefore, results will be considered in two categories: Operational Benefits and Energy Savings.

Operational Benefits are derived from the system because it allows staff to both control and monitor the environment on an ongoing basis. This ensures that the museum is in a position to take direct steps to meet its mission. Problems can be quickly identified and corrective action may follow before irreparable damage is done to collections. In addition, long term studies of the environment can be done to validate the quality of the facility. These studies ensure the safety of current collections. They may also be used with those managing traveling exhibitions of artworks, and thus offer local patrons the opportunity to view these works. In essence, the operational benefits are much like an insurance policy. The true benefit is extremely difficult to measure. Though some might propose that the benefit is equal to the full monetary and artistic value of museum collections, that are protected by the system.

Energy Savings are the more traditional evaluation of a DDC system. Yet in this case, energy savings are an added benefit to the museum. All facility operations must carefully control their energy costs, and this may be more true for public agencies with limited access to additional capital. Like any other business, a museum wants to direct as much of its budget to the mission. If additional money must be diverted to energy costs the museum is less successful in achieving its mission. Yet many of the typical DDC control sequences that result in savings cannot be implemented in a museum. For example, night setback/up, optimal start/stop, etc. or demand limiting are not likely to be feasible in this application. These limitations severely impact energy and cost savings.

Another key issue with these museums is that they did not have a full-time on-site energy specialist like ETHICON. However, they did have the benefit of a relationship with the Energy Conservation and Management Division (ECMD), which was used effectively. This relationship provided project management for an EMS maintenance program to evaluate system performance and make recommendations for corrective action.

The project was initiated in 1988. By this time it was apparent that there were HVAC systems' control problems, as well as increasing energy costs, that needed to be investigated. EMS maintenance was conducted in 1989 and 1990, as part of a $0.5 million HVAC maintenance program funded by EMNRD. The $23,264 EMS maintenance project for the Museum of Indian Arts and Culture realized $3,925 in avoided energy costs during the July 1989 through June 1990 fiscal year period, which is a simple payback of 5.9 years. Similar cost avoidance was achieved in the following fiscal year (1990-91), compared to the July 1988 through June 1989 pre-maintenance period.

A before-and-after monthly comparison of energy costs for the building is shown in Figure 20-3. Note that avoidance first shows up in September 1989, which coincides with substantial completion of the work. Utility billing histories for electricity and natural gas were normalized for weather and billing period variables to present an accurate picture of the avoided costs.

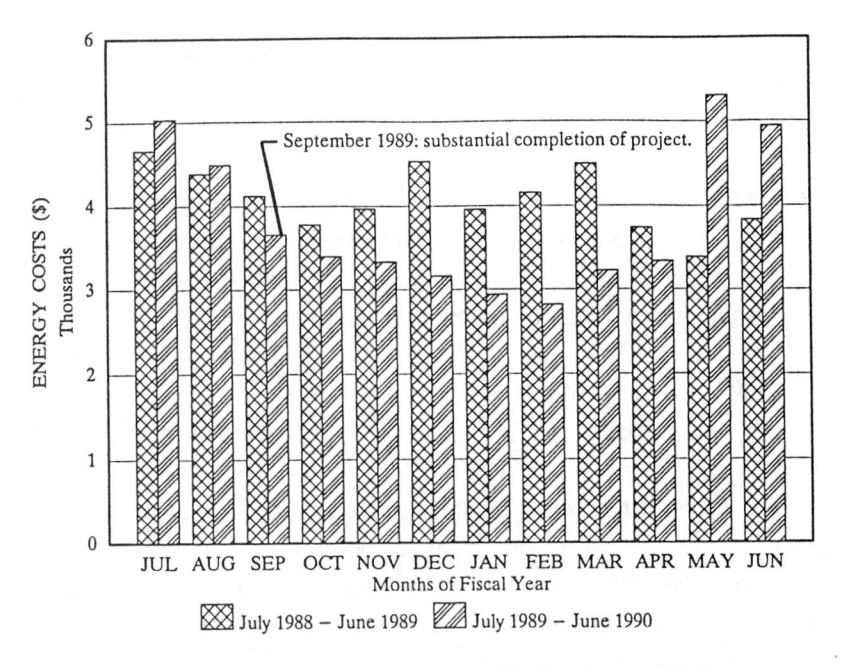

DDC Commercial Building Case Study:
The Museum of New Mexico and Energy Management

Figure 20-3 Energy Costs, Museum of Indian Arts and Culture

The Museum of Fine Arts did not fair as well in cost reduction, though operational benefits were achieved. The $17,918 spent for its EMS maintenance did not achieve a positive payback through the same type of normalized utility billing analysis. This is due to the variation in activities held at the building, additional electricity load from office equipment, and varying operating policies for HVAC systems. In addition, analysis of utility billing data of the more recent 1991-92 fiscal year shows that the Museum of Indian Arts and Culture is no longer achieving a positive dollar-based payback. The mixed results show that EMS projects cannot be justified on avoided costs. First, ECMD's experience with EMS indicates that maintenance should not be periodic, but is needed on a regular daily basis to keep systems, computer hardware, and energy costs in line. Second, operational benefits are real and must be considered. In both buildings, EMS maintenance funding was used for

- HVAC control programming
- replacement and repair of EMS components, as necessary

- limited replacement and repair of HVAC components
- training.

Previous HVAC programming was revised or totally replaced. Museum administrators and maintenance staff, who had worked with the buildings' operations for several years, were consulted. Consensus was reached on how to go about the programming. Main points followed were

- programming should be simplified or application specific
- maintain constant indoor conditions, 24 hours a day, in all spaces
- give low EMS user levels to MNM staff and high levels to EMNRD staff and their contractor.

Concerning the last point, higher user levels were eventually given to MNM staff.

It was found that many EMS components had failed or degraded severely since the EMS was installed. Sensors, relays, and computer hardware were replaced. HVAC components closely related to EMS operation, such as control valves, were also repaired or replaced.

The activities carried out on this one-time basis were consistent with those discussed in Chapter 17. Daily attention to this level of system management is recommended.

Training was conducted for MNM maintenance and administrative staff, to get them more familiar with EMS and to achieve energy savings through more extensive use of EMS capability. Careful consideration of the following issues is critical for EMS training to provide the payback anticipated.

- The entire maintenance staff should not be provided with training. Instead, reserve it for specific personnel
- Maintenance staff are not familiar with the computer interface, typically
- Incentives are desirable for maintenance staff to pursue new skills
- Training quality by the contractor should be excellent.

EMS training should be reserved for selected individuals of the maintenance staff that have already shown knowledge and leadership in operation of HVAC systems. During the training, document that the selected maintenance staff are truly interested. Disinterest might be attributed to, simply, the inability to type on the computer keyboard. In addition, the wages for State government maintenance personnel are low, compared to the private sector, so this labor force may be difficult to retain.

Contractors should not be allowed to treat training as an afterthought. Contract payment should be withheld so that they do not have to be pushed and reminded to satisfy the training requirements listed in their agreement. Hypothetical situations should not be used to lead maintenance staff through the training. Current HVAC problems at the site provide excellent training opportunities. Extensive one-on-one training during the real life of the building's HVAC is beneficial to the maintenance staff, perhaps more than anything else.

HVAC Test and Balance

During the EMS maintenance at the Museum of Indian Arts and Culture, a HVAC test and balance procedure was recommended. There appeared to be severe HVAC equipment problems in the newer of the two buildings, because computer programming revisions and EMS hardware maintenance by the contractor were not effecting the expected improvements in indoor environment conditions. An amendment to the existing contract was approved and $5,300 worth of diagnostic testing was performed. The in-place performance of the heating and cooling systems was documented[3].

It was also found that several terminal units had terminal piping that was smaller than specified on the construction drawings. This meant that fluid flow to the units, in cooling mode or heating mode, was not sufficient. From performance data, it was determined that about five terminal piping runs should be replaced, due to actual poor performance on the cooling side.

Energy Modeling

At about the same time as the test and balance procedure, a $5,000 project was initiated to quantify the energy performance of the Museum of Indian Arts and Culture. With EMS and HVAC systems issues it appeared that documenting energy consumption with the DOE-2.1d simulation would be an excellent tool for current and future reference. This was combined with the need for ECMD to upgrade its building energy

simulation capabilities, so the necessary software was purchased and installed. Some of the more challenging modeling tasks turned out to be
- HVAC equipment
- self-shading effects on the building
- envelope
- scheduling.

The model came to within 30 percent of annual energy consumption of actual utility billings for the building[4]. MNM has shown interest in using the model to explore energy implications of expansion options for the building.

CONTINUING ON

Future Strategies

The future will see MNM conservators advocating better exhibition conditions, MNM and OCA administration planning for facilities' expansion, ECMD pushing for responsiveness to the State Energy Policy, an increasing number of visitors to the museums, and increased variety of museum activities. Energy consumption and costs will increase as a result. However, there are innovative strategies available that may help avoid costs and maintain the necessary indoor conditions to protect museum collections.

Temperature can be allowed to float off of setpoint to accommodate humidity requirements: Indoor conditioning for people is temperature-based; for collections, it should be humidity-based. When humidity is high and HVAC systems reach their dehumidification limit on a hot, but wet, summer day, allow the HVAC system to float and the temperature to move up[5]. These conditions do occur during the monsoon season in July and August. A look at the psychometric chart for 5,000 feet elevation[6] (Santa Fe is at 7,000 feet) shows that an unacceptable 44% RH at 72° F would become an acceptable 35% RH at 78° F, based on MNM's acceptable range of 30 to 40% RH.

The collections can accept the temperature variation more than humidity variation. An operational benefit of the EMS is the ability to program the necessary subroutines into the existing EMS. HVAC systems can be controlled to produce this result. For storage areas that are usually not occupied with people, this strategy has great utility. This strategy can reduce energy costs and can be an interim solution before systems are upgraded.

Hire a HVAC technician – This person would conduct the types of duties outlined in Chapter 17. These include:
- management building automation systems
- consult with MNM administrative and maintenance personnel on HVAC equipment, systems, control, and indoor conditioning issues
- review specifications, design, cost estimates, scheduling, and procurement of equipment and services for automated building control systems.

Green Lights for MNM – A significant portion of MNM energy costs is for lighting. Many of MNM's lighting systems must be specialized to enhance presentation of exhibitions, while protecting them from heat and radiation damage. This often intensive use of lighting should be reviewed by ECMD and proposals developed, with MNM input, on how to reduce the energy use, but maintain or upgrade the quality of the light. There are conventional lighting systems throughout the museums, in administrative, public, and storage areas, where energy efficiency can be improved.

ECMD is actively investigating how the State of New Mexico can be a Green Lights Partner with the U.S. Environmental Protection Agency. On becoming a Partner, ECMD will work with MNM and OCA to achieve Green Lights goals of improved energy efficiency in existing and new lighting systems.

Energy Management Plan for MNM – The Governor's Executive Order on energy, based on the 1991 State Energy Policy, directs each State agency to develop and carry out an energy management plan. An energy management goal has been established that, between 1992 and 1995, State government should reduce energy consumption in State facilities and transportation fleets by 10% over the projected State utility and fuel expenditures for 1995. ECMD will assist MNM and OCA to prepare a plan and meet the energy management goal.

Communication and Collaboration

Two key occurrences in late 1991 that will spur MNM and ECMD on to further joint efforts are
- release of the 1991 State Energy Policy
- serious investigation of indoor environment conditions by MNM conservators.

These two activities were developed independently, but MNM conservators inquired about ECMD technical assistance, which then brought the two groups together for serious negotiations. MNM is working with an art preservation consultant to present recommendations to OCA management. ECMD is offering technical assistance through the Policy. Roles are being defined so that MNM and ECMD can proceed with a reasonable program.

Communication is now much improved. Previously, in the administration of preserving the arts and culture for New Mexico, certain responsibilities were not defined. This led to
- communication gaps
- short-term solutions
- damaged art.

Among MNM administration, OCA administration, MNM maintenance, EMNRD, and EMNRD's contractors, there were many communication gaps. There are different ways in which people see maintenance needs, technical needs, budget allocation, budget planning, and new MNM programs. One perspective of EMNRD is: if maintenance and repair budgets are funded adequately, indoor environment conditions can then be offered for exhibition of prestigious collections, while continuing to display and store MNM's own wonderful collections. Energy efficiency improvements on MNM facilities can help fund the maintenance and repair budgets.

In the preservation of cultural collections at MNM it is now defined as the first priority that indoor environment conditions should be based on humidity requirements of the collections and not on the temperature-based requirements of people.

The challenge will be for MNM to achieve optimum humidity and temperature in all of its facilities. ECMD is looking forward to providing technical assistance to do this at the lowest operational expense feasible.

ACKNOWLEDGEMENTS

Numerous consultations have been held with Ron Vigil, Deputy Director, OCA, on energy efficiency of MNM facilities. Claire Munzenrider, Chief Conservator, MNM, has communicated the needs of art conservation to EMNRD. John J. McGowan, CEM, former Director, ECMD, contributed significantly to development of the relationship between OCA and EMNRD. Throughout his work as a consultant to OCA, Steven Weintraub, Art Preservation Services, has raised issues concerning preservation of collections, the building environment, and control of humidity. Key ideas of the above mentioned people, developed during the normal course of State activities, are contained in this work.

REFERENCES

1. Munzendider, C., personal communication, Chief Conservator, Museum of New Mexico, Office of Cultural Affairs, Santa Fe NM, May 1992.
2. Thomson, G., The Museum Environment, Second Edition, Butterworths, London, 1986.
3. Sedillo, P. M., "Mechanical System Study for the Museum of Indian Arts", report prepared for Energy, Minerals and Natural Resources Department, Contract No. 77-521.33-134A, Miller Metal Co., Albuquerque NM, July 1990.
4. Jones, R. W., "Installation and Verification of Building Energy Simulation Software: DOE2 Model of Museum of Indian Arts and Culture", report prepared for Energy, Minerals and Natural Resources Department, Contract No. 78-521.03-272, Balcomb Solar Associates, Santa Fe NM, February 1991.
5. Weintraub, S., personal communication, Conservation Consultant, Art Preservation Services, New York NY, November 1991.
6. ASHRAE Psychometric Chart No. 4, American Society of Heating, Refrigeration, and Air-Conditioning Engineers, Atlanta, GA, 1965.

Chapter 21

DDC Case Study: Monitoring and Control of Large Computers and Support Systems

By John P. Cilia

PREFACE

The case studies in this book have been drawn from a variety of applications for DDC systems. Our last case study was a commercial building in the institutional sector. Commercial buildings in general account for a major portion of the traditional DDC marketplace. The Industrial application considered first has also been a common application for DDC. Yet the Industrial environment also has seen a good deal of integration. As you know by now, Integration in this context means combining several systems that provide different functions; fire, security, HVAC or process control. Finally, this case study addresses a totally new application for DDC, control of large computers and support systems.

John P Cilia is an Advisory Engineer with the International Business Machines (IBM) Corporation. This chapter was reprinted with permission from the Association of Energy Engineers. It was originally published in the proceedings of the World Energy Engineering Congress, Atlanta, Georgia in October of 1992.

To achieve effective control in one of the most sophisticated operations found in most buildings requires a significant effort. The efforts to integrate multiple systems, and allow one system to make equipment start/stop decisions reaches a new level of complexity when you are stopping a main frame computer!

System Integration is an exciting topic, and is also a logical progression for the control industry. Yet many hurdles must be cleared to make it a reality. As we discussed in Chapter 16 multiple system integration is a challenge that requires significant cooperation and technology sharing. This Case Study is an example of just that sort of activity to accomplish integration in a new area, Data Processing.

There are a number of technologies discussed in this paper, and a brief note on these is warranted. The DDC System is referred to here as a Distributed DDC (DDDC) or Building Automation System. As noted earlier in this book, these terms are synonymous for our purposes because they all employ distributed intelligent control modules. There are also a number of networking terms, some of which were introduced in earlier chapters. Be careful to note use of the terms "Host" and "Enterprise". There are used here in the computer industry context, and refer to data processing equipment and systems. If there is an interest in understanding this technology further the reader is encourage to do additional research. Sources for such information include Data Processing Industry books and Journals, and also "Networking for Building Automation and Control Systems" by John J. Mc Gowan, published by The Fairmont Press.

Finally, the readers attention is drawn to the Data Processing environment. This is the most complex process in most buildings, and traditionally has been totally isolated from the rest of the building. In many cases separate electrical service was provided for power, often with independent emergency backup. Integration of Data Processing with other parts of the organization, beyond the PC or data terminal, was not considered. It is a tremendous testament to DDC technology that this new control application is now a reality. DDC Systems are viewed today as highly reliable. They are able to carry out complex control sequences and can respond quickly to changes in condition. These are among the critical features that make it possible to include monitoring and control of large computer systems among DDC system applications.

Monitoring and Control of Large Computers and Support Systems

By J. P. Cilia

ABSTRACT

Distributed Direct Digital Control (DDDC) for building and facility automation systems has been implemented and operational at IBM for many years. This DDDC system monitors and controls chillers, boilers, lighting, heating, ventilating and air-conditioning. Finding new ways to monitor and control untapped high potential users of energy and support unmanned areas for optimum operation and support is a major concern for IBM Corporation. The desire to remotely monitor and control unmanned computer raised floor areas resulted in the development of the System View Site Manager Services (SSMS) offering.

Besides powering on-off computers and peripherals, SSMS can communicate environmental information such as room temperature, relative humidity, fire, smoke and security system status, and alarms through the same network.

This chapter describes the steps required to design and implement a system for a computer raised floor area. This unique installation will remotely monitor and control large frame computers, peripherals, and environmental systems simultaneously. Expansion of SSMS using NetView Automation throughout the plant and the enterprise is also covered.

UNDERSTANDING YOUR SYSTEMS AND OPERATIONS

The same requirements to understand your mechanical and electrical systems apply to Data Process (DP) equipment. These basic requirements are necessary for the monitoring and control of DP equipment residing inside large buildings, hospitals, college campuses, or plants.

In order to reduce operating costs, there has been a trend in the 90s to reduce staff personnel within all fields of a business. The DP field is no exception. Reducing staff personnel implies that the existing operation has too many bodies to perform certain tasks or, in most cases today, cost reductions are imposed in every area of the business in order to be competitive in the world market. This plan of action requires some kind of automated systems to replace most of the tasks performed by your DP personnel.

This section is intended to provide the reader with a basic picture of the major DP equipment and operation required to operate, manage, and support any large buildings, hospitals, college campuses, or plants.

Hardware

Today, most companies use a large number of computers and peripherals to operate, manage and support their business. Typical examples are office automation and telecommunication.

To show the varieties of hardware in the DP environment, a partial list of potential hardware located in your site are identified.

- Workstations
  - Personal computers (PC's)
  - Printers, plotters and displays
- Mid-Range and Large Systems
  - Processors
    - Control units
    - Printers, plotters, terminals
    - Magnetic Media
      ◊ Direct access storage devices (DASD's) and tapes
- Telecommunications Systems
  - Modems
  - Communication controllers

Operation

Today, integration of office automation, computers, manufacturing automation and peripherals is accomplished by one or more networks which transfer data from one system to another.

This connectivity provides the door to the host computer using expert system software and artificial intelligence to achieve the highest productivity at the minimum cost. Existing automation packages are available from many manufacturers and software companies to provide such support. The operation, management, and support of a site is done through the host computer. The host computer determines which systems, nodes and control units should be operational at different times of the day.

You can see that an understanding of how your DP equipment operates is absolutely required if you want to control any of your DP equipment. This task usually requires a team of people working in different areas of your DP environment. Many meetings will be necessary to gather all relevant data in order to gain a true understanding of the operation of your DP equipment.

REQUIREMENT GATHERING

Now that you understand how your DP equipment operates, the next step is to specify necessary requirements to decide what you can turn on or off and when. A major problem not yet covered should be coming to your mind. How can you make sure that a computer is not working, i.e., idling, before you shut it down? This question will be discussed and answered later under the Process Data Monitoring section of this chapter.

Equipment Survey

This task requires that an expert(s), on your DP equipment, checks whether the hardware you chose to monitor and control has existing connections for the power-control interface (PCI). Many computer manufacturers like IBM, have provided PCI for some of their hardware since 1966. One of the most common interfaces is the "standard 6-wire interface". In the event that some of the DP equipment you select does not have a PCI, you can go back to the unit manufacturer and request such a connection to be installed. This PCI provides the necessary connection to remotely power up and down (control function) a device, and to confirm whether the unit is up or down (monitor function).

Environment monitoring and control is also required to facilitate the operation of your DP room. This will allow you to reduce operating costs by shutting down air-conditioning units (ACU's), as well as, monitoring the room temperature and humidity. Other information such as security and fire/smoke alarm can be used to provide a local and/or remote alert to enhance the operation of your DP area, and reduce the chance of plant shut-down due to a system malfunction. Additional energy management strategies can be used to control peripheral equipment, such as terminals and printers, to reduce energy consumption during scheduled operation shutdown.

Process Data Monitoring

The final and most challenging task is the monitoring of your application programs. A simple scenario is used to explain this process. Let's assume for a moment that you can power off an office automation system configured as follows: one computer, one control unit, and several attached DASD's and terminals. Your site is unoccupied after 8:00pm every Friday until 6:00am Monday. You plant to shut down your office automation system by Friday, 10:00pm, for the weekend. Before turning off the equipment and its support systems, such as ACU's, you want to make sure that nobody is "LOGGED ON" the network. This requires a software package capable of reading all the attached devices on the network, i.e., nodes, and determines if each node is activated or de-activated. The application program will read the actual time of day (TOD) and day of week (DOW), determine it is Friday, 10:00pm, and initiate a pre-defined strategy. This strategy will start a program and simulate key strokes on the screen, as if a person were typing on the keyboard. The program reads the status of the nodes on the screen and, if all identified nodes are logged off, the program proceeds to shut down the DP equipment and its support systems in a pre-defined sequence. If one or more nodes are logged on, the program will hold the shutdown process for another hour and try again. This strategy is just one of many possible ways that can be utilized to schedule a weekend shutdown of your DP equipment. Let's assume that the system tried again two hours later and one node is still logged on. It is midnight and usually the building is empty. There are at least two reasons that can be derived by this moni-

DDC Case Study:
Monitoring and Control of Large Computers and Support Systems

toring process. First, the person forgot to log off his terminal before leaving office. The second one is potentially a serious problem; the person is still in the office, unable to move for help. Your automation system can display an alert on a terminal and simultaneously send a message on the pager of your security guard at the front entrance of the plant. The message on the pager will identify the building and room number where the node is installed. If the guard does not acknowledge such alert within a pre-defined time by dialing back the computer with a specific code to clear the alert, the system can escalate and call one or more predefined pages until confirmation is acknowledged by the system.

The process requires a software program to allow the user to interface and communicate with the office automation system, as well as, provide the necessary functions to monitor and control the equipment. One approach used in IBM has been to utilize a Personal System/2 (PS/2) to act at the gateway between DP equipment, application software, and the host computer. The host computer will have an automated expert program package which will provide the necessary interfaces with all major application programs running your enterprise. This automated expert program will monitor computer systems, networks and software. In the event of a problem, an alert and possible corrective action can be handled. Automated operations products such as IBM NetView and Automated Operations Control (AOC) are available to predict system outages and re-route workload to improve system availability. Figure 21-1 describes how an IBM Enterprise Systems Architecture/390 (ESA/390) host computer is connected to remote sites using Enterprise Systems Connection (ESCON) products with SSMS.

HARDWARE AND SOFTWARE COMPONENTS

To accomplish the control and process data monitoring functions described requires the hardware and software components listed below.

PS/2

- PS/2 Model 8580, 20 Mhz, 110 MB hard disk, 8 MB
- Color Display Model 85XX
- Mouse Model 8770
- Proprinter Model 42XX

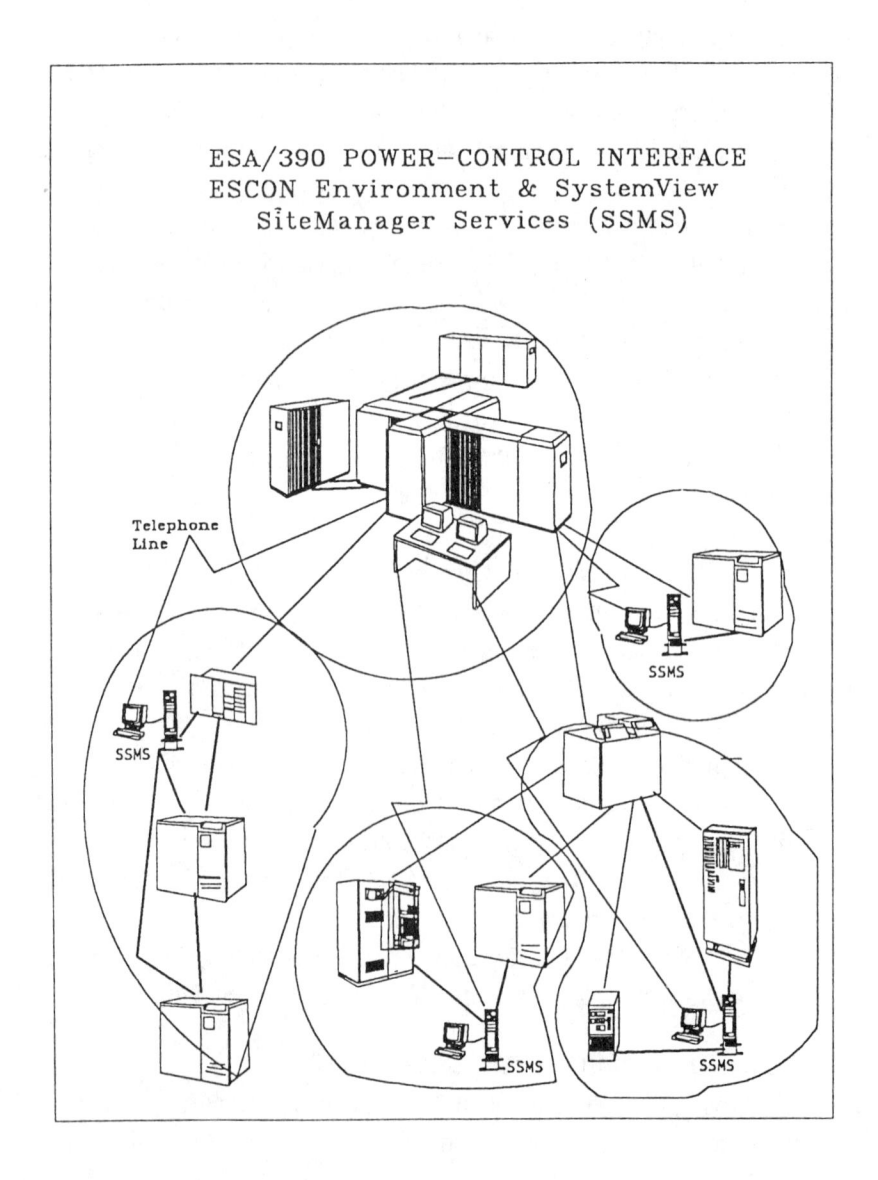

Figure 21-1 Power-Control Interface for ESCON Environment and SSMS

DDC Case Study:
Monitoring and Control of Large Computers and Support Systems

Communication cards are used to communicate to the host computer, to other computers and terminals for monitoring application programs.

Communications
- Dual Async Adapter/A (RS232)
- Synchronous Data Link Control (SDLC) Adapter
- 3270 Emulation
- 5250 Emulation
- X.25 Adapter
- IBM PS/2 300/1200/2400 bits/second Internal Modem Local Area Network (LAN)
- IBM Token-Ring Network 16/4 Mbit/second Adapter/A
- Ethernet Adapter
- IBM PC Network Broadband Adapter II/A
- IBM PC Network Baseband Adapter/A

Environmental Monitoring and Control Architecture

Different components are used to interface with the computers, peripherals and environment. The same approach used for standard DDC or Building Automation System (BAS) is applied for the DP equipment.

The PS/2 is connected to an Environmental Master Controller (EMC) which acts as a policeman between the PS/2 and attached DP equipment and sensors. The EMC is linked to the PS/2 through a standard RS-232 port. The EMC has two RS-485 communication trunk lines running at 51.2 KBaud to a maximum of 4000 feet for each trunk line. Each trunk line can support up to 31 Environmental Local Controllers (ELC's). EMC has a 3-cell rechargeable battery to back up memory for a minimum of 72 hours in case of a power failure.

ELC has an input/output capacity to handle 11 analog and/or digital inputs for monitoring and 8 digital and/or analog outputs for controlling devices. The communication between the ELC on the trunk line is done via a two-wire bus.

Connection between field sensors and the ELC are direct plug-in with a DB-9 connector. Input and output connectors are color coded to facilitate installation. Normally open (NO) and normally closed (NC) contacts as well as 2000 Ohms RTD, 4-20mA and 1-5Vdc are supported.

Connection from the ELC to the computers and peripherals is done through the PCI 6-wire, PCI 2-wire and PcI-RPC (Remote power control). Each PCI provides connection for power control and monitoring. Different PCI's voltage and current load are available to handle most common DP equipment.

Software Base Components

The software used to run the PS/2 and to monitor and control DP equipment is integrated in two packages: PS/2 operating system (OS/2 version 2.0), and SSMS. OS/2 provides the different interfaces necessary to handle the hand checking between the PS/2 components and the communication to the outside world. The extended services (ES) of OS/2 includes the communication manager providing the necessary drivers to interface with the host computer, LAN's and other computers. SSMS interacts directly with external systems by using the existing protocols supported under the ES communication manager capabilities of OS/2. SSMS provides the sensors and controllable configuration lists of all sensors and devices attached to the system as well as the drivers to communicate with EMC and the pagers. Other screens to select which alert and messages to be forwarded to different pager numbers and devices that can be powered on or off are also available.

IMPLEMENTATION

Figure 21-2 shows a typical application between DP equipment, host computers, devices, and sensors.

Connectivity and application monitoring software (dialog) are integrated under the communication manager of OS/2 extended services.

PCI's and environmental monitoring and control are provided by the ELC's and the PCI's devices.

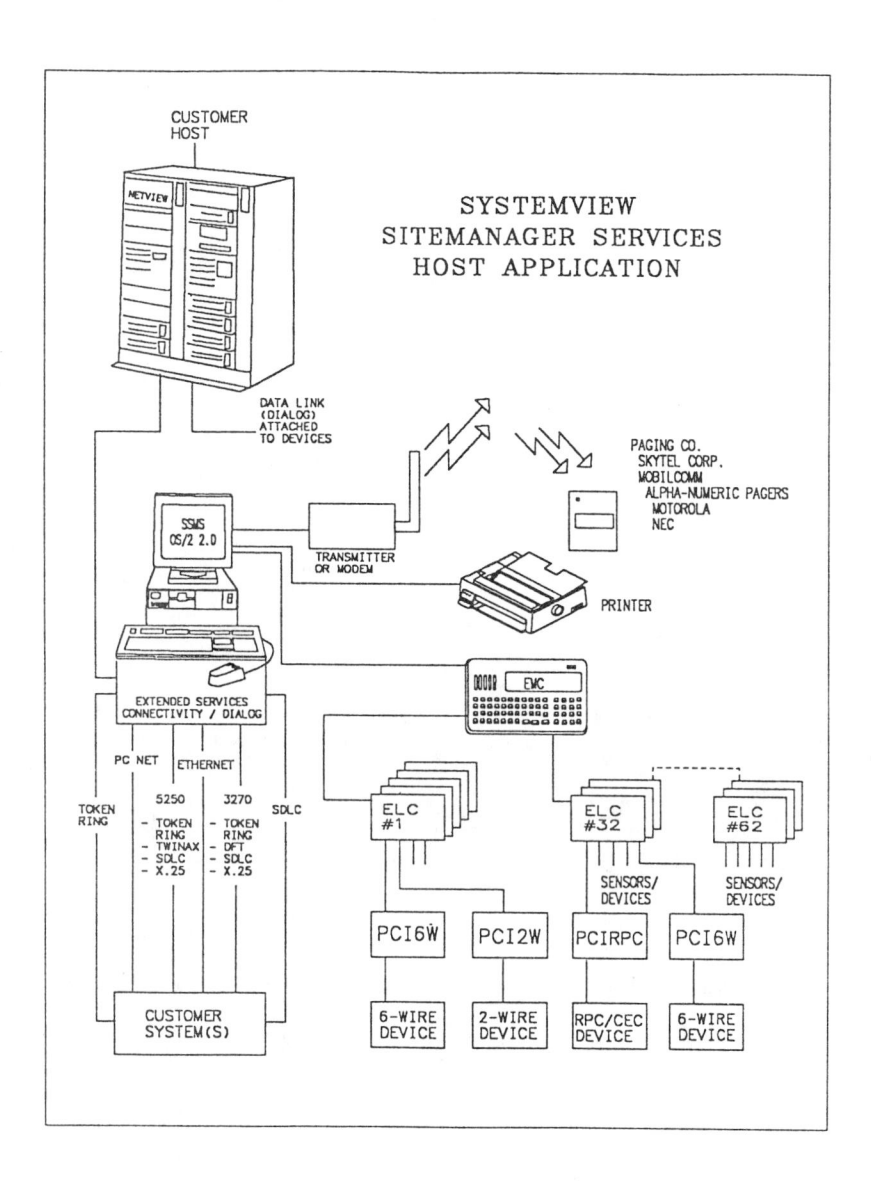

DDC Case Study:
Monitoring and Control of Large Computers and Support Systems

Figure 21-2 Typical application utilizing SSMS

Appendix A: Glossary

80286/80386/80486 – Microprocessors developed by INTEL.

Actuator – A component that is used to electrically or pneumatically regulate a process. It may do so in various stages or continuously, such as water flow control.

AHU – Abbreviation for air handling unit.

Alarm handling – The treatment and processing of alarms.

Analog input – Input from a continuously varying signal, e.g., a temperature sensor.

Analog output – Output for a continuously varying signal, e.g., for a water flow valve.

ANSI – American National Standards Institute – an industry supported standards organization that establishes U.S. industrial standards and their correspondence to those set by the ISO.

Architecture – Industry term used to describe a wiring configuration which connects controllers, field devices and network equipment. In some cases architecture may involve electrical wiring. Of more relevance to this topic, however, are those that provide a vehicle for controller to controller communication, typically over a twisted pair media.

ASCII – American Standard Code for Information Interchange. This coded character set is used internally by the IBM PC and most other microprocessors.

ASHRAE – American Society of Heating Refrigeration and Air Conditioning Engineers, a leading engineering organization in the controls industry. ASHRAE supports research, and has historically established standards within our industry in a wide variety of areas. ASHRAE has taken the lead in establishing a control industry standard through establishing a standards committee, 135P.

ATC – Automatic Temperature Control

BACnet – Building Automation Control network – an emerging communication standard for building control components, driven by an ASHRAE (American Society of Heating, Refrigeration, and Air conditioning Engineers) committee consisting of building facility operations personnel and building control industry supplier personnel. The intent is to create a communication standard that encompasses all facets of the ISO Reference Model plus an application layer that will enable multiple vendor products to intercommunicate and interoperate with one another. The final definition and availability of this standard has not yet been determined.

Bandwidth – The capacity of a communications channel.

Baud – Per Thomas Madron, baud "is a unit of transmission speed equal to the number of discreet conditions or signal events per second" A discreet condition or signal event refers to computer data that is represented by an electrical signal. Generally each event consists of the portion of that signal that correlates with one bit of data being sent on the communication path. Madron further states that "baud is the same a 'bits per second' only if each signal event represents exactly one bit, although the terms are often used interchangeably."

Bit Per Second – See Baud

Bus – A path for electrical signals; also a type of network topology with a single cable onto which all devices attach. All network stations receive the same message through the cable as the same time.

Central Operator Interface – A PC based software system designed to provide a full range of interaction with control systems. Interaction will

Appendix A: Glossary 469

include at a minimum: Control system Commissioning, Monitoring, status and Alarm dial in/out.

Central/Hand-held Operator – Hardware and software which provides local and remote network communication with distributed controllers. The interface may reside in a hand-held device, a controller or in a separate piece of hardware, such as a PC.

Channel – A path which communication may be sent.

Closed Loop – A method of control in which feedback is used to link a controlled process back to the original command signal.

Commission – A term used to define the process of programming a control panel directly out of the carton. Commissioning of a DDC controller requires point by point definition, control strategy definition and setpoints. Finally the commissioning process must also include addressing controllers to the network and establishing any parameter necessary for remote communication with the network.

Control sequence – A control sequence establishes relationships between the various data points (inputs and outputs). The number of points required depends on the type of control strategy.

Control System – One or more components which perform a complete set of control functions for load control, HVAC control or both. The term system is defined as including all software or hardware components necessary to achieve the full range of building functions desired. The control system will provide a variety of functions as defined for a particular building. Among the functions which may be included are: Time of Day and Holiday scheduling, Possibly Event Initiated schedules, Load Control, HVAC control, Integrated HVAC control strategies, Demand limit control and Local/Remote Alarming.

Control, Adaptive – A control algorithm or technique where the controller changes its control parameters and performance characteristics in response to its environment and experience. See self-adapting and self-learning.

Control, Cascade – An algorithm or method of control in which various control units are linked in sequence, each control unit regulating the operation of the next control unit.

Controller, DDC – A computer that uses internal logic to look at real-time variables such as hardware inputs, makes decisions, and provides an output.

Controller – An individual component, usually designed in modular form, which performs load or HVAC control functions.

CPU – Central Processing Unit. The microprocessor that provides processing capability, or intelligence, for a microcomputer.

CRT – Abbreviation for cathode ray tube. CRT is a term used to indicate the system monitor/display.

CSMA/CD – A data link control photocol applicable to a broadcast network in which all stations can receive all messages. It finds its widest application in LANs and is generally referred to as Ethernet.

DDC – Abbreviation for direct digital control. See Direct digital control.

DDC Controller – Direct Digital Control (DDC) is implemented with a microprocessor based controller which performs control loop functions via sophisticated (P+I or P+I+D) algorithms and strategies.

De facto – Refers to standards that are widely used even though they may not be sanctioned by any official standards bodies.

Dead Band – A temperature range over which no heating or cooling energy is supplied. Opposite of overlap.

Dedicated – Reserved for a single function. A dedicated network server, for example, cannot be used for any purpose other than network serving.

Digital input – Input for discrete electrical values, e.g., an impulse transducer for flow measurement totalization.

Digital output – Output with discrete electrical values, e.g., a pump switching status (on/off)

Appendix A: Glossary

Direct Digital Control (DDC) – Direct Digital Control (DDC) is implemented with a microprocessor based controller which performs control loop functions via sophisticated (P+I or P+I+D) algorithms and strategies. A controller technology that handles inputs and outputs electronically with software providing the logic.

Disk, Hard – Disk memory that uses rigid disks rather than flexible disks as the storage medium. Hard-disk devices can generally store more information and access it faster.

Distributed Control System – DDC Control for a complete building via microprocessor based distributed controllers, also referred to in this text as Digital Distribution.

Distributed Controller – DDC Controller designed to be installed in close proximity to controlled equipment (i.e.: in equipment control panel or in electrical room). This type of equipment is distinguished by the requirement for network communication to provide full access to the distributed microprocessors.

Distributed processing – A technology in which processing occurs in multiple machines on the network. Distributed processing is typical on most local area networks.

EEPROM – Electrically Erasable Programmable Read Only Memory

EMS – Energy Management System.

EPROM – An integrated circuit for data storage.

Event Initiated Program – A program which issues specified commands to any number of system points based on a system event such as a hardware change-of-status, absolute or elapsed time, an operator command or a command from another program. Called EIP.

Execution Time – The time required for the CPU to decode and execute a computer instruction.

Fiber Optics – A communication technique where information is transmitted in the form of light over a transparent material (fiber) such as a strand of glass. Advantages are noise free communication not susceptible to electromagnetic interference.

Gateway – A device that connects two networks with dissimilar protocols.

Global data point – A system input or output that can be addressed by all connected controllers.

Hardware – All computer components that can be touched, such as the monitor, keyboard, central processing unit, etc.

Heating Ventilation and Air Conditioning (HVAC) – Building Equipment providing space conditioning or HVAC is one of the core areas of concern for any control system. HVAC systems may be divided into two categories: Central station and Packaged. Central station systems provide a plant approach to space conditioning, and are normally applied in large facilities. Due to the extensive requirements for heating and cooling, boilers, chillers and large air handling equipment are applied. Packaged equipment combines the heating, cooling and air handling in the same piece of equipment, hence the name packaged. These self-contained systems usually provide constant volume air to a single occupied zone. For more explanation of this complex topic additional research is recommended.

Holiday – A day on which a special daily program, not described in the weekly program, is valid as an exception.

HVAC Control – HVAC equipment requires that a series of control loops be established. These loops define a sequence consisting of a sensed value in a conditioned media, a controller, and a driven device which interacts with the controlled media. A control set is an assembly of independent control loops which collectively perform a required sequence of operation, usually for a single unit.

I/O – Input/output.

IAQ – Indoor air quality.

If-Then-Else – A program statement often used in modern high-level language to provide a closed branch in the program. When the IF statement is true, the THEN procedure is executed. Otherwise the ELSE procedure is executed.

Initialization, System – The process of loading the operating system into the computer and defining the processing environment.

Initialize – To originate or establish the basic conditions or startup state.

Input – Information or data supplied to a computer for processing.

Integrated Control – Integrated control requires a network and includes commands that consist of control decisions made in one controller, but carried out by one or more other distributed controllers.

Integration – Combining or networking HVAC, fire, security, and other systems into one that allows multiplexed data exchange between systems and data display from any system on the same PC screen in a similar format.

Intelligent Buildings Institute (IBI) – The IBI has had extensive involvement in the open protocol issue. IbI has sponsored a series of symposiums to expand industry knowledge of this issue and to clarify positions of various industry members. The IBI has also conducted a series of end user focus groups to clarify requirements of individuals actually using control systems.

Interface – The point at which systems communicate with each other. An interface may be hardware-to-software or software-to-software.

Interoperability – Also called internetworking; implies that computer systems and/or software between different vendors will work together in some complementary fashion.

ISO – International Standards Organization, Paris, which developed the Open System Interconnection (OSI) model.

KB – Kilobyte.

LAN (Local Area Network) – A group of computer systems (nodes) tied together by a high-speed network within a confined geographical area, such as a building.

MA – Milliamp

Modem Interface – Provides direct dial in/out communication over telephone lines with a DDC system and distributed controllers or a network interface.

Modem – Modulator/Demodulator

Mouse – A device used to move the cursor to select items or make entries on the computer display. A mouse generally has one or two buttons.

MS DOS – A common operating system for IBM PCs and other compatible computers. It is a product of Microsoft Corporation.

MS Windows® – A graphical user interface for PCs, developed by Microsoft Corporation.

Multiple access – The condition in which multiple users can open the same file simultaneously. Same as concurrent access.

Multiplexing – Supporting simultaneous multiple transmissions on one medium.

Multitasking – The concurrent handling of multiple jobs by one CPU. On a network server, multitasking usually means that the server can be used in the foreground as a local workstation while network tasks are carried out in the background.

Multiuser – A system in which multiple users share one CPU. Multiuser commonly indicates a host-to-terminal system.

Network Architecture – As defined under distributed network architecture, a media which supports communication between controllers. Network architectures to date have employed proprietary protocols for communication and interface.

Network interface card (NIC) – A circuit board that permits direct connection of a microcomputer to a network cable.

Night Cycle – An energy management application program which maintains specified levels of temperature in the controlled space during unoccupied periods.

Night Purge – An energy management application program which causes the cool night air to be used to reduce the temperature in a building

thereby reducing energy requirements during morning occupancy cool down.

NIST (National Institute for Standards and Technology) – Formerly called the National Bureau of Standards (NBS), the NIST is a U.S. government standards body that defines computer and networking specifications as standards. Vendors who want to sell products to the government must build their products in compliance with such standards.

Node – A physical address on the network, that is able to communicate is called a node. There are variations in the capabilities of nodes based upon internal processing power and the programming of the device. In control industry networks nodes are normally Distributed Controllers of some type.

Office automation – The practice of using technology, usually in the form of a computer system, to perform tasks that previously were handled manually.

One step distribution – A term for the distribution system whereby a manufacturer sells directly to the contractor.

Open Protocol – A protocol which has been developed for use by a specific manufacturer or another entity for use in network or remote communication with control systems. The term open is used when the manufacturer publishes the protocol source code. With published source code any manufacturer, or software house, can develop interfaces to allow controllers using that protocol to be integrated with a network of controllers using another manufacturers protocol.

Open Systems Interconnection (OSI) – A model developed by the International Standards Organization describing the network communications process. Current plans underway in the ASHRAE standards committee include using the OSI model in development of a control industry protocol.

Open systems – Computing devices that are mutually compatible because of their reliance on industry-standard technologies and their conformance with industry-accepted standards.

OSI (Organization for International Standards) – Originally a European consortium defining standards for networking. Defines standard formats for messages sent across system boundaries.

Parallel interface – A socket on the rear of a PC where printers are normally connected. With a parallel interface, data is transmitted simultaneously over several wires.

Password – A security tool used to identify authorized network users and to define their privileges with the network.

Password level – Indicates which commands an authorized user can use.

PBX – Public Branch exchange. A central control unit for telephone systems.

PC – Abbreviation for personal computer.

Peer-to-peer – A network scheme that assumes that connected devices have local processing power and can pass data between one another without the presence of a host computer.

Peripheral – A device, such as a printer or disk drive, that is connected to and controlled by a computer.

PI – Proportional plus integral control logic.

PID – Proportional plus integral plus derivative control logic.

Plug and play – A term used to describe a central operator interface's ability to operate on a Bus without any PC engineering. Data is read directly from the connected Bus devices.

Proprietary Protocol – A protocol developed for use by a specific manufacturer in network or remote communication with control systems. The term proprietary is used when the manufacturer does not publish the protocol source code. In this way controllers using this protocol cannot be integrated with a network of controllers using another manufacturers protocol.

Protocol – A specification that describes the rules and procedures that products should follow to perform activities on a network, such as transmitting data.

Pseudo – A variable value that is calculated by the controller as part of the program sequence.

Pseudo data point – A system variable; it is not accessible externally.

RAM – Random access memory.

Real-time Clock – A program-accessible clock which indicates the passage of actual time. The clock may be updated by hardware or software.

Remote Communication – A function which allows users with a central OI to meet basic requirements for interface with control systems at a remote geographic location. This requires a modem interface at the site and the OI, and generally includes, at a minimum, these features: Control system Commissioning, Monitoring, status and Alarm dial in/out.

Reversible actuator – A bi-directional motor to position a control device.

RS-232-C – A common protocol for connecting microcomputer system components.

RTD – Resistance Temperature Detector

Serial interface – Socket on the rear of the personal computer. Used for transmitting data one line after another. The data transmission is therefore slower than with a parallel interface.

Serial transmission – A data transmission method in which each bit in a byte is sent sequentially, one at a time. When the data arrives at its destination, the bits are reassembled into 8-bit bytes.

Server – A hardware and software device that acts as an interface between a local area network and a peripheral device. The server receives requests for peripheral services and manages the requests so that they are answered in an orderly, sequential manner.

Setpoint – The preset value that must be achieved or maintained, such as room temperature = 72 degrees F.

Software – The instruction programs that tell a computer what to do.

Standards – Products or procedures that have become the most common, or most frequently used, within the industry. Standards are not specific to one company, though they may be owned and licensed from one source.

Time program – A program to control the building that is dependent on the calendar. In time programs, switching conditions can be defined for each day of the year.

TOD – Abbreviation for time of day.

Topology – The physical layout of a local area network. Some common topologies are Bus, Ring, and Star.

Totalizer – A counter that measures the time since the counter was last reset to zero.

Traffic – The volume of messages sent over a shared medium.

Transducer – A device that sends information, such as a digital impulse emitter.

Trend Log – Saves changes in the input or output variable logged with user address, value (or status), data and time for up to 20 data points.

Twisted-Pair Wiring – Cable comprised of two copper wires twisted together at 6 turns per inch to provide electrical self-shielding. Telephone wire is often twisted-pair, this is also the most common media implemented in control networks.

Two step distribution – A term for the distribution system whereby Honeywell sells to a contractor/distributor who then sells to the building owner.

Two-position actuator – An actuator with two positions (Open/Closed)

Upload – Transfer of data or programs from a controller onto the hard disk of the PC.

User address – A name for a data point that is determined by the user.

User level – Indicates the commands that an authorized user can use (see password level).

VAV – Variable air volume.

WAN (Wide Area Network) – A network that extends beyond an area served by the dedicated communications lines of a LAN. Typically employs telephone lines for communications over long distances.

Wild card function – Permits the search for data points in accordance with texts that partially agree; the parts not in agreement are replaced with wild card characters.

Appendix B: Listing of DDC Manufacturers

The listing below is not intended to be all inclusive, but does cover most of the control system manufacturers. Inclusion in this listing does not mean these are truly distributed DDC offerings. As recommended throughout this book, any manufacturer, distributor or contractor of DDC systems should be carefully evaluated to ensure they meet the customer's requirements and needs.

A.E.T. Systems, Inc.
77 Accord Executive Park Dr., Norwell, MA 02061

Alerton Technologies
2475 140th Ave., N.E., Bldg. E., Bellevue, MA 98005

American Auto-Matrix
One Technology Dr., Export, PA, 15632

Andover Controls
300 Brickstone Square, Andover, MA, 01810

Automated Logic Corp.
1283 Kennestone Circle, Marietta, GA, 30066

Broadmoore Electric
1947 Republic Ave., San Leandro, CA, 94577

Carrier Corp.
Carrier Parkway, P.O. Box 4808, Syracuse, NY 13221

CSI Control Systems Intl.
1625 W. Crosby Rd., Carrollton, TX 75006

Delta Controls, Inc.
13520 78th Ave., Surrey, BC, Canada V3W 8J6

Dencor, Inc.
1450 W. Evans Ave., Denver, CO 80223

EDA Controls Corp.
6645 Singletree Dr., Columbus, OH 43229

Elemco Building Controls
324 Motor Pkwy., Hauppauge, NY 11788

Energy Control Systems
2940 Cole Court, Norcross, GA 30071

Functional Devices, Inc.
310 S. Union St., Russiaville, IN 46979

GC Controls Inc.
P.O. Box 461, Greene, NY 13778

Honeywell Inc.
Home and Building Control, Honeywell Plaza, P.O. Box 524, Minneapolis, MN 55408

Inncom International, Inc.
P.O. Box 966, Old Lyme, CT 06371

Integrated Energy Controls
5150 W. 76 St., Minneapolis, MN 55439

ISI Wireless Inc.
3000-D South Highland Dr., Las Vegas, NV 89109

Johnson Cntrls, Inc.
507 E. Michigan St., P.O. Box 423, Milwaukee, WI 53201

Kreuter Mfg. Co.
19476 New Paris Industrial Dr., New Paris, IN 46553

Landis & Gyr Powers, Inc.
1000 Deerfield Parkway, Buffalo Grove, IL 60089

Novar Cntrls. Corp.
24 Brown St., Barberton, OH 44203

Siebe Environ. Cntrls.
1354 Clifford Ave., Rockford, IL 61132

SnyderGeneral, Inc.
13600 Industrial Park Blvd., Minneapolis, MN 55440

Staefa Control Systems Inc.
8515 Miralani Dr., San Diego, CA 92126

Teletrol Systems Inc.
324 Commercial St., Manchester, NH 03101

Tour & Andersson, Inc.
1001 Briggs Rd., Suite #205, Mt. Laurel, NJ, 08054

The Trane Company
20 Yorkton Court, St. Paul, MN 55117

Unity Systems, Inc.
2606 Spring St., Redwood City, CA 94063

Xencom Systems
12015 Shiloh Rd., Suite #155, Dallas, TX 75228

York International
631 S. Richland Ave., York, PA 17405

Index

Analog Input/Output, 26, 198
Architecture, 21, 27, 116, 118, 177, 180, 189, 474
ASHRAE, 44, 113, 130, 220, 258, 275, 340, 468
Association of Energy Engineers, 130, 416
BACnet – Building Automation Control Network, 44, 139, 258, 276, 468
Building Wide Control, 10, 105, 116, 128, 145, 176, 184, 196
Central Operator Interface, 118, 124, 179, 216, 246, 251, 263, 279, 283, 289, 296, 322, 398, 468
Central Processing Unit (CPU), 107, 109
Commission, 179, 186, 224, 249, 299, 338, 469
Control loops/sequence/strategy/ algorithm, 192, 195, 208, 210, 212, 240, 469
Control Integration – See Integrated Control
Data Sharing – See Shared Data
DDC, 6, 113, 191, 205, 211, 261, 340, 363, 397
Demand limit control, 146, 150, 209, 221, 248
Demand side management, 2, 115, 261, 326, 333, 417
Distributed Controller, 10, 120, 189
Distributed Direct Digital Control System, 1, 9, 21, 64, 127, 131, 180, 223, 256, 268, 422
Distributed processing, 15, 234
Duty cycling, 146, 161

Economizer cycle, 146, 154, 213
Energy Management System, 8, 38, 64, 280, 341, 363, 417, 426
EMS – See Energy Management System
Equipment/General Purpose Control, 189, 196, 209
Equipment Level Controller (ELC), 26, 105
EUI (energy use index), 100
Fine-tuning, 397, 406
Firmware, 202, 207
Flash memory, 187
Global Data, 183, 200
Graphics, 337, 341, 350, 372
Heating Ventilation and Air Conditioning (HVAC), 15, 16, 106, 222, 261, 307, 340, 428, 437, 472
Indoor Air Quality, 115, 215, 216, 219, 397
Information management, 10, 411
Installation, 224, 249, 370
Integrated Control, 15, 141, 176, 180, 185, 219, 249, 455, 473
Integration, 15
Intelligent Buildings Institute (IBI), 17, 473
Inter-operability, 238, 251, 267, 270, 473
Local Area Control Network, 113, 131
Monitoring, 152, 163, 179
Network Management, 180
Night Set back/set up, 153
Node, 27, 117, 182, 289, 475
Object Orientation, 144, 184, 294, 372
Open Protocol, 28, 256, 475

Open Systems Interconnection (OSI), 269, 475
Operator Interface, 105, 223, 227, 251, 279
Optimal start/stop, 111, 146, 148
Optimizing functions, 130, 146, 157, 214, 398
Peer-to-peer network, 128, 139, 182, 190, 204, 476
Performance Contracting, 422
Plug and Play, 184, 300, 312, 476
Point schedule, 165, 176
Programming, 179, 186, 297
Proportional, Integral, Derivative (PID), 114, 134, 137, 212, 476
Proprietary protocol, 24, 204, 269, 476
Protocol, 27, 180, 204, 268, 322, 476
Remote Communication, 164, 179, 182, 204, 223, 249, 264, 289, 310, 476
Reset temperature, 156, 212
RS-232, 181

Safety/alarm, 146, 152, 208, 214, 217, 246, 307
Scheduling (Time of Day), 147, 209
Sensors, 194, 199, 370
Shared Data, 141, 146, 162, 178, 200, 263
Smoke and fire management system, 15
Specification, 300, 337, 363
Standard Protocol/Communication, 184, 204, 260
Start/stop optimization – See Optimal Start/Stop
System architecture, 105, 176
Training, 65, 415
Transducers, 199, 478
Trend Log, 152, 179, 217, 246, 306
Trouble diagnosis, 164
Twisted-pair wiring, 224, 227, 478
Zone Control, 11, 105, 120, 190, 196, 232